FROM IDEA TO APPLICATION
Some selected nuclear techniques
in research and development

The following States are Members of the International Atomic Energy Agency:

AFGHANISTAN	HOLY SEE	PHILIPPINES
ALBANIA	HUNGARY	POLAND
ALGERIA	ICELAND	PORTUGAL
ARGENTINA	INDIA	QATAR
AUSTRALIA	INDONESIA	ROMANIA
AUSTRIA	IRAN	SAUDI ARABIA
BANGLADESH	IRAQ	SENEGAL
BELGIUM	IRELAND	SIERRA LEONE
BOLIVIA	ISRAEL	SINGAPORE
BRAZIL	ITALY	SOUTH AFRICA
BULGARIA	IVORY COAST	SPAIN
BURMA	JAMAICA	SRI LANKA
BYELORUSSIAN SOVIET	JAPAN	SUDAN
SOCIALIST REPUBLIC	JORDAN	SWEDEN
CANADA	KENYA	SWITZERLAND
CHILE	KOREA, REPUBLIC OF	SYRIAN ARAB REPUBLIC
COLOMBIA	KUWAIT	THAILAND
COSTA RICA	LEBANON	TUNISIA
CUBA	LIBERIA	TURKEY
CYPRUS	LIBYAN ARAB JAMAHIRIYA	UGANDA
CZECHOSLOVAKIA	LIECHTENSTEIN	UKRAINIAN SOVIET SOCIALIST
DEMOCRATIC KAMPUCHEA	LUXEMBOURG	REPUBLIC
DEMOCRATIC PEOPLE'S	MADAGASCAR	UNION OF SOVIET SOCIALIST
REPUBLIC OF KOREA	MALAYSIA	REPUBLICS
DENMARK	MALI	UNITED ARAB EMIRATES
DOMINICAN REPUBLIC	MAURITIUS	UNITED KINGDOM OF GREAT
ECUADOR	MEXICO	BRITAIN AND NORTHERN
EGYPT	MONACO	IRELAND
EL SALVADOR	MONGOLIA	UNITED REPUBLIC OF
ETHIOPIA	MOROCCO	CAMEROON
FINLAND	NETHERLANDS	UNITED REPUBLIC OF
FRANCE	NEW ZEALAND	TANZANIA
GABON	NICARAGUA	UNITED STATES OF AMERICA
GERMAN DEMOCRATIC REPUBLIC	NIGER	URUGUAY
GERMANY, FEDERAL REPUBLIC OF	NIGERIA	VENEZUELA
GHANA	NORWAY	VIET NAM
GREECE	PAKISTAN	YUGOSLAVIA
GUATEMALA	PANAMA	ZAIRE
HAITI	PARAGUAY	ZAMBIA
	PERU	

The Agency's Statute was approved on 23 October 1956 by the Conference on the Statute of the IAEA held at United Nations Headquarters, New York; it entered into force on 29 July 1957. The Headquarters of the Agency are situated in Vienna. Its principal objective is "to accelerate and enlarge the contribution of atomic energy to peace, health and prosperity throughout the world".

Printed by the IAEA in Austria
February 1978

PANEL PROCEEDINGS SERIES

FROM IDEA TO APPLICATION
Some selected nuclear techniques
in research and development

PROCEEDINGS OF AN ADVISORY GROUP MEETING ON
APPLIED NUCLEAR PHYSICS
ORGANIZED BY THE
INTERNATIONAL ATOMIC ENERGY AGENCY
AND HELD IN SAN JOSÉ, COSTA RICA, 9–13 MAY 1977

INTERNATIONAL ATOMIC ENERGY AGENCY
VIENNA, 1978

FROM IDEA TO APPLICATION: SOME SELECTED
NUCLEAR TECHNIQUES IN RESEARCH AND DEVELOPMENT
IAEA, VIENNA, 1978
STI/PUB/476
ISBN 92–0–131078–1

FOREWORD

In May 1977, the International Atomic Energy Agency organized an Advisory Group meeting on applied nuclear physics. At the kind invitation of the Government of Costa Rica the meeting took place at the University of Costa Rica, in San José. The aim of the Advisory Group was to study some selected nuclear techniques that have emerged from fundamental research in atomic and nuclear physics and have found their way into applied research and technological processes.

An important aspect of the Group's work was to examine these techniques for their suitability for application in modest laboratories, especially in developing countries. The contributions prepared by the participants for the meeting consisted of review papers endeavouring to trace the evolution of the described techniques from the time when they were used exclusively for fundamental studies to the present when many of them are firmly established as valuable tools in applied research and technological development. The papers pay particular attention to work and activities where these techniques are seen to have been successful or even uniquely effective. It is hoped that this approach will arouse the reader's interest and prove particularly valuable to those scientists intending to introduce a certain nuclear technique or looking for a suitable method to solve a particular problem they may be faced with.

The list of techniques included in these proceedings is of course not complete. The subjects of beam-foil spectroscopy and synchrotron radiations were discussed at the meeting but are not included here, since they require extensive equipment or multimillion dollar accelerators and are thus not of 'primary' interest to modest laboratories. Mössbauer spectroscopy is certainly one of the most popular inexpensive techniques; the subject played an important part in the meeting, but it was felt that a review paper was not needed at present. Amongst other useful books available, the reader may be referred to 'Mössbauer Spectroscopy and its Applications' (STI/PUB/304), published in 1972 by the IAEA in its Panel Proceedings Series.

The whole family of techniques based on nuclear or electron resonances, such as NMR (nuclear magnetic resonance), ENDOR (electron nuclear double resonance), EPR (electron paramagnetic resonance) and NQR (nuclear quadrupole resonance), was not represented at the meeting. They were the subject of an IAEA Panel in 1973, and a series of review papers, in a form much similar to the present set, was published in Atomic Energy Review, 1974, Vol.12, No.4.

Among the techniques represented here is 'nuclear track formation', despite the fact that a book was recently published on this topic. It is felt that this cheap and widely applicable technique still does not receive the attention it deserves. Neutron scattering is represented because the observation has been made that scientists at many research reactors where neutron diffractometers and spectrometers are in operation would welcome advice on how to utilize their instrumentation for applied studies.

CONTENTS

Charged-particle activation analysis 1
    E.A. Schweikert
Spatially sensitive analytical techniques 33
    T.B. Pierce
Detection of characteristic X-rays: methods and applications 49
    V. Valković
Neutron absorption physics in the development and practice of
    activation analysis 81
    S.S. Nargolwalla
Recent analytical applications of neutron-capture gamma-ray
    spectroscopy 125
    J.A. Lubkowitz, M. Heurtebise, H. Buenafama
Applications of positron annihilation 151
    P. Hautojärvi, A. Vehanen
Applications of thermal neutron scattering 183
    G. Kostorz
Developments in ion implantation 215
    D.W. Palmer
Track formation: principles and applications 261
    M. Monnin
List of participants 299

CHARGED-PARTICLE ACTIVATION ANALYSIS

E.A. SCHWEIKERT
Center for Trace Characterization,
Texas A & M University,
College Station, Texas,
United States of America

Abstract

CHARGED-PARTICLE ACTIVATION ANALYSIS.
The paper discusses the methodology and application of nuclear activation with ion beams ($1 \leqslant Z \leqslant 5$) of energies in the range of 2 to 10 MeV/amu as a tool for chemical characterization. With these excitation parameters, radioactive products are obtained from virtually all stable elements.

Light element analysis. Oxygen, carbon and nitrogen can be determined at levels as low as a few pp10^9 via $^{16}O(^3He,p)^{18}F$, $^{12}C(^3He,\alpha)^{11}C$ and $^{14}N(p,\alpha)^{11}C$ respectively. Recently, triton activation has been shown to be inherently still superior to ^{3}He activation for the determination of oxygen [$^{16}O(^3H,n)^{18}F$]. Lithium, boron, carbon and sulphur can be detected rapidly, nondestructively and with high sensitivity (~ 0.25 ppm for Li and B) via "quasi-prompt" activation based on the detection of short-lived, high-energy beta emitters (10 ms $\leqslant T_{\frac{1}{2}} \leqslant$ 1 s). A new development is heavy ion activation, where hydrogen is detected via $^1H(^7Li,n)^7Be$ for example.

Nondestructive multielement analysis. Proton activation has the inherent potential for meeting requirements of broad elemental coverage, sensitivity (ppm and sub-ppm range) and selectivity. Up to 30 elements have been determined in Al, Co, Ag, Nb, Rh, Ta and biological samples, using 12-MeV proton activation followed by gamma-ray spectrometry. These capabilities are further enhanced with the counting of X-ray emitters, 28 elements ($26 \leqslant Z \leqslant 83$) have been determined in different matrices by nondispersive X-ray counting.

Heavy element analysis. Elemental abundances of Tl, Pb and Bi can be measured with high sensitivity (~ 1 pp10^9) and accuracy using proton activation. $^{204}Pb/^{206}Pb$ ratios can also be determined with a relative precision of a few per cent. Although charged-particle activation analysis is a well-established trace analysis technique, broad potential capabilities remain to be explored, e.g. those arising from ultrashort-lived nuclides, heavy ion interactions and the combination of delayed and prompt methods.

1.0 INTRODUCTION

As early as 1938, Seaborg et al. [1] described a procedure for determining trace amounts of gallium in high purity iron based on deuteron activation. Despite this early example of charged particle activation analysis (CPAA), this chemical analysis technique did not receive much attention until the early sixties. The renewal in interest was largely due to the pioneering work of Sue [2], Albert [3], Gill [4] and Saito [5], showing the unique potential of this activation mode for determining light elements (e.g. boron, carbon, oxygen) at the sub-ppm level. These prospects were considerably enhanced with the introduction of ^{3}He activation by Markowitz in 1962 [6]. An additional factor essential to the development of this field has been the increasing availability of suitable sources of charged particles in recent years (cyclotrons and Van de Graaff accelerators). Thus, the scope of CPAA has been broadening rapidly during the past decade encompassing now elemental analysis procedures with protons, deuterons, tritons, ^{3}He, α and most recently even Li and B ions. With this variety of bombarding particles and given the energy range of interest here (2-10 MeV/amu), a great number of nuclear reactions can be induced, yielding radioisotopes with suitable decay characteristics. Indeed for virtually all stable elements a charged particle activation procedure can be devised which will feature high sensitivity (ppm to ppb level) and specificity.

The present paper discusses the methodology and applications of the charged particle technique based on delayed counting. Related and complementary prompt techniques, elastic and inelastic scattering and particle induced x-ray emission are treated elsewhere [7, 8]. The following topics are examined below: selection of activation reactions; experimental conditions and quantitative aspects; applications as a single and a multielement technique for low, medium and high Z elements; current status of CPAA and anticipated developments.

2.0 METHODOLOGY

2.1 Selection of Activation Reactions

The activation reactions must meet the analytical requirements of sensitivity, accuracy and experimental feasibility.

The considerations pertinent to the analytical sensitivity are based on the rate equations governing nuclear transformations and the decay of activation products. For charged particle induced reactions, it can be shown that the disintegration rate of a radionuclide in the target at the end of irradiation is:

$$D(t) = \frac{mN°f}{A} \int_0^{x_1} b_x \sigma_x dx (1 - e^{-\lambda t}) \qquad (1)$$

where:

- D: disintegrations per second
- m: weight of trace element
- $N°$: Avogadro's number
- f: isotopic abundance of the target nuclide
- x_1: sample thickness where the incident particle energy reaches a value equal to the threshold of the nuclear reaction
- b_x: intensity of particle beam at depth x in the target

σ_x: reaction cross section at x in the target for an infinitely thin target dx
λ: decay constant of the nuclide
t: irradiation time
A: atomic weight of the trace element

Charged particle beam intensities are usually expressed in terms of the beam current, I, in microamperes. For singly charged particles, the relationship between b and I is as follows:

$$b = kI = 6.2 \times 10^{12} I \qquad (2)$$

Considering eq. 1, it follows that for an activation reaction to be of interest, it must feature a high yield, which depends upon the magnitude of the cross section and the isotopic abundance of the target element. The excitation function, i.e. the dependance of the cross section on the energy of the incident particle must be taken into account. In principle, the bombarding energy should be as high as the accelerator can deliver to obtain as large an integrated cross section as possible, with limitations set by the occurrence of interfering reactions (see below). Energies widely used in CPAA range from 5-20 MeV for protons, deuterons and He-3 ions and from 20-40 MeV for He-4 ions. Extensive information on excitation functions, reaction thresholds and Q-values is available in the literature, a useful recent compilation is that of Lange [9].

Several factors affect the accuracy. The radioactive product nuclide should have distinctive decay characteristics (type and energy of emissions, half life). This is a necessity in the case of purely instrumental analysis. Useful compilations of the data necessary for nuclide identification are those of Lederer [10], Seelmann-Eggebert [11], Erdtmann [12], the Nuclear Data Group of Oak Ridge [13], Ajzenberg-Selove [14] and Endt [15]. A review on the nuclear data requirements for activation analysis and the state of pertinent compilations has recently been made by Krivan [16]. The above requirement is much less stringent if the radionuclide of interest is separated chemically after irradiation. Then the only inherent limitation is that imposed by the half life which must be sufficient to carry out the separation.

Equally important from the standpoint of accuracy is the requirement that the reaction of interest have no or a minimum of primary and secondary interfering reactions [16]. Reaction thresholds or Q-values and Coulomb barriers must be considered in a survey of possible interferences.

Further, the experimental feasibility can be affected by the overall level of matrix activity induced by an activation procedure. In examining Q-values, it becomes apparent that proton and He-3 activation will be preferred modes of activation. Indeed discriminations of impurity vs. matrix activation is possible in many cases, based on the differences in Q-values for proton-induced reactions. In He-3 activation the thresholds for (^{3}He,n), (^{3}He,p), (^{3}He,α) are uniformly low; however, advantage can be taken of the

Coulomb barrier for determining low Z elements in high Z matrices [6]. No discrimination of impurity vs. matrix activation is possible with deuteron induced reactions which have low thresholds for too many elements, and with He-4 ions where all reactions have high thresholds. Activation reactions may also be rejected because of the impractical half life of the product nuclide. Other limitations common to all CPAA procedures are discussed below.

2.2 Experimental Considerations

Beam Monitoring. The items which require attention are the beam energy, intensity and homogeneity. Energies quoted for CPAA done with cyclotrons should be considered "nominal" only. Actual energies and energy resolution from the same accelerator may vary depending on the location in the beam transport system. A much more critical parameter, however, is the beam intensity as it translates into heat dissipation in the sample. Due to the heat generated by charged particle irradiation, the application of CPAA is severly limited or impossible for specimens with low melting points, poor thermal conductivity or high volatility. Steps for avoiding sample destruction due to excessive temperatures include: converting samples into targets of greater thermal stability (e.g. ashing of biological specimens, preparation of samples as thin targets), cooling and/or rotation of samples during irradiation and, for samples with large surface areas, sweeping or defocusing of beams. The homogeneity of the beam must also be considered along with its intensity. Great care must be taken to obtain a beam of uniform beam density. Given a typical 1 cm^2 sample area to be irradiated, normal focusing equipment should be sufficient to produce a uniform 1 cm^2 beam spot. A simple way of testing the beam density consists of subjecting a thin plastic sheet (e.g. mylar) to a very short irradiation, beam inhomogeneities that could not be tolerated are revealed by damage in plastic foil. Other approaches for achieving homogeneous beams rely on the use of diffuser foils or magnetic sweeping of a tightly focused beam. With a beam of uniform density and proper sample cooling, beam intensities of several $\mu A/cm^2$ have been applied. Upper limits for beam intensities vary depending on the nature and energy of the projectile and the thermal conductivity of the target [17]. Beam intensities can be estimated based on the activities accumulated in thin foils placed on top of the sample. These might simply serve as flux monitors or might be the standards used for quantitative calculations (see below). Instrumentation for real time monitoring of beam intensity and homogeneity should include a Faraday cup and an additional device capable of sensing at least gross beam inhomogeneities that can develop during bombardment. A collimator made of four electrically isolated segments was found adequate in our work [18].

Samples and Their Handling. The physical characteristics of samples, their cleaning and the preservation of their integrity need to be considered. In the case of solids, the area to be irradiated should be flat and polished. As already indicated, a typical dimension for the beam spot is 1 cm^2. For powders, pelletizing is recommended; graphite and boric acid are excellent binding agents [19]. Liquids can be encapsulated in thin-window cells and can, with proper precaution, be irradiated in vacuum or in air [20, 21]. Gases have also been activated using flow-through irradiation cells [22]. In the case of solids, one of the significant advantages of activation analysis is the possibility of removing surface contamination with a post-irradiation mechanical or chemical etch deep enough to remove the activated surface layer including recoil activity. The etching procedure must be carefully designed in the case of sub-ppm determinations particularly of such elements as oxygen or carbon to avoid errors due to "grinding in" of contaminants by mechanical abrasion or readsorption of activity onto the sample surface in the chemical etch [23]. Several authors have found that radiation damage can result in differences in the etching rate. Even if only a very small fraction of the sample surface remains incompletely cleaned, this can significantly affect the results. The accuracy of the analysis is critically dependent on the preservation of the sample integrity during particle bombardment. Partial or total losses of the species of interest can occur due to sample heating, diffusion, recoil. Conversely, additional impurities are driven into a sample by recoil during bombardment. "Anomalous" diffusion of some trace elements in certain metal under charged particle bombardment has also been observed [24].

Irradiation Set-up. The hardware components needed at the irradiation site are summarized in Figure 1. This schematic diagram shows the simple set-up for a single target [25]. The sample holder may be replaced with a rotating target wheel accomodating multiple samples, or a thin window end plate may be substituted for irradiations in air. This latter arrangement is particularly well suited for work where a fast sample transfer system is needed [26, 27], or for in-situ counting of very short-lived radioisotopes [28]. Finally the target holder may be replaced with a target chamber (Figure 2) when prompt radiation measurements are coupled with nuclear activation [29]. The exploitation of very short-lived nuclides ($T_{\frac{1}{2}}$ < 1 sec) requires pulsing of the charged particle beam to achieve maximum build-up of the nuclides of interest while minimizing the production of longer lived species. Clearly, irradiations must be carried out under precisely reproducible conditions. Several methods have been described for Van de Graaff accelerators [30, 31], for cyclotrons the pulsing of the Dee voltage was found to be very satisfactory [28].

Counting. For γ-ray spectrometry Ge(Li) detectors are used, NaI(Tl) are preferred for measuring β^+ annihilation radiation by γ-γ coincidence counting. X-ray counting is carried out with thin Ge(Li) and Si(Li) detectors [19]. Special counting techniques have been described for high energy β-emitters, using Cerenkov detectors [32] or thin plastic scintillators [33]. For data reduction, numerous γ-ray spectral analysis programs are available (e.g. 34, 35), for decay curve analysis the "CLSQ" program of Cumming is widely used [36].

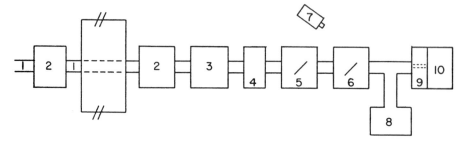

1. 4 INCH ID ALUMINUM PIPE FOR BEAM TRANSPORT
2. QUADROPOLE FOCUSING MAGNET
3. HIGH VACUUM TURBINE PUMP
4. ISOLATION VALVE
5. GRAPHITE BEAM MONITOR FOR TOTAL BEAM CURRENT MEASUREMENT
6. ALUMINUM OXIDE BEAM VIEWER
7. TELEVISION CAMERA
8. ROUGHING PUMP
9. GRAPHITE COLLIMATOR
10. SAMPLE HOLDER

FIG.1. Experimental configuration of cyclotron terminal for activation analysis [25].

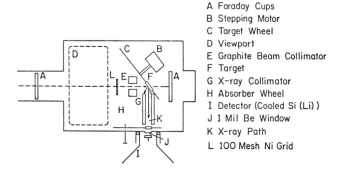

A Faraday Cups
B Stepping Motor
C Target Wheel
D Viewport
E Graphite Beam Collimator
F Target
G X-ray Collimator
H Absorber Wheel
I Detector (Cooled Si (Li))
J 1 Mil Be Window
K X-ray Path
L 100 Mesh Ni Grid

FIG.2. Set-up for combined nuclear and atomic activation [29].

2.3 Quantitative Aspects

Different methods have been devised for quantitative calculations applicable to thick targets. Among these is the average cross section ($\bar{\sigma}$) concept of Ricci et al. [37]. $\bar{\sigma}$ is defined as follows:

$$\bar{\sigma} = \frac{\int_0^R \sigma_x dx}{\int_0^R dx} = \frac{1}{R} \int_0^R \sigma_x dx \qquad (3)$$

where σ_x is the value of the cross section at depth x in the target and R is the range of the particles in that target. For activation at saturation, one obtains:

$$D = bn\bar{\sigma}R \tag{4}$$

where D is the number of disintegrations per second, b is the intensity of the particle beam, n is the number of target atoms in the irradiated sample. Obviously $\bar{\sigma}$ is constant for a given reaction at a given incident particle energy. A comparison of the disintegration rates of a sample, u, and a standard, s, both irradiated under the same conditions gives then:

$$\frac{D_u}{D_s} = \frac{b_u n_u R_u}{b_s n_s R_s} \tag{5}$$

Equation 5 indicates also that the simple relative comparison method can be used in CPAA provided an appropriate standard is available and range corrections are made. Although inherently simple, the comparison method can only be applied if the trace element is known to be homogeneously distributed in the sample used as standard. The internal standard method, subject to the same requirement, can also be used [38]. Another approach was developed by Engelmann [39]. It is based on the concept of an equivalent thickness, e, over which the specific activity of the trace species would be constant. In this case, referring to eq. 1 given before, the term

$$\int_0^{x_1} b_x \sigma_x dx$$

may be expressed as

$$eb_0 \sigma_0$$

where b_0 is the intensity of the incident particle beam and σ_0 the cross section for an infinitely thin target (dx) irradiated at the incident energy. It can be readily shown that when the sample weight is expressed as a function of e, a direct comparison between the weighed activities induced in a thin monitor foil of known composition and in the sample can be made. The values for σ_x, σ_0 are derived from an experimentally measured activation curve (i.e. relative excitation function) for a given nuclear reaction in a given matrix. An important feature of this procedure is to minimize systematic errors due to differences in beam characteristics for different accelerators and irradiation positions [39-42]. Rook et al. have shown that the equivalent thickness method is readily usable in practice since an activation curve determined experimentally in a matrix can be transformed into the corresponding activation curve in any other matrix by using the differential range-energy relationship [43]. Values for ranges can be obtained from the tables of Williamson et al. [44].

An important aspect of quantitative work concerns the validity of the numbers obtained. As far as the accuracy is concerned, there are two sources of absolute errors: interfering reactions and nonhomogeneous impurity distribution. The methods of quantitation used assume a uniform impurity distribution. CPAA is however inherently very sensitive to concentration gradients [45]. A number of different factors affect the precision, these will only be briefly listed here:

- Integrity of the standards
- Resolution of the activation or excitation functions (well defined curves are necessary for accurate integration and comparison with thin monitor foils)
- Resolution of the beam energy and current measurements
- Quality of the post-irradiation etch
- Measurement errors in the sample thickness before and after etching
- Counting statistics and decay curve analysis. A selection has to be made among a number of components, which could be present, to retain only those which are effectively major contributors of the observed gross activity. (It might be noted here that excellent fits can be obtained for any decay curve provided a large enough number of components are considered even if those components have wrong half-life values). In summary, satisfactory accuracy and precision may be achieved with repeated analyses using whenever possible different nuclear reactions.

TABLE I. ACTIVATION REACTIONS

Activation Reaction	Interfering Reactions	Maximum "Safe" Energy
$^{18}O(p,n)^{18}F$	$^{19}F(p,pn)^{18}F$	10 MeV
$^{16}O(^{3}He,p)^{18}F$ $^{16}O(^{3}He,n)^{18}Ne \longrightarrow ^{18}F$	$^{19}F(^{3}He,\alpha)^{18}F$, $^{20}Ne(^{3}He,\alpha n)^{18}Ne \longrightarrow ^{18}F$, $^{23}Na(^{3}He,2\alpha)^{18}F$, $^{24}Mg(^{3}He,2\alpha p)^{18}F$, $^{27}Al(^{3}He,3\alpha)^{18}F$, $^{29}Si(^{3}He,^{14}N)^{18}F$	15 MeV
$^{16}O(\alpha,pn)^{18}F$	$^{15}N(\alpha,n)^{18}F$, $^{19}F(\alpha,\alpha n)^{18}F$, $^{20}Ne(\alpha,\alpha d)^{18}F$, $^{23}Na(\alpha,\alpha 2n)^{18}F$, $^{24}Mg(\alpha,2\alpha d)^{18}F$, $^{27}Al(\alpha,3\alpha n)^{18}F$, $^{28}Si(\alpha,^{14}N)^{18}F$, $^{29}Si(\alpha,^{15}N)^{18}F$	< 35 MeV

3.0 SCOPE

As noted before, the major thrust of CPAA has been and still is in the field of light element analysis [17, 46], but growing attention has been given to the trace detection of medium and high Z elements, from calcium to lead. No exhaustive review can be given here, useful further information can be found in references [47-57]. As far as applications are concerned, CPAA has been found suitable for solving many problems in different fields. A number of reports deal with the characterization of semiconductor materials, more specifically the determination of traces of lithium, boron, carbon, nitrogen or oxygen in silicon [40, 58-76], in germanium [63, 77-80], in gallium phosphide [80, 81] or in chalcogenide glasses [82]. Other work has dealt with the assay of environmental samples [55], biological specimens [83], geological materials [84], petroleum products [20, 21] and many high purity metals [19, 52-54, 57, 85-87]. Some applications of CPAA studied in our laboratory are further outlined below, followed by a comparison with related techniques.

3.1 Single Element Analysis

The flexibility in experimental parameters (i.e. type and energy of bombarding particles) can be taken advantage of to optimize an analysis procedure for a given problem. The examples chosen illustrate the different requirements that accompany different applications: accurate ultratrace analysis (< ppm) in support of solid state research, trace analysis (> ppm) applicable to a large number of varied samples, isotopic ratio measurements.

3.1.1 Determination of Oxygen

The activation reactions successfully used in practice are summarized in Table I. Although the number of possible interfering reactions appears overwhelming, these are in almost all cases negligible for bombarding energies up to the maximum "safe" energies indicated. Reasons follow:

- Fluorine, neon, sodium, magnesium and aluminum are unlikely to be present or at much lower concentration levels than oxygen [17, 69].
- These interferences are further minimized in the case of ^{3}He activation, where the ^{18}F yields from these elements are one or several orders of magnitude smaller than that resulting from $^{16}O(^3He,p)^{18}F$ [81].
- The relative effects of interfering reactions vary widely with different activating particles. The possibility of errors due to interferences can thus also be checked a posteriori by comparing results obtained using different activation modes. The agreement on the data obtained, in the case of

TABLE II. OXYGEN IN SILICON 10–15 ppm LEVEL, σ AT THE 95% CONFIDENCE LEVEL

Sample	Particle	Energy (MeV)	PPM	σ
1	^{4}He	34	12.0	
1	^{4}He	33	13.7	
1	^{1}H	12	11.9	
AVERAGE 1			12.5	± 1.0
2	^{3}He	12	15.4	
	^{1}H	10	14.0	
AVERAGE 2			14.7	± 0.8

TABLE III. OXYGEN IN SILICON 1–10 ppm LEVEL

Sample	34 MeV α Activation (ppm)	Infrared Spectroscopy* (ppm)
622368	10.9	9.4
623364	8.3	7.4
623363	5.6	5.7
623352	4.7	4.8

*Data supplied by J. A. Baker, Dow Corning Corp., Hemlock, MI., USA

silicon (see results below), confirms again that these errors are negligible at the oxygen levels measured thus far (≥ 10 ppb).
- In silicon, the effect of the "fission" reaction Si(^{3}He,x)^{18}F is negligible for ^{3}He energies up to 15 MeV for oxygen levels at least as low as ~ 1 ppb [17, 69]. When α activation, (E_d > 35 MeV), is applied on silicon, Si(α,x)^{18}F precludes oxygen measurements below 0.05 ppm [17].

A post-irradiation etch is absolutely necessary for trace determinations of oxygen, carbon or nitrogen. This step and other items to be considered in sample handling have been described earlier. In the case of silicon, the CPAA procedure is nondestructive due to (a) the short-lived matrix activities ($T_{\frac{1}{2}}$ ≤ 2.5 min) allowing the detection of trace species with $T_{\frac{1}{2}}$ ≥ 20 min (e.g. ^{18}F), and (b) its high purity. For many other materials (e.g. Ge, Al, chalcogenide glasses), sub-ppm determinations cannot be carried out nondestructively. A post-irradiation chemical separation of ^{18}F from other activities generated in the target is needed. Several authors have described such procedures (e.g. 72, 77).

TABLE IV. EXPERIMENTAL DETECTION LIMITS FOR THE DETERMINATION OF OXYGEN IN SILICON (IRRADIATION OF 2 h AND COUNT RATE OF 25 counts/min AT t_0).

Reaction	Beam Current $\mu A/cm^2$	Max. Irradiation Energy (MeV)	Detection Limit (ppm)
$^{18}O(p,n)^{18}F$	10	12	0.06
$^{16}O(\alpha,pn)^{18}F$	4	34	0.05
$^{16}O(^3He,p)^{18}F$	4	15	0.005

Once a method is developed (including counting and quantitation steps), its successful application depends on a careful evaluation of the accuracy, precision and detection limit that can be achieved under experimental conditions. The case of oxygen analysis in silicon can be used as an illustration. On samples containing oxygen at the 10 ppm level, comparisons of results from different activation modes and from the totally independent IR absorption spectroscopy are possible. Data from duplicate analyses on some samples are given in Tables II and III. Results in Table II verify that for the concentration range considered, errors due to interfering reactions are minimized. The agreement between the results by α activation and by IR validates the activation method as a whole at the 10 ppm level. For oxygen concentrations below 0.5 ppm only CPAA can be applied. With repeated determinations on each sample, average deviations of + 30% to + 50% were obtained for oxygen levels ranging from 1 to 0.05 ppm [17]. For concentrations at or below the 10 ppm level (Table IV), the average deviation may reach + 100% due to the poor counting statistics. It must be emphasized that the CPAA data reflects accurately the ultratrace levels, improvements in the precision can be anticipated as more analyses on such ultrapure materials are made.

3.1.2 Quasi-prompt Activation

Ideally, an analysis technique should be sensitive, selective, rapid and require a minimum of sample handling. Quasi-prompt activation, i.e. the analytical exploitation of short-lived nuclides ($T_{\frac{1}{2}}$ < 1 sec) is, a priori, an attractive approach for meeting these requirements: only short irradiations are required, matrix activities should be minimized (many elements yield little or no very short-lived activity with charged particle beams), the samples remain intact, maximum sensitivity can be obtained with repetitive irradiation-counting cycles, the cost per sample analyzed should be low due to short beam time required and the elimination of any post-irradiation sample handling. Several light elements yield very short lived species under charged particle bombardment. Among these, only those that are products of interference-free reactions with adequate cross sections are of interest. An additional requirement

TABLE V. QUASI-PROMPT ACTIVATION[a]

Reaction	$T_{1/2}$ msec	$E_{\beta max}$	Interference-free Detection Limits ppm[b]	Experimental Detection Limits ppm[c]
$^{12}C(p,n)^{12}N$	11	16.4	0.25	5 in iron
$^{14}N(p,2n)^{13}O$	8.7	16	2000	
$^{24}Mg(p,n)^{24m}Al$	129	13.3	350	
$^{28}Si(p,n)^{28}P$	270	11.5	25	
$^{32}S(p,n)^{32}Cl$	297	9.4	30	75 in petroleum products
$^{46}Ti(p,n)^{46}V$	486	6.6	300	
$^{50}Cr(p,n)^{50}Mn$	288	6.1	450	
$^{54}Fe(p,n)^{54}Co$	194	7.3	500	
$^{7}Li(d,p)^{8}Li$	850	13.0	0.15	0.5 in germanium
$^{11}B(d,p)^{12}B$	20	13.4	0.25	See Table VI
$^{40}Ca(d,n)^{41}Sc$	600	5.5	1000	

a) Incident particle energy 40 MeV for $^{12}C(p,n)^{12}N$ and $^{14}N(p,2n)^{13}O$ reactions. All other reactions studied at 20 MeV.
b) Based on 10 repetitive irradiations at 2 μA and 100 counts for a count time equal to $T_{1/2}$ of the product nuclide.
c) Same irradiation conditions as for b).

TABLE VI. RESULTS OF BORON DETERMINATIONS AND EXPERIMENTAL LIMITS OF DETECTION WITH DEUTERON ACTIVATION [28]

Sample	Amount found (ppm)		Experimental detection limit (ppm)
	Quasi-prompt	Other methods	
NBS glass SRM 610	352 ± 7[a]	351[b]	2 – 3
NBS glass SRM 612	38 ± 5[a]	32[b]	2 – 3
Si IX 225	73 ± 5[a]	77[c]	0.5
Si IX 252	11 ± 2[a]	11[c]	0.5
Si IX 253	2.3 ± 0.7[a]	1 – 2[c]	0.5
NBS SRM 1571 (orchard leaves)	36 ± 5	33 ± 3[b]	0.75

[a] Average deviation based on 10 determinations.
[b] Non-certified value, National Bureau of Standards.
[c] Data supplied by F.A. Thrumbore, Bell Laboratories, Murray Hill, NJ, USA.

is that the radionuclide must emit characteristic γ-rays or be a high energy β-emitter. This is necessary for its selective identification which is based on the type and energy of the emission and the half-life. A corollary is that no other nuclides of similar or close decay characteristics should be produced concurrently. A survey of quasi-prompt activation reactions yielding nuclides with $10 \leq T_{\frac{1}{2}} \leq 1000$ msec and meeting the above criteria is given in Table V. Based on the thick target yields, proton or deuteron activation appears feasible for measuring lithium, boron, carbon, silicon or sulfur. The relative excitation functions for $^7Li(d,p)^8Li$ and $^{11}B(d,p)^{11}B$ show low threshold and peak energies, thus these reactions can also be used with lower energy accelerators (E_d = 5 MeV for ex.). Special experimental requirements, beam pulsing and selective and efficient detection of high energy β particles must be met. These have been addressed earlier. A series of results of boron determinations and experimental detection limits are given in Table VI. Widely different kinds of samples were assayed and because of the short beam time needed for each sample, large numbers of specimens can be processed readily. A typical decay curve from the analysis of boron in silicon is given in Figure 3. It shows that little matrix activity is produced in silicon with a 20 msec bombardment with 20 MeV deuterons. In the case of lithium, ppm sensitivity can also be achieved in very different matrices. Quasi-prompt procedures for carbon, sulfur and silicon are less sensitive with interferences between sulfur and silicon. Among the applications reported are the determination of carbon in steel [28] and of sulfur in petroleum products [20, 21].

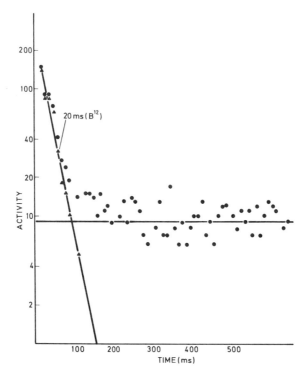

FIG.3. Quasi-prompt activation. Decay curve for boron in silicon (11.2 ppm B).

3.1.3 Isotopic Ratio Measurements

Among the elements with varying natural isotopic abundances, lead has received particular attention. Based on lead isotope ratios age determinations in geological specimens can be made [56], patterns of environmental pollution can be established with lead "fingerprints" [55]. This discussion reviews only procedures for determining $^{204}Pb/^{206}Pb$. Different activation reactions may be used for this purpose [57]. Examples of some relevant proton and deuteron reactions are listed in Table VII. The activation method cannot match the precision of mass spectrometry. But for applications where a precision of a few percent is sufficient, CPAA is preferred because of its simplicity (little or no sample preparation) and inherent accuracy (freedom from reagent blanks). Stable tracer and environmental studies are among such applications [55]. It should be noted that many charged particle activation reactions are available for total lead determinations, detection limits as low as 1 ppb have been reported [88]. Like in the case of the low Z elements, CPAA provides for some of the heaviest species, Tl, Pb, Bi, a unique combination of sensitivity, selectivity and applicability to many different kinds of samples [18].

TABLE VII. REACTIONS USED FOR LEAD ISOTOPE RATIO MEASUREMENTS [55]

Reaction	Threshold energy (MeV)	Half-life	Principal gamma rays (keV)	Interfering reactions
^{204}Pb(p,n)^{204}Bi	~10	11.2 h	376	^{206}Pb(p,3n)^{204}Bi if $E_p \geqslant 20$ MeV
^{206}Pb(p,2n)^{205}Bi	~12	15.3 d	703	^{207}Pb(p,3n)^{205}Bi if $E_p \geqslant 18.5$ MeV
^{204}Pb(d,2n)^{204}Bi	~8	11.2 h	376	^{206}Pb(d,4n)^{204}Bi if $E_d \geqslant 22$ MeV
^{206}Pb(d,2n)^{206}Bi	~8	6.24 d	803, 880	^{207}Pb(d,3n)^{206}Bi if $E_d \geqslant 13.5$ MeV

3.2 Nondestructive Multielement Analysis

Proton activation has the inherent potential for meeting the requirements of multielement analysis: broad elemental coverage, sensitivity (ppm and sub-ppm range) and selectivity. Moreover for nondestructive analysis, discrimination of impurity vs. matrix activation is possible in many cases, based on the differences in Q-values for proton induced reactions. Protons of 10 to 15 MeV energy are most advantageous, yielding high specific activities for (p,n) reactions. Possible interfering reactions [(p,α), (p,2n), (p,pn), (p,d)] have low cross sections at these proton energies and are thus (usually) negligible sources of errors. Multielement determinations can also be accomplished by ^{3}He activation [49]. Further, mention should be made of procedures utilizing fast neutrons produced by a high energy deuteron beam impinging on a beryllium target [89].

The feasibility of nondestructive multielement analysis must be evaluated on a case by case basis. Several items affect the performance of CPAA: (a) Activities due to major, minor and/or trace components of the matrix impose the actual detection limits that can be achieved with γ-ray or x-ray spectrometry. These are also dependent on the number of countings, i.e. on the extent that detection conditions can be optimized given the half lives of the different radionuclides involved. (b) Gamma-ray interferences may occur [50, 52]. Errors can be avoided by identifying peaks via decay curve analysis and/or by using several γ-rays to measure the same nuclide. (c) The magnitude of nuclear interferences can be readily assessed experimentally. Three examples of applications are described below, they illustrate the diversity of trace analysis problems that can be handled by CPAA.

TABLE VIII. NONDESTRUCTIVE TANTALUM ANALYSIS [50][a]

Activation Data			Gamma-ray Interferences[b]				Experimental Detection Limit in Ta, ppm[c]
Activation Reaction	$T_{1/2}$	$E\gamma$, MeV	$E\gamma$, MeV	Nuclide	$T_{1/2}$	Mode of Production	
$^{48}Ti(p,n)^{48}V$	16.0 d	0.983 1.312					0.02
$^{56}Fe(p,n)^{56}Co$	77.3 d	0.847 1.238 1.771 2.598	0.850	^{96}Tc	4.3 d	$^{96}Mo(p,n)^{96}Tc$	0.1
$^{90}Zr(p,n)^{90}Nb$	14.6 h	1.129 2.319 0.141	1.127	^{96}Tc	4.3 d	$^{96}Mo(p,n)^{96}Tc$	0.005
$^{93}Nb(p,n)^{93m}Mo$	6.9 h	0.263 0.685 1.477	0.265	^{182}Ta	115 d	$^{181}Ta(n,\gamma)^{182}Ta$	0.3
$^{94}Mo(p,n)^{94}Tc$	4.8 h	0.850 0.702 0.871	0.847	^{56}Co	77.3 d	$^{56}Fe(p,n)^{56}Co$	not measured
$^{95}Mo(p,n)^{96}Tc$	4.3 d	0.778 0.850 0.812	0.847	^{56}Co	77.3 d	$^{56}Fe(p,n)^{56}Co$	0.2
$^{182}W(p,n)^{182}Re$	13 h	1.122 1.222	1.122 1.222	^{182}Ta ^{182}Ta	115 d 115 d	$^{181}Ta(n,\gamma)^{182}Ta$	not meaningful
$^{182}W(p,2n)^{181}Re$	19 h	0.366					0.5

[a] Irradiation with 15 MeV protons.
[b] For Ep = 15 MeV, nuclear interferences were found to be negligible in practice.
[c] 2.5 hour bombarding time with a beam intensity of 2 μA, followed by a cooling period of 10 to 50 hours.

3.2.1 Determination of Some Trace Elements in High Purity Tantalum and Niobium

In tantalum, the impurities of interest in connection with metallurgical and physical studies are titanium, iron, zirconium, niobium, molybdenum and tungsten. It has been shown that proton activation is the only nondestructive technique available to date capable of detecting these elements in high purity tantalum [52]. Pertinent data on this analysis is given in Table VIII. For analyses, irradiations of 2 to 4 hours (15 MeV, 1 to 2 µA beam) were sufficient to determine the six trace elements of interest with a precision of better than 20% even at the 0.1 ppm level.

Niobium has several special properties which are, or may be, utilized in various scientific and industrial fields. Of particular interest is the superconductivity exhibited by niobium and its compounds. In this context, the purity grade of niobium plays an important role. Impurities of interest include: tantalum, tungsten, molybdenum, zirconium, iron, titanium, vanadium, chromium and hafnium. These nine trace elements can be determined simultaneously using 12 MeV proton activation [(p,n) reactions] followed by γ-ray and x-ray counting of the radionuclides. Except for tantalum, sub-ppm detection limits were obtained in all cases [87]. It is interesting to note that proton activation not only provides a nondestructive multielement capability but in the case of niobium, it is, excluding mass spectrometry, the most sensitive technique for the determination of titanium, vanadium, iron, zirconium and molybdenum.

3.2.2 Application of Proton Activation in Biological Samples

The importance of trace elements in biological systems has been recognized for many years. Some elements are known to play an essential role in the life process, while others are known to be toxic and therefore their concentration in biological materials, including food, are of great importance. In this context the analytical task is formidable, considering the multiplicity of trace elements of known or potential interest, the varying levels of concentration and the number of specimens to be assayed to validate any findings. These requirements cannot be fully met by any current analytical technique including proton activation, but it can provide a useful analysis capability. A survey of 12 MeV proton activation in botanical specimens and animal tissues has shown that 10 elements can be determined at the ppm level with a 1 hour 0.5 µA irradiation [83]. In this nondestructive approach only nuclides with half-lives longer than a few hours are useful as analytical signals. A decay time of at least 10 hours is required due to the high levels of matrix activities, the most important of these being ^{18}F ($T_{\frac{1}{2}}$: 109.7 min) which is formed by the reaction $^{18}O(p,n)^{18}F$. The major limitation for biological samples is the heat generation in the sample which prevents the use of high beam intensities (e.g. 1 µA) and thus limits sensitivities. Despite this, the detection limits (Table IX) are adequate for several essential trace elements (Cu, Fe,

TABLE IX. 12-MeV PROTON ACTIVATION APPLIED IN BIOLOGICAL SAMPLES [81]

Reaction	Half-life	γ-rays Used, MeV	Experimental Detection in Orchard Leaves ppm[a]
$^{75}As(p,n)^{75}Se$	120 d	0.280; 0.264	3
$^{44}Ca(p,n)^{44m}Sc$	2.44 d	0.271	500
$^{65}Cu(p,n)^{65}Zn$	243 d	1.115	3
$^{56}Fe(p,n)^{56}Co$	77.3 d	1.240; 0.849	6
$^{96}Mo(p,n)^{96}Tc$	4.3 d	0.813; 0.778	2
$^{206}Pb(p,n)^{206}Bi$	6.2 d	0.881; 0.803	10
$^{88}Sr(p,n)^{88}Y$	107 d	1.836	10
$^{48}Ti(p,n)^{48}V$	16 d	1.311; 0.980	0.2
$^{66}Zn(p,n)^{66}Ga$	9.5 h	2.760; 0.839	2
$^{92}Zr(p,n)^{90}Nb$	14.6 h	2.319; 1.129	2.5

[a] One hour irradiation of 0.6 μA followed by a cooling period of 10 h.

Mo and Zn, see 3.3 below). These possibilities could be substantially enlarged with chemical separations, lead and strontium would be among the elements which could be detected with great sensitivity.

An alternate approach would consist of using only very short irradiations and basing the analysis on short lived radioisotopes. Debrun et al. [27] have investigated such possibilities based on the measurement of γ-ray emitters with half lives ranging from 1 sec. to 1 min. One or several of the following elements, Se, Br, Y, Zr, La, Pr, Dy and Nd, could be detected with 10 to 17 MeV proton irradiations of a few seconds. The respective detection limits vary with the major, minor and trace component make-up of the matrix. In the case of biological materials, they are estimated to be in the tens to hundreds of ppm.

3.2.3 Multielement Assays With X-ray Counting

The possibilities of CPAA and γ-ray spectrometry can be expanded with delayed x-ray counting. Indeed most of the radionuclides of medium and high Z elements, produced by charged particle reactions, decay principally by internal conversion or electron capture and are thus predominantly x-ray emitters. The main features of nuclear activation followed by nondispersive x-ray spectrometry (using a Si(Li) or thin Ge(Li) detector) are: the direct relationship between the x-ray energy and the atomic number of the pertinent element; the relatively simple structure of x-ray spectra (in comparison with γ-ray spectra); the availability of

TABLE X. 20-MeV PROTON ACTIVATION WITH X-RAY COUNTING [19]

Reaction	Half-life	Q-Value, MeV	X-ray detected (energy in keV)	Interference-free[a] detection limits, μg
^{56}Fe(p,n)^{56}Co	78.5 days	−5.4	Fe Kα (6.4)	1.0×10^{-2}
^{69}Ga(p,n)^{69}Ge	120 days	−1.7	Ga Kα (9.2)	4.0×10^{-1}
^{75}As(p,n)^{75}Se	19.5 hr	−5.4	As Kα (10.5)	1.9×10^{-2}
^{76}Se(p,n)^{76}Br	4.4 hr	−2.7	Se Kα (11.2)	2.6×10^{-2}
^{85}Rb(p,n)^{85m}Sr	67.7 min	−2.0	Sr Kα (14.1)	4.4×10^{-3}
^{86}Sr(p,γ)^{87}Y		>0		
^{87}Sr(p,n)^{87}Y	80.3 hr	−2.5	Sr Kα (14.1)	6.4×10^{-4}
^{86}Sr(p,n)^{86m}Y	2.6 hr	−6.1	Y Kα (14.9)	3.8×10^{-2}
^{89}Y(p,n)^{89}Zr	78.5 hr	−3.6	Y Kα (14.9)	7.4×10^{-4}
^{92}Mo(p,γ)^{93}Tc		>0		
^{94}Mo(p,2n)^{93}Tc	2.7 hr	−13.6	Mo Kα (17.5)	4.2×10^{-3}
^{94}Mo(p,γ)^{95}Tc		>0		
^{95}Mo(p,n)^{95}Tc	20.0 hr	−2.4	Mo Kα (17.5)	1.0×10^{-4}
^{103}Rh(p,p')^{103m}Rh	57.0 min	0	Rh Kα (20.1)	3.4
^{102}Pd(p,γ)^{103}Ag	67 min	>0		
^{104}Pd(p,2n)^{103}Ag		−13.4	Pd Kα (21.1)	2.8×10^{-1}
^{104}Pd(p,n)^{104}Ag	68 min	−5.1		
^{107}Ag(p,n)^{107}Cd	6.5 hr	−2.2	Ag Kα (22.1)	1.1×10^{-2}
^{110}Cd(p,n)^{110}In	4.9 hr	−4.7	Cd Kα (23.1)	1.0×10^{-1}
^{110}Cd(p,γ)^{111}In		>0		
^{111}Cd(p,n)^{111}In	2.8 days	−2.8	Cd Kα (23.1)	5.8×10^{-4}
^{112}Cd(p,2n)^{111}In		−11.2		
^{113}In(p,n)^{113m}Sn	20.0 min	−1.8	Sn Kα (25.1)	8.6×10^{-2}
^{115}In(p,p')^{115m}In	4.4 hr	0.0	In Kα (24.1)	2.3×10^{-2}
^{116}Sn(p,γ)^{117}Sb		>0		
^{117}Sn(p,n)^{117}Sb	2.8 hr	−2.6	Sn Kα (25.1)	8.4×10^{-4}
^{118}Sn(p,2n)^{117}Sb		−11.9		
^{118}Sn(p,γ)^{119}Sb		>0		
^{119}Sn(p,n)^{119}Sb	38.0 hr	−1.4	Sn Kα (25.1)	3.2×10^{-4}
^{120}Sn(p,2n)^{119}Sb		−7.8		
^{121}Sb(p,n)^{121}Te	17 days	−2.1	Sb Kα (26.3)	2.4×10^{-1}
^{133}Cs(p,n)^{133m}Ba	38.9 hr	−1.3	Cs Kα (30.8)	4.5×10^{-1}
^{139}La(p,n)^{139}Ce	137 days	−1.1	La Kα (33.3)	5.8×10^{-1}
^{140}Ce(p,2n)^{139}Pr	4.4 hr	11.8	Ce Kα (34.5)	2.2×10^{-3}
^{142}Ce(p,n)^{142}Pr	19.6 hr	−1.7	Ce Kα (34.5)	7.8×10^{-4}
^{141}Pr(p,2n)^{140}Nd	3.37 days	10.5	Pr Kα (35.8)	1.5×10^{-4}
^{141}Pr(p,n)^{141}Nd	2.5 hr	−2.6	Pr Kα (35.8)	9.7×10^{-4}
^{151}Eu(p,n)^{151}Gd	120 days	−1.2	Eu Kα (41.3)	6.6×10^{-4}
^{153}Eu(p,p')^{153m}Eu	96 min	0	Eu Kα (41.3)	8.6×10^{-3}
^{155}Gd(p,n)^{155}Tb	5.1 days	−1.7	Gd K$\alpha_{1,2}$ (42.3; 43.0)	
^{156}Gd(p,n)^{156}Tb	5.3 days	−3.2	Gd K$\alpha_{1,2}$ (42.3; 43.0)	2.1×10^{-4}
^{160}Dy(p,n)^{160m}Ho	4.8 hr	−4.1	Ho K$\alpha_{1,2}$ (46.7; 47.5)	4.3×10^{-3}
^{161}Dy(p,2n)^{160m}Ho		−10.5		
^{165}Ho(p,pn)^{164m}Ho	37 min	8.04	Ho K$\alpha_{1,2}$ (46.7; 47.5)	3.5×10^{-2}
^{165}Ho(p,n)^{165}Er	10.3 hr	−1.2	Ho K$\alpha_{1,2}$ (46.7; 47.5)	3.0×10^{-4}
^{164}Er(p,2n)^{163}Tm	1.8 hr	11.8	Er K$\alpha_{1,2}$ (48.2; 49.1)	1.5×10^{-3}
^{166}Er(p,γ)^{167}Tm		>0		
^{167}Er(p,n)^{167}Tm	0.3 days	−1.7	Er K$\alpha_{1,2}$ (48.2; 49.1)	1.5×10^{-4}
^{168}Er(p,2n)^{167}Tm		−9.5		
^{171}Yb(p,n)^{171}Lu	8.2 days	−3.7	Yb K$\alpha_{1,2}$ (51.3; 52.3)	1.2×10^{-1}
^{172}Yb(p,2n)^{171}Lu		−10.5		
^{175}Lu(p,n)^{175}Hf	70 days	−1.7	Lu K$\alpha_{1,2}$ (52.9; 54.1)	3.3×10^{-4}
^{177}Hf(p,n)^{177}Ta	56.4 hr	−1.9	Hf K$\alpha_{1,2}$ (54.6; 55.8)	2.1×10^{-4}
^{178}Hf(p,n)^{178}Ta	2.2 hr	−2.7	Hf K$\alpha_{1,2}$ (54.6; 55.8)	1.3×10^{-3}
^{182}W(p,n)^{182}Re	64.0 hr	−3.7	W K$\alpha_{1,2}$ (57.9; 59.3)	1.6×10^{-4}
^{187}Os(p,γ)^{188}Ir		>0		
^{188}Os(p,n)^{188}Ir	41.0 hr	−3.6	Os K$\alpha_{1,2}$ (61.4; 63.0)	1.9×10^{-3}
^{190}Os(p,n)^{190}Ir	3.2 hr	−2.8	Os K$\alpha_{1,2}$ (61.4; 63.0)	5.6×10^{-4}
^{191}Ir(p,n)^{191}Pt	3.0 days	−1.5	Ir K$\alpha_{1,2}$ (63.3; 64.9)	1.6×10^{-3}
^{191}Ir(p,pn)^{190}Ir	3.1 hr	−7.6	Ir K$\alpha_{1,2}$ (63.3; 64.9)	8.2×10^{-4}
^{194}Pt(p,n)^{194}Au	39.0 hr	−3.3	Pt K$\alpha_{1,2}$ (65.1; 66.8)	1.4×10^{-3}
^{197}Au(p,n)^{197}Hg	64.1 hr	−1.6	Au K$\alpha_{1,2}$ (67.0; 68.8)	7.7×10^{-3}
^{203}Tl(p,n)^{203}Pb	52.1 days	−1.6	Tl K$\alpha_{1,2}$ (70.8; 72.8)	7.3×10^{-3}
^{204}Pb(p,p')^{204m}Pb	66.9 min	0	Pb K$\alpha_{1,2}$ (72.8; 74.9)	2.6×10^{-2}
^{206}Pb(p,n)^{206}Bi	6.2 days	−4.4	Pb K$\alpha_{1,2}$ (72.8; 74.9)	8.4×10^{-4}

[a] Based on a 3 μA irradiation for 3 hours or one half-life of the product nuclide (whichever is shorter) and a count time of one half-life of the product nuclide.

radioactive decay rates as an additional criterion of identification. These advantages must be weighed against possible limitations arising from self-absorption and enhancement effects common to all x-ray techniques. An additional limitation proper to nondispersive x-ray counting of radioactive samples arises when β activity is present, which results in increased background. This bremsstrahlung background can be reduced with absorbers (accompanied by a corresponding loss in low energy x-rays) or eliminated by magnetic deflection of the β particles [90]. 20 MeV proton activation shows broad elemental coverage and capability: 33 elements can be measured with detection limits from 10^{-4} µg to 1 µg (Table X). These values are for interference-free conditions. Nuclear interferences are, in most cases, small (< 10%) even with bombarding energies of 20 MeV, many can be totally eliminated by reducing the bombarding energy to 10 MeV. X-ray interferences can occur, i.e. overlaps among x-rays from adjacent elements or between K and L x-rays of similar energies. However, the half-lives of the nuclides involved are different and proper assignments can be made via decay curve analysis. Possible limitations due to β-activity and remedies have already been mentioned.

The applicability of this technique for multielement trace analysis has been tested on NBS glass samples. These glasses are doped with 61 trace elements and their major constituents (SiO_2, CaO, Na_2O and Al_2O_3) yield intense, long-lived β and γ-activity following proton bombardment. They can thus be viewed as fairly representative of a variety of difficult matrices including geological and biological specimens. In the case of a glass with an average trace element concentration of ~ 40 ppm (NBS SRM 612), 26 elements were determined by x-ray spectrometry following a 30-min. irradiation at 0.5 µA. Spectra from a glass with trace element concentrations in the 300 to 400 ppm range (NBS SRM 610) obtained after different decay times are given in Figure 4. They illustrate the combination of x-ray spectrometry and decay rates, which yields quantitative data on a truly large number of elements. The applications best suited for this procedure would include the determination of medium and high Z impurities in samples with major constituents composed of low Z elements ($1 \leq Z \leq 14$, e.g. B, Be, C, Al, Si, their carbides, nitrides, oxides, rocks, organic compounds, biological material).

3.3 Comparison of CPAA With Other Techniques

As CPAA applications are developed, they must be validated and justified on a performance basis. Some comments on how CPAA does and might fit into the broad arsenal of trace analysis tools follow.

CPAA has received most attention because of its high sensitivity and accuracy for light elements, particularly B, C, N, O. A further distinctive trait of CPAA is its applicability as an in-depth (bulk) characterization method. "Classical" methods adequate also for bulk probing and applicable to B, C, N or O include infrared spectroscopy, vacuum fusion, mass spectrometry and electrical resistivity

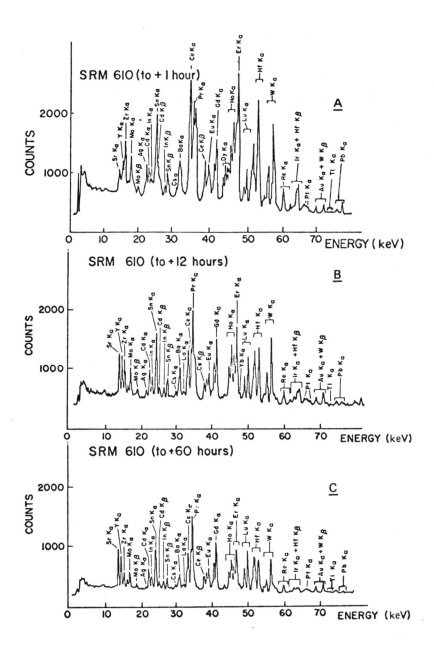

FIG.4. *X-ray spectra from NBS glass SRM 610 at various times after a 15-min 0.5-μA irradiation with 20-MeV protons. Count times were 20 min* [19].

techniques. As a general rule these are adequate for light element concentrations as low as 10 ppm. From that point on they are hampered by increasing difficulties as trace levels decrease (blanks, errors due to surface adsorption, limitations in signal-to-noise ratios). At sub-ppm levels, photon activation competes well with CPAA for carbon and nitrogen [91]. Both methods are free from reagent blanks and errors due to surface contaminants. CPAA is inherently more sensitive, but there are more interferences than with photon activation [72]. For oxygen, ^{3}He, and probably triton activation (see below), feature superior sensitivity. Photon activation cannot compete at levels below 0.5 ppm due to the short half-life of ^{15}O ($T_{\frac{1}{2}}$: 2.03 min.) produced by $^{16}O(\gamma,n)^{15}O$, and due to the necessity of a radiochemical separation of ^{15}O for proper identification [92]. In summary it is felt that CPAA has currently a unique capability for the in-depth determination of sub-ppm levels of oxygen and boron. Hydrogen might be added if heavy ion activation is included (see below). At levels above the ppm, CPAA is well suited (accuracy, thorough selective activation, freedom from reagent blanks and surface contamination, wide dynamic range) as a reference technique for calibrating other methods.

When appraising CPAA as a multielement technique, its features should be compared with those of other methods also capable of determining simultaneously several medium and/or high Z elements, e.g.: photon or particle induced x-ray emission analysis, emission spectrography, mass spectrometry, neutron or photon activation analysis. Such comparisons must be made and are only valid for a given analytical problem. Depending on the requirements of the analysis, the importance of various criteria of evaluation varies, e.g. importance of reagent blanks, speed of analysis, sample size needed. As an example, the findings of a comparison of instrumental proton and neutron activation analysis applied to biological samples are summarized here [81]. At the outset, activation conditions had to be defined. These were set as follows: a one hour irradiation of 0.6 µA/cm^2 with 12 MeV protons and a one minute and a 14 hour irradiations with 4×10^{12} n/cm^2/sec on a 500 mg sample. These conditions were arbitrary but deemed representative of what can be achieved in practice. The data obtained on orchard leaves samples (NBS SRM 1571) is presented in Figure 5, giving sensitivities for various elements. Clearly, neutron activation can determine a greater number of elements with high sensitivity. (Moreover improvements in its performance by using longer irradiations, higher fluxes, fast or resonance neutrons can be readily envisioned). On the other hand, with proton activation, elements which are not accessible by the neutron technique can be measured: titanium, strontium and lead. For others like iron, copper and zinc, better detection limits are obtained even with a relatively short activation. (One might note also, that beyond proton activation, other charged particle beams may be advantageously used for the detection of specific trace elements, e.g. Li, B, Ca, P, Pb). A more quantitative comparison can still be attempted by computing a figure of merit for each element. This figure of merit was defined

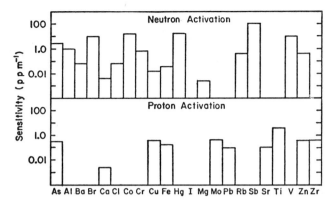

FIG.5. Comparison of proton and neutron activation. Sensitivities (expressed as −log of detection limits in ppm) obtained for orchard leaves.

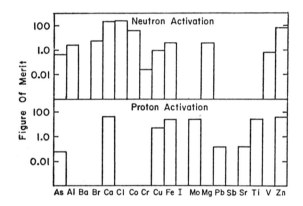

FIG.6. Comparison of proton and neutron activation. Figures of merit obtained for different elements.

as the ratio of the average concentration of a given element in the human body divided by its detection limit. A large numerical value would thus indicate that the element in question can readily be measured by the method selected. Conversely, a figure of merit below unity would indicate that the respective element could probably in many tissue samples not be detected. With neutron activation, 9 elements had figures of merit greater than unity, with proton activation this was the case for 4 elements. This can readily be understood when considering the short irradiation time used in proton activation. A graphical illustration of these figures of merit is given in Figure 6.

An interesting observation deals with the comparative capabilities for the determination of known essential elements. Figures of merit equal or greater than unity were obtained with neutron activation for cobalt, iron and zinc;

with proton activation for copper, iron and zinc. This would indicate that both activation modes are approximately equivalent as far as the number of essential elements that can be detected is concerned. It should be emphasized again that these conclusions are based on arbitrary experimental conditions. The above considerations exemplify the need for an in-depth evaluation as basis for comparisons.

4.0 CURRENT TRENDS AND PROSPECTS

The progress made in CPAA in recent years is leading to numerous new opportunities for further studies and developments. Among the possibilities ready to be explored and applied are those arising from heavy ion interactions, triton activation, ultrashort half-lives, the combination of delayed and prompt methods and the refinement of precision and accuracy.

4.1 Novel and Improved Capabilities

4.1.1 Heavy Ion Activation Analysis

The increasing availability of heavy ion beams opens new possibilities. In particular the determination of hydrogen, deuterium and helium appears feasible via "inverse" nuclear reactions. This approach can be briefly described with the following example: instead of measuring the ^{11}B content in a sample via $^{11}B(d,p)^{12}B$, deuterium would be measured via the "inverse" reaction $^{2}H(^{11}B,p)^{12}B$. The merits of this approach termed HIAA (Heavy Ion Activation Analysis) can be seen by considering a few interesting points: (a) there are a number of well known "forward" reactions, e.g. (p,n), (d,p) which could be "inverted"; there are only a few methods available for the determination of hydrogen, deuterium or helium and these are of limited capability (sensitivity, size of sample analyzed); (c) the reaction products are radioactive species, thus, except for very short-lived nuclides, the interference of surface contamination can be avoided by post-irradiation etching. A study of HIAA has been undertaken by McGinley et al. [93], information on the reactions studied to date is presented in Table XI. Among these ^{7}Li and ^{10}B activation have been applied for the determination of hydrogen (Figure 7). Clearly the HIAA technique should be readily applicable also to the determination of helium. Further possibilities might exist for other very low Z nuclides.

4.1.2 Triton Activation

The favorable characteristics of $^{16}O(^{3}H,n)^{18}F$, high cross section, complete freedom from interferences, have been pointed out by several authors [94-96]. Recently, Borderie et al. [95] have measured the specific activities induced by 3.5 MeV triton activation on 20 elements with Z < 34. Sensitive (< ppm) and selective determinations of

TABLE XI. REACTIONS OF INTEREST FOR HEAVY ION ACTIVATION ANALYSIS

Reaction	$T_{\frac{1}{2}}$	$E_{\beta max}$ (MeV)	E_γ (keV)	Thick target yield (cps/μg)[a]	Interference-free detection limit (ppm)
$^1H(^7Li,n)^7Be$	53 d	—	480	0.051	0.1[b]
$^2H(^7Li,p)^8Li$	0.85 s	13.0	—	5.79 × 10³	0.1[c]
$^1H(^{10}B,n)^{10}C$	19 s	1.9	720	—	
$^1H(^{10}B,\alpha)^7Be$	53 d	—	480	0.0261	0.5[b]
$^1H(^{11}B,n)^{11}C$	20 min	0.96	511 (β^+)	—	
$^2H(^{11}B,p)^{12}B$	0.02 s	13.0	4400	6.9 × 10⁴	0.1[c]
$^1H(^{16}O,\alpha)^{13}N$	10 min	1.19	511 (β^+)	—	
$^1H(^{19}F,n)^{19}Ne$	18 s	2.23	511 (β^+)	—	
$^2H(^{19}F,p)^{20}F$	11 s	5.4	1630	1.24 × 10²	1.2[c]
$^2H(^{22}Ne,p)^{23}Ne$	38 s	4.4	440	—	

[a] Irradiated at 1 μA for 1 hour or 1 half-life, whichever is shorter.
[b] Irradiated at 3 μA for 3 hours, counted for 12 hours following a 1-week delay. $E(^7Li) = 74$ MeV and $E(^{10}B) = 64$ MeV.
[c] 20 repetitive irradiations at 1 μA, count time equal to the half-life. $E(^{11}B) = 60$ MeV and $E(^{19}F) = 109$ MeV.

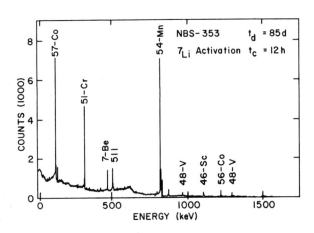

FIG. 7. Gamma-ray spectrum of titanium sample (NBS 353) irradiated with 40-MeV ^{10}B ions (1000 sec, 1 μA bombardment). Hydrogen is determined via $^1H(^{10}B,\alpha)^7Be$.

B, N, O, F, Na, Mg, S, Cl, V and Mn are possible nondestructively in medium and high Z targets. The work so far has been limited by the low triton energies. Further studies with higher energy beams should be of great interest.

4.1.3 Quasi-prompt Activation

The possibilities of quasi-prompt activation with very short-lived γ-ray emitters ($T_{\frac{1}{2}} \leq 1$ msec) has not yet been explored. It can be pointed out that an accelerator is ideally suited for such work given the inherent simplicity of producing short beam bursts and given that detectors can be placed directly at the irradiation site. A series of isomeric states are known with half lives from 1 μsec to 1 msec. Cyclic activation procedures with γ-ray spectrometry can thus be envisioned.

4.2 Combination of Nuclear and Atomic Activation

Heavy ion beams are of interest not only for nuclear activation of the hydrogen isotopes and helium but also as a means for exciting characteristic x-rays from medium and high Z elements via ion-atom collisions (atomic activation). Studies carried out by Zeisler [98] and Cross [99] have shown that high energy ($E \geq 0.2$ MeV/amu), heavy ion ($Z \geq 3$) induced x-ray emission features high sensitivity (10^{-10}-10^{-11} g on sample sizes of 10^{-5}-10^{-4} g) and selectivity as well as applicability to a wide range of elements. With a properly selected ion beam, one can thus envision a unique multi-element analysis capability, determination of ^1H, ^2H, ^4He via nuclear activation and of medium and high Z elements via their characteristic K and L x-rays emitted during bombardment. One can anticipate substantial further developments maximizing the analytical information (prompt and delayed emissions) that can be derived from ion bombardment.

4.3 Improvements in Precision and Accuracy

The wide ranging and in certain cases unique capabilities of CPAA remain to be fully evaluated with respect to sources of errors. The aim should be to bring CPAA procedures to the level of performance of a reference method, particularly for those elements where possibilities with other techniques are restricted. Some items in need of consideration include:

- γ-ray interferences. Typical are those arising from nuclides produced concurrently by proton and fast neutron reactions. They are difficult or impossible to detect when occurring at levels close to the experimental detection limits and are thus likely to go unnoticed. Little data seems to be available on this topic in the format needed for CPAA.
- Preservation of sample integrity during irradiation.
- Lack of adequate standards at the sub-ppm level.

The topics outlined above are difficult to handle experimentally: the effects to be investigated, although important at the level of trace analysis, are very small in absolute terms.

5.0 CONCLUSIONS

CPAA has a broad range of applications either as a high performance single element technique or as a tool for multi-element assays. Among the advantages are the sensitivity (10^{-6}-10^{-10} g), the selectivity (many activation reactions are interference-free) and the applicability to virtually all stable elements. Limitations arise from the hardware requirements (i.e. ion beams of several MeV/amu are needed) and the generation of heat in thick samples. CPAA detects particularly well the very light and very heavy elements and thus complements the other trace analysis methods for which the preferred domain of application is in the medium Z region. The capabilities of CPAA are far from being fully assessed. Novel analytical procedures with significant problem solving abilities can be envisioned, based on ultrashort-lived nuclides, heavy ion interactions or combinations of delayed and prompt methods.

ACKNOWLEDGEMENTS

It is a pleasure and a privilege to acknowledge the numerous helpful discussions, ideas and observations of my co-workers present and past: J. N. Beck, T. D. Burton, J. B. Cross, S. A. Dabney, J.-L. Debrun, G. Francis, S. M. Kormali, V. Krivan, J. R. McGinley, L. R. Novak, D. C. Riddle, H. L. Rook, G. J. Stock, D. L. Swindle, J.-P. Thomas, R. Zeisler, and L. Zikovsky.

Thanks are also due to the cyclotron operations personnel, Texas A&M University.

The financial support by the National Science Foundation for the studies carried out at Texas A&M University is gratefully acknowledged.

REFERENCES

[1] SEABORG, G. T., LIVINGOOD, J. J., J. Am. Chem. Soc. 60 (1938) 1784.
[2] SUE, P., C. R. Acad. Sci. Paris C 242 (1956) 770.
[3] ALBERT, Ph., Modern Trends in Activation Analysis (Proc. 1961 Int. Conf. College Station), Texas A&M University (1961) 78.
[4] GILL, R. A., Proton Activation Analysis in the Determiniation of Submicrogram Amounts of Boron in Silicon, AERE Report C/R 2758 (1958).
[5] SAITO, K., NOZAKI, T., TANAKA, S., FURUKAWA, M., CHENG, H., Int. J. Appl. Radiat, Isot. 14 (1963) 357.
[6] MARKOWITZ, S. S., MAHONY, J. D., Anal. Chem. 34 (1962) 329.

[7] PIERCE, T. B., Spatially Sensitive Analytical Techniques, these proceedings.
[8] VALKOVIC, V., X-ray Fluorescence, these proceedings.
[9] LANGE, J., MUNZEL, H., KELLER, K. A., PFENNIG, G., Excitation Functions for Charged Particle Induced Reactions, Landolt-Börnstein, Neue Serie, Gruppe I, Band 5b, Nuclear Reactions, Springer, Berlin (1973).
[10] LEDERER, C. M., HOLLANDER, J. M., PERLMAN, I., Table of Isotopes, 6th Ed., Wiley, New York (1968).
[11] SEELMANN-EGGEBERT, W., PFENNIG, G., MUNZEL, H., Chart of Nuclides, 4th Ed., Gersbach u. Sohn, München (1974).
[12] ERDTMANN, G., SOYKA, W., Die γ-Linien der Radionuklide, Report Jül-1003-AC (1973).
[13] Nuclear Data, Sections A and B, Nuclear Data Group, Oak Ridge, Academic Press, New York (1965-present).
[14] AJZENBERG-SELOVE, F., Nucl. Phys. A190 (1972) 1.
[15] ENDT, P. M., VAN DER LEUN, C., Nucl. Phys. A105 (1967) 1.
[16] KRIVAN, V., Applications of Nuclear Data in Science and Technology (Proc. Symp., Paris 1973), IAEA, Vienna (19).
[17] SCHWEIKERT, E. A., McGINELY, J. R., FRANCIS, G., SWINDLE, D. L., J. Radioanal. Chem. 19 (1974) 89.
[18] RIDDLE, D.C., On the Determination of Some Medium and High Z Elements Using Charged Particle Activation Analysis, Ph.D. Dissertation, Texas A&M University, College Station (1973).
[19] McGINELY, J. R., SCHWEIKERT, E. A., Anal. Chem. 48 (1976) 429.
[20] THOMAS, J. P., SCHWEIKERT, E. A., Nucl. Instr. Meth. 99 (1972) 461.
[21] BURTON, T. D., SWINDLE, D. L., SCHWEIKERT, E. A., Radiochem. Radioanal. Lett. 13 (1973) 191.
[22] DeMICHELE, D. W., FARES, Y., GOESCHL, J. D., BALTUSKONIS, D. A., Plant Physiology (1977) in press.
[23] ROOK, H. L., SCHWEIKERT, E. A., WAINERDI, R. E., Anal. Chem. 40 (1968) 1194.
[24] ALBERT, Ph., personal communication (1976).
[25] ROOK. H. L., The Determination of Trace Levels of Oxygen and Carbon by Charged Particle Activation Analysis, Ph.D. Dissertation, Texas A&M University, College Station (1969).
[26] LEE, D. M., LAMB, J. F., MARKOWITZ, S. S., Anal. Chem. 43 (1971) 542.
[27] DEBRUN, J.-L., RIDDLE, D. C., SCHWEIKERT, E. A., Anal. Chem. 44 (1972) 1386.
[28] McGINELY, J. R., SCHWEIKERT, E. A., Anal. Chem. 47 (1975) 2403.
[29] McGINLEY, J. R., STOCK, G. J., SCHWEIKERT, E. A., CROSS, J. B., ZEISLER, R., ZIKOVSKY, L., J. Radioanal. Chem. (1977) in press.
[30] McCARTHY, A. L., COHEN, B. L., GOLDMAN, H. L., Phys. Rev. 137 (1965) B250.
[31] BLACK, J. L., MAHIEUX, J., Nucl. Instr. Meth. 58 (1968) 93.
[32] THOMAS, J. P., SCHWEIKERT, E. A., Radiochem. Radioanal. Lett. 9 (1972) 155.

[33] McGINLEY, J. R., SCHWEIKERT, E. A., Radiochem. Radioanal. Lett. 25 (1976) 1.
[34] YULE, H. P., (Proc. 1968 Int. Conference, Gaithersburg) NBS Spec. Publ. 312, Vol. II, (1969) 1155.
[35] QUITTNER, P., Gamma-Ray Spectroscopy, Adam Hilger Ltd., London (1972).
[36] CUMMING, J. B., Report BNL-6470 (1962).
[37] RICCI, E., HAHN, R. L., Anal. Chem. 39 (1967) 794.
[38] ALBERT, Ph., Ann. de Chimie, 13e serie (1956) 827.
[39] ENGELMANN, Ch., Radiochemical Methods of Analysis (Proc. Symp. Salzburg 1964), Vol. I, IAEA, Vienna (1965) 405.
[40] THOMAS, J.-P., Report Lycen 7016 (1970).
[41] CHEVARIER, N., GIROUX, J., TRAN, M. D., TOUSSET, J., Bull. Soc. Chim. France 8 (1967) 2893.
[42] DEBRUN, J.-L., Etude du dosage de quelques elements par activation dans les particules chargees et les photons de 35 MeV, Dissertation Univ. of Paris (1969).
[43] ROOK. H. L., SCHWEIKERT, E. A., WAINERDI, R. E., Modern Trends in Activation Analysis (Proc. 1968 Int. Conference, Gaithersburg), NBS Spec. Publ. 312, Vol. II (1969) 768.
[44] WILLIAMSON, C. F., BOUJOT, J. P. PICARD, J., Table of Ranges and Stopping Powers of Chemical Elements for Charged Particles of Energy 0.05 to 500 MeV, CEA Report R3042 (1966).
[45] ROOK, H. L., SCHWEIKERT, E. A., Radiochem. Radioanal. Lett. 3 (1970) 239.
[46] NOZAKI, T., YATSURUGI, Y., AKIYAMA, N., ENDO, Y., MAKIDE, Y., J. Radioanal. Chem. 19 (1974) 109.
[47] KORMALI, S. M., SCHWEIKERT, E. A., J. Radioanal. Chem. 31 (1976) 413.
[48] KORMALI, S. M., SWINDLE, D. L., SCHWEIKERT, E. A., J. Radioanal. Chem. 31 (1976) 437.
[49] SASTRI, C. S., PETRI, H., ERDTMANN, G., Modern Trends in Activation Analysis (Proc. 1976 Int. Conf. Munchen) to appear in J. Radioanal. Chem. (1977).
[50] DEBRUN, J.-L., BARRANDON, N., BENABEN, P., Anal. Chem. 48 (1976) 167.
[51] PRETORIUS, R., SCHWEIKERT, E. A., J. Radioanal. Chem. 7 (1971) 319.
[52] KRIVAN, V., SWINDLE, D. L., SCHWEIKERT, E. A., Anal. Chem. 46 (1974) 1626.
[53] SWINDLE, D. L., SCHWEIKERT, E. A., Anal. Chem. 45 (1973) 2111.
[54] SWINDLE, D. L., NOVAK, L. R., SCHWEIKERT, E. A., Anal. Chem. 46 (1974) 655.
[55] DESAEDELEER, G., Anal. Chem. 48 (1976) 572.
[56] COBB, J. C., J. Geophys. Res. 69 (1964) 1895.
[57] RIDDLE, D. C., SCHWEIKERT, E. A., Anal. Chem. 46 (1974) 395.
[58] SUE, P., ALBERT, Ph., C. R. Acad. Sci. Paris, 242C (1956) 2461.
[59] ENGELMANN, Ch., C. R. Acad. Sci. Paris, 285C (1964) 4279.
[60] NOZAKI, T., TANAKA, S., FURUKAWA, M., SAITO, K., Nature 190 (1961) 39.
[61] ROMMEL, H., Anal. Chim. Acta 34 (1966) 427.
[62] VIALATTE, B., REVEL, G., J. Radioanal. Chem. 12 (1972) 371.

[63] VANDECASTEELE, C., ADAMS, F., HOSTE, J., Anal. Chim. Acta 71 (1974) 67.
[64] SCHUSTER, E., WOHLLEBEN, K., Z. Anal. Chem. 240 (1968) 175.
[65] BUSCH, G., SCHADE, H., GOBBI., A., MARMIER, P., J. Phys. Chem. Solids 23 (1962) 513.
[66] KUIN, P. K., Practical Aspects of Activation Analysis with Charged Particles (Proc. 2nd Int. Conf. Liege), Euratom Report EUR 3896 d-f-e (1968) 31.
[67] ENGELMANN, Ch., FRITZ, B., GOSSET, J., GRAEF, P., LOEUILLET, M., ibid. 319.
[68] ENGELMANN, Ch., GOSSET, J., LOEUILLET, M., MARSCHAL, A., OSSART, P., BOISSIER, M., Modern Trends in Activation Analysis (Proc. 1968 Int. Conf. Gaithersburg), NBS Spec. Publ. 312 (1969), Vol. II, 819.
[69] ENDO, Y., YATSURURGI, Y., AKIYAMA, N., NOZAKI, T., Anal. Chem. 44 (1972) 2258.
[70] LAMB, J. F., LEE, D. M., MARKOWTIZ, S. S., Anal. Chem. 42 (1970) 212.
[71] GIROUX, J., TALVAT, M., THOMAS, J.-P., TOUSSET, J., Radioanal. Chem. 6 (1970) 423.
[72] ENGELMANN, Ch., Contribution a l'etude de l'analyse par activation an moyen des particules chargees et des photons gamma, dissertation Univ. of Paris (1970).
[73] LIGEON, E., BONTEMPS, A., J. Radioanal. Chem. 12 (1972) 335.
[74] GIROUX, J., TALVAT, M., THOMAS, J.-P., TOUSSET, J., Bull. Soc. Chim. France 2 (1971) 706.
[75] ENGELMANN, Ch., MARSCHAL, A., Radiochem. Radioanal. Lett. 6 (1971) 195.
[76] VALLADON, M., DEBRUN, J.-L., Modern Trends in Activation Analysis (Proc. 1976 Int. Conf. Munchen), to appear in J. Radioanal. Chem. (1977).
[77] ALEKSANDROVA, G. I., DEMIDOV, A. M., KOTELNIKOV, G. A., PLESHAKOVA, G. P., SUKHOV, G. V., CHOPOROV, D. Y., SHMANENKOV, G. I., Soviet Atomic Energy 23 (1967) 787.
[78] HOLM, D. M., BRISCOE, W., PARKER, J., SANDERS, W. M., LASL Report DC-8866 (1966).
[79] KIM, C. K., Radiochem. Radioanal. Lett. 2 (1969) 53.
[80] KIM, C. K., Anal. Chim. Acta 54 (1971) 407.
[81] KRASNOV, N. N., Soviet Atomic Energy 26 (1969) 284.
[82] SWINDLE, D. L., SCHWEIKERT, E. A., Talanta 22 (1975) 84.
[83] ZIKOVSKY, L., SCHWEIKERT, E. A., J. Radioanal. Chem. 39 (1977) in press.
[84] SIPPEL, R. F., GLOVER, E. D., Nucl. Instr. Meth. 9 (1960) 37.
[85] KRIVAN, V., Anal. Chem. 47 (1975) 469.
[86] KORMALI, S. M., SCHWEIKERT, E. A., J. Radioanal. Chem. 22 (1974) 139.
[87] KRIVAN, V., J. Radioanal. Chem. 26 (1975) 151.
[88] PARSA, B., MARKOWITZ, S. S., Anal. Chem. 46 (1974) 186.
[89] KRIVAN, V., MUNZEL, H., J. Radioanal. Chem. 15 (1973) 575.
[90] AMIEL, S., MANTEL, M., ALFASSI, Z. B., Modern Trends in Activation Analysis (Proc. 1976 Int. Conf. Munchen) to appear in J. Radioanal. Chem. (1977).

[91] ENGLEMANN, Ch., Radiochemical Methods of Analysis (Proc. Symp. Salzburg 1964), Vol. I., IAEA, Vienna (1965) 341.
[92] HISLOP, J. S., STEVENS, J. R., WOOD, D. A., Modern Trends in Activation Analysis (Proc. 1976 Int. Conf. München), to appear in J. Radioanal. Chem. (1977).
[93] McGINLEY, J. R., ZIKOVSKY, L., SCHWEIKERT, E. A., Modern Trends in Activation Analysis (Proc. 1976 Int. Conf. München), to appear in J. Radioanal. Chem. (1977).
[94] SMALES, A. A., Ann. Rep. Progr. Chem. $\underline{46}$ (1949) 290.
[95] BARRANDON, N., ALBERT, Ph., Modern Trends in Activation Analysis (Proc. 1968 Int. Conf. Gaithersburg), NBS Spec. Publ. 312, Vol. II (1969) 794.
[96] REVEL, G., LINCK, I., DA CUNHA BELO, M., PASTOL, J. L., KRAUS, L., Radiochem. Radioanal. Lett. $\underline{27}$ (1976) 191.
[97] BORDERIE, B., BARRANDON, J. N., DEBRUN, J.-L., Modern Trends in Activation Analysis (Proc. 1976 Int. Conf. München), to appear in J. Radioanal. Chem. (1977).
[98] ZEISLER, R., CROSS, J. B., SCHWEIKERT, E. A., Anal. Chem. $\underline{48}$ (1976) 2124.
[99] CROSS, J. B., ZEISLER, R., SCHWEIKERT, E. A., Nucl. Instr. Meth. $\underline{142}$ (1977) 111.

SPATIALLY SENSITIVE ANALYTICAL TECHNIQUES

T.B. PIERCE
Atomic Energy Research Establishment,
Harwell, Oxfordshire,
United Kingdom

Abstract

SPATIALLY SENSITIVE ANALYTICAL TECHNIQUES.
The properties of many materials are dependent not only upon the overall composition, but also upon the way the elemental constituents are distributed throughout the sample. The paper reviews the theoretical background and the experimental requirements of analytical methods which exploit the manoeuvrability and focusing ability of the ion beams in particle accelerators to induce nuclear reactions in the selected part of the sample. By detecting and analysing the products of a nuclear reaction, quantitative results on the sample constituents and their spatial distribution are obtained. Examples of profitable practical applications of the techniques are presented.

1. INTRODUCTION

The growing appreciation of the importance of the effect of location as well as concentration on the properties of many materials has stimulated interest in spatially sensitive analytical techniques. For materials sensitive to elemental distributions, a gross analysis of the composition of a bulk sample often does not provide adequate information for subsequent interpretation and more detailed measurements are required. In order to meet this growing need, substantial attention has been allocated to the development of spatially sensitive analytical techniques and the most possible methods of measurement, both novel and established, have been examined to assess whether some degree of spatial sensitivity can be conferred upon them. Two types of approach have been most widely investigated. One of these relies on general illumination of the sample surface with radiation and exploits a position-sensitive detector to identify the location of the source of some product emitted as a result of the interaction of the sample with the primary radiation beam; autoradiography and certain types of microscopy fall into this category. The other approach consists of interrogating the sample with a beam of radiation of carefully controlled geometry which interacts only with that particular part of the sample that is of interest and monitoring the products of interaction with a detector which, in this case, need not necessarily be spatially sensitive. This microprobe-type approach has exploited irradiation with electrons, low-energy ions and photons and the products measured have included secondary ions, X-rays and lines in the visible spectrum [1]. Best known of the microprobes is the electron microprobe, which has now been very widely applied, and sophisticated commercial equipment has been developed possessing substantial analytical and image-processing capabilities. Until comparatively recently, however, nuclear-analytical techniques have made only a relatively minor contribution to this important and expanding area of analytical science, reported applications being restricted mainly to autoradiography and track-counting methods. One requirement which existing microanalytical techniques do not always meet satisfactorily, however, is the quantitative determination of the distribution of light elements since these are often difficult to determine by existing methods. Over recent years, considerable attention has been devoted to evaluating and applying analytical techniques, exploiting measurement of the prompt radiation emitted as a result of nuclear interaction rather than the induced radioactivity produced as a result of nuclear decay. Such prompt-radiation techniques are well applied to the determination of light elements since prompt particles or de-excitation gamma radiation can be produced from light elements at incident ion energies which are substantially lower than those required to penetrate the Coulomb barrier of intermediate or heavy elements. Reactions can therefore be preferentially generated with light elements without exciting radiation from heavier elements which might complicate the determination. Prompt-radiation measurement can be exploited to permit correlation of analytical data with location by two routes. Firstly, the characteristics of nuclear interaction can be used to give information about sub-surface compositions and, secondly, the incident ion-beam can be reduced in size by collimation or by focussing to provide a microprobe which is the nuclear analogue of the electron microprobe; this paper is concerned solely with the latter approach. In general, determinations with the nuclear microprobe require measurements of prompt particles or photons resulting from nuclear reaction processes, particles produced as a result of elastic scattering or X-rays. The characteristics of analytical procedures exploiting each of these sources of radiation will be considered briefly in this paper before methods of production of small-diameter ion beams and applications are discussed.

2. ANALYTICAL METHODS

The most familiar analytical techniques exploiting sample irradiation with neutrons, charged particles or gamma rays are based almost exclusively upon counting the induced radioactivity formed in the sample as a result of nuclear reaction. These activation techniques have been extensively investigated by

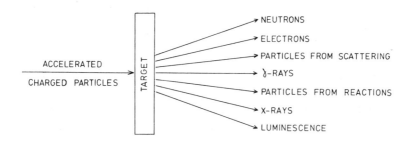

FIG.1. Diagrammatic representation of principles of microprobe operation.

many workers following the demonstration of neutron activation by Hevesy [2] and charged-particle techniques by Seaborg [3]. Neutron activation analysis now has an established place as a high-sensitivity analytical method while charged-particle irradiation has been primarily reserved for the determination of those light elements to which neutron activation can not be conveniently applied; in almost all applications, the purpose of the experimental procedure has been to provide analytical data representative of the bulk composition of the sample. In extending nuclear-analytical techniques to provide information with good spatial resolution, a primary requirement is for high intrinsic sensitivity of the analytical method since, in the end, the volume of sample from which relevant analytical information is available is likely to be very small. Fortunately, many of the prompt-radiation techniques, in common with conventional activation methods, do offer good "best-case" sensitivity although the actual lowest concentration that can actually be achieved when investigating a "real" sample will be less and will be a function of composition. Nevertheless, the several techniques do have their own particular characteristics and these are summarized in turn below while the basic microprobe procedure is summarized diagrammatically in Fig.1. These emitted radiations included in Fig.1 but not discussed in the paper are excluded only because they have not so far provided the basis for major microprobe applications.

(a) Gamma radiation

Excited nuclear states are produced as a result of nuclear reaction and routes may be open for these states to lose their excitation energy by gamma-ray emission. In many cases, photons are emitted within picoseconds of the initial interaction of the incident ion so that a detector must be in position and counting during the irradiation for them to be measured. These gamma rays are of a specific energy and therefore can be considered to be characteristic of a particular nucleus; they therefore offer a means of identifying individual elements and, in appropriate cases, permit multi-element analysis of complex samples. Further, since such lines can result from stable-to-stable transitions, a radioactive product is no longer needed for measurement. This greatly extends the range of methods available for light-element determination over those offered by conventional charged-particle activation analysis, which generally produces positron emitters that cannot be distinguished from each other by established methods of gamma-ray spectroscopy. As a specific example of the production of characteristic gamma radiation from stable nuclei, a 3.09-MeV gamma ray is emitted from carbon-13 produced by the reaction $^{12}C(d,p)^{13}C$ when a carbon-containing material is irradiated with deuterons of about 1 MeV, although both carbon-12 and carbon-13 are stable.

Measurement of carbon has figured largely in analytical applications of prompt-radiation techniques and various methods have been developed to meet particular requirements. Proton irradiation provides a 2.13-MeV gamma ray for measurement [4] by the $^{12}C(p,\gamma)^{13}N$ reaction. Higher sensitivity can be attained by means of the $^{12}C(d,p)^{13}C$ reaction as described above [5] and other reactions available for carbon include those induced by helium-3 ions [6] and inelastic scattering reactions. Where a number of alternative routes for generating prompt gamma radiation are available, the analyst has some opportunity to choose lines which appear in convenient regions of the gamma-ray spectrum and are away from those energies which are likely to be subject to interference due to other target elements. However, the potential of this approach tends to be restricted by the sensitivity and convenience of the various reactions.

Analytical techniques have been described for most of the light elements and a number of examples are given in Table I. Gamma-ray measurement has the attraction that spectra do not experience the same sensitivity to sample thickness as, for example, particle peak shape, and can be interpreted by conventional techniques of gamma-ray spectroscopy. Gamma-ray detection close to an accelerator often suffers from a high background which tends to limit the sensitivity with which particular elements can be determined. The available sensitivity may be degraded still further by the usual limitations of gamma-ray spectroscopy if spectra are complex and wanted lines occur in regions of Compton events. Further, nuclear interference, familiar in activation analysis, can also be experienced during prompt gamma-ray measurement when a particular excited level whose de-excitation radiation is being counted can be formed by two routes, one from the wanted element and one from a near neighbour in the periodic table.

TABLE I. EXAMPLE OF ANALYTICAL DETERMINATIONS EXPLOITING THE MEASUREMENT OF PROMPT GAMMA RADIATION

Element	Reaction	Ref.
Lithium	$^7\text{Li}(p,\gamma)^8\text{Be}$	[7]
Beryllium	$^9\text{Be}(p,\gamma)^{10}\text{B}$	[7]
	$^9\text{Be}(\alpha,n)^{12}\text{C}$	[8]
Boron	$^{11}\text{B}(p,\gamma)^{12}\text{C}$	[7]
Carbon	$^{12}\text{C}(d,p)^{13}\text{C}$	[5]
	$^{12}\text{C}(p,\gamma)^{13}\text{N}$	[4]
Nitrogen	$^{14}\text{N}(d,p)^{15}\text{N}$	[9]
	$^{15}\text{N}(p,\alpha)^{12}\text{C}$	[10]
Oxygen	$^{16}\text{O}(d,p)^{17}\text{O}$	[9]
	$^{18}\text{O}(p,\gamma)^{19}\text{F}$	[11]
Fluorine	$^{19}\text{F}(p,\alpha)^{16}\text{O}$	[7, 8, 12]
Magnesium	$^{24}\text{Mg}(p,p')^{24}\text{Mg}$	[13]
Aluminium	$^{27}\text{Al}(p,\gamma)^{28}\text{Si}$	[14]
	$^{27}\text{Al}(p,p')^{27}\text{Al}$	[13]
Silicon	$^{28}\text{Si}(p,p')^{28}\text{Si}$	[15]
Sulphur	$^{32}\text{S}(p,p')^{32}\text{S}$	[16]

(b) **Particles from nuclear reactions**

Charged particles produced by nuclear reactions such as (d,p), (α,p) and (p,α) provide a powerful basis for analytical measurement. These particles are emitted as a result of particular nuclear mechanisms and have energies which can be related to the transition occurring. Energy analysis of individual particles therefore provides a route to identifying the reaction which has taken place and also offers a means of distinguishing particles emitted from different nuclei. A major difference between charged particles and gamma photons arises from the energy loss experienced by charged particles as they pass through matter with the result that particle peak-shapes are critically dependent on sample thickness. Whilst this can complicate interpretation of spectral features, it also offers a means of examining the depth beneath the surface at which the reaction has occurred and so can provide the basis for "sub-surface" analysis. Particle measurement also possesses two further advantages which are not associated with gamma-

TABLE II. SOME ANALYTICAL METHODS EXPLOITING CHARGED-PARTICLE DETECTION

Element	Reaction	Ref.
Lithium	$^6\text{Li}(d,\alpha)^4\text{He}$	[19]
	$^7\text{Li}(d,\alpha)^5\text{He}$	
Beryllium	$^9\text{Be}(d,p)^{10}\text{Be}$	[20]
Boron	$^{10}\text{B}(d,p)^{11}\text{B}$	[21]
Carbon	$^{12}\text{C}(d,p)^{13}\text{C}$	[17, 22]
Nitrogen	$^{14}\text{N}(d,p)^{15}\text{N}$	[22]
Oxygen	$^{16}\text{O}(d,p)^{17}\text{O}$	[17, 22, 23]
	$^{16}\text{O}(d,\alpha)^{14}\text{N}$	[24]
	$^{18}\text{O}(p,\alpha)^{15}\text{N}$	[25]
Fluorine	$^{19}\text{F}(p,\alpha)^{16}\text{O}$	[19]
Sulphur	$^{32}\text{S}(d,p)^{33}\text{S}$	[26]

ray measurement: (a) the natural particle background is very low and (b) surface barrier detectors used for particle measurement are relatively insensitive to beta and gamma radiation normally emitted during radioactive decay. This latter characteristic enables the microprobe to be applied to highly radioactive samples which can be of value when examining materials used in nuclear reactors particularly after irradiation. Spectral interpreation can, of course, only be achieved within the constraints imposed by spectral composition, and a clear signal is needed from the component of interest for reliable quantitative interpretation to be feasible. A large number of reactions producing particles is available for light-element determinations and these have been used for a range of analytical measurements [17]. Such reactions not only offer a means of measuring the presence of individual elements in a variety of matrices but also permit the movement of particular elements to be traced by doping with reactants enriched in a particular stable isotope [18]. Table II includes examples of analytical techniques which are based on particle measurement.

Whilst the reactions given in Table II exploit counting of charged particles, in principle, detection of neutrons produced by, for example, the (d,n) reaction could also be used for analytical determination. However, the more limited capability of existing neutron-detection techniques has restricted neutron counting to a few applications where the particular advantages are worth while; an example is the detection of beryllium [20].

(c) Elastic scattering

Elastic scattering of charged particles offers an alternative means of analysing surfaces and provides information which is primarily a function of the mass of the target nucleus. Interactions which occur during the scattering process are analogous to a "billiard-ball" type collision and result when a positively charged projectile is scattered from a positive target nucleus. The amount of energy retained by the projectile after scattering is dependent upon the quantity of energy lost by imparting recoil to the scattering nucleus and, since energy loss is greater the lighter the target nucleus, the method provides a technique of mass analysis with the lowest energy ions scattered from the light elements. The ratio of the energies of the incident particle (mass m) before (E_0) and after (E_1) collision with a target nucleus (mass M) is given by

$$\frac{E_1}{E_0} = \left(\frac{m \cos \theta + \sqrt{M^2 - m^2 \sin^2 \theta}}{m + M} \right)$$

where θ = the scattering angle.

When target thicknesses become appreciable compared with the range of penetration of the particle in the sample matrix, the spread of energies of the scattered particles will increase since some particles lose energy moving through the sample to and from the scattering site. Thus, for a thick target, the spectrum will take the form of the traditional Rutherford plateau with an edge characteristic of scattering material. Interpretation of distribution of energies of particles scattered from thick targets can be valuable since it may provide a means of deriving information about sub-surface elemental concentrations. However, when a thin-surface layer is being examined in the presence of a thick substrate, difficulties can arise if the substrate is of greater mass than the surface film since the peak from the surface component will be present on the plateau from the substrate. Elastic scattering is therefore best applied to the analysis of a thin film of a heavy element present on a lighter substrate. Providing that the mass of the surface film is sufficiently different from that of the substrate, a very sensitive analytical method is available.

Since the differential cross-section for particle scattering is a function of the square of the atomic number of the scattering nucleus, the gross particle yield from a thick target can be utilized to provide information about the mean atomic number of the target area being examined. This enables particle scattering to be used to provide a means of rapid examination of sample surfaces before more detailed examination of selected regions of particular interest [27]. The technique provides a valuable complement to other applications offering more specific analytical information since it enables semi-quantitative information to be collected rapidly from large areas.

When samples can be prepared which are sufficiently thin to avoid appreciable energy loss in the sample, the Rutherford plateau is absent and a mass analysis of the target composition can sometimes be carried out, particularly at low masses where the best mass separation is achieved. The technique is valuable for the examination of biological samples since established methods of specimen preparation enable thin targets to be produced.

(d) X-ray generation

The passage of charged particles through matter leads to the generation of X-rays which offers yet another means of obtaining characteristic information about the elemental content of individual samples. Since X-rays are emitted simultaneously with products of nuclear interaction, the two types of radiation can be measured simultaneously to obtain more detailed information about composition if required. Analytical measurements requiring only X-ray counting following charged-particle irradiation with the nuclear microprobe are often hard to justify since the majority of such determinations are likely to be able to be carried out more cheaply and more effectively with the electron microprobe, but the use of X-ray measurement to supplement the information obtained from nuclear reactions can be very valuable. The potential of X-ray methods is discussed elsewhere in these Proceedings and therefore is not considered in any detail here, but a number of aspects of the approach relevant to the microprobe are worth summarizing. Adequate access to the detector for radiation emitted from the target is needed in spite of the limited space available around the point of interaction of the ion beam with the sample, and this can be difficult to achieve, particularly if high collection efficiency is required. In general, energy dispersive detectors have been preferred for X-ray counting, since their high efficiency permits good count-rates to be achieved from many materials even when beam currents are low. However, the limitations imposed by the relatively poor resolution of solid-state detectors results in the need for crystal spectrometers for particular applications and access to both types of detector is required from time to time.

The overall capability of X-ray measurement has been summarized in proceedings of conferences which have considered the topic in depth [28]. Many of the applications reported have centred on the examination of the composition of either thin targets or of samples possessing matrices of low Z-numbers. Whilst this configuration is desirable to ensure a low Bremsstrahlung component in the background, in practice, many samples do not conform to this type of composition. Application of X-ray measurement to less satisfactory matrices complicates detection and reduces the overall sensitivity of the determination. However, measurement of the high yield of X-rays from major elements in metal samples permits the discontinuity between the mounting material and the sample to be clearly identified and complements optical assessment of sample position which is not always adequate to permit accurate alignment of the ion beam with a particular region of interest. X-ray results when correlated with information from light elements permit the presence of specific compounds to be identified in certain types of metallurgical sample (e.g. chromium carbide or metal oxides).

3. PRODUCTION OF SMALL-DIAMETER ION BEAMS

Small-diameter ion beams were initially produced for the microprobe by collimation [29]. Beam collimator systems are comparatively inexpensive and are designed to produce ion beams with diameters of 20–30 μm without too much difficulty. Collimation in fact may be the preferred means of microbeam production if smallest diameters are not required. Examination of the sample with a smaller beam than warranted by the definition required is not advisable, since beam current is often restricted, time taken for the analysis long and data generated unnecessarily detailed. More than one collimating stop is usually needed to remove the edge-scattered component and adjustable stops and stop-positions aid alignment procedures which can be tedious if the collimating system contains two or three fixed stops pierced with holes a few tens of micrometres in diameter. Focussing the ion beams by means of suitable magnets does of course provide a more flexible equipment and permits more efficient use of the available ion beam. Various ways have now been reported to achieve small focussed particle beams and diameters of 2 or 3 μm can be attained. The relatively high beam densities that can be obtained by focussing permit these beams to be usefully employed for analysis and the stability of beam position (in spite of the relatively long flight-path between the beam defining stops and target) enables reproducible measurements to be carried out. The system currently in use at AERE, Harwell has now been routinely employed for a number of years and exploits a system of lenses in the form of a Russian Quadruplet [30] to provide beam diameters of the order of 3 μm. The system was constructed from four specially designed quadrupole lenses, the distances between opposite poles differing by ± 0.013 mm while the variation in the shortest distance between neighbouring poles on a single magnet was only ± 0.045 mm in 14.25 mm. Quadrupoles were aligned on the same geometrical axis with an accuracy of about ± 50 μm and magnetic centres of the four quadrupoles coincided with the geometrical centres to within about $\pm 130 \mu$m [31]. In many of the applications examined with this microprobe, samples are moved mechanically relative to the position of the ion beam, but in certain cases, particularly those where rapid scanning of the surface is needed to provide a visual image or to reduce the heat-damage in the sample, electrostatic deflection is employed. Aberrations caused by beam movement must be kept to a minimum to preserve resolution and for the

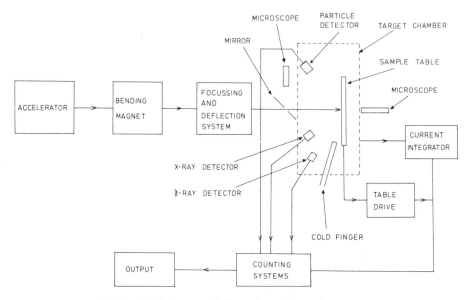

FIG.2. Block diagram of the nuclear microprobe.

IBIS system they were found to be about 5.5% of the lateral displacement in the X direction and 6.7% in the Y direction when a potential was applied to a single deflector plate, but when equal and opposite potentials were applied to both Y plates, aberration was reduced to 2.7% of the displacement.

4. EXPERIMENTAL CONFIGURATION OF THE MICROPROBE

The component parts of the microprobe include (a) a source of energetic charged particles, (b) a beam handling and focussing system to direct the ion beam on the target, (c) a target chamber with associated sample handling facilities and (d) radiation detectors and the associated counting, control and data processing equipment needed for collection and interpretation of the analytical signals. The experimental arrangement of the nuclear microprobe has been described in detail elsewhere [32] and consequently only the main components of the system are considered here; an outline diagram is given in Fig.2. The equipment itself, although in use for several years, must still be regarded as experimental since it has not benefited from the level of instrumental development that has been devoted to other microanalytical techniques, for example the electron microprobe. Consequently, facilities available are rudimentary by microanalytical standards and correspond much more closely to a research configuration than to a fully developed piece of analytical instrumentation. The equipment is centred on the 3-MV electrostatic generator IBIS and the microbeam line is secured to a substantial steel supporting beam to ensure the absolute rigidity of line the components need to avoid movement during use. An adjustable beam-defining stop acts as the objective for the ion-optics and the ion beam is focussed down on the plane of the sample-table in the chamber some 12 ft (3.66 m) away. The table can be moved in three dimensions by stepping motors under remote control from outside the chamber. In practice, mechanical movement of the table is preferred for line-scans, the table being moved discontinuously with irradiations of each point of the sample surface being controlled automatically to a given ion dose and the sample being moved to a new position relative to the ion beam between irradiations. The ion beam can also be deflected electrostatically, and this mode is preferred when a high-speed raster scan is needed to generate a photographic image or when the heat dissipated in the sample is to be spread over a large area to reduce damage. The sample table itself has space for two sample discs up to $1\frac{1}{8}$ in (2.86 cm) in diameter as well as a position for a quartz disc used during beam alignment procedures. A microscope views the point of interaction of the ion beam with the sample and enables samples to be positioned initially so that the region of interest is irradiated. A second microscope at the rear of the target chamber permits the quartz disc to be examined during initial setting-up procedures so that both the shape and size of the beam can be examined. Gamma-ray, particle and X-ray detectors view the point of interaction of the ion beam and permit the appropriate radiations to be counted. Detector output is fed to a pulse-height analysing system after conventional amplification and counting systems are usually kept as simple as possible to reduce the amount of data handling needed. Pulses are normally fed to one or more single-channel analysers and counts are collected for a known ion dose falling on the sample. Analyser output can also be fed to an oscilloscope operating in sequence with electrostatic deflection of the ion beam to provide a visual representation of elemental distributions.

An important part of the microprobe technique is the preparation of samples in a suitable form for irradiation and preferred procedures generally follow a standard metallographic technique. The materials of interest are placed in some mounting material and are polished with successively finer grades of abrasive until a good finish is achieved. Samples are normally held in Woods Metal to reduce the total quantity of light elements present in the mount and to reduce the chance of contamination. Since samples of fairly well-defined compositions have generally been examined with the microprobe, the preferred method of calibration has been with standards and for this purpose several standard materials, analysed by established analytical techniques, have been prepared in the same mount and irradiated to permit appropriate calibration curves to be generated.

5. APPLICATIONS

The applications that have been reported for the nuclear microprobe have been restricted by the limited number of installations so far in regular use. The potential of the technique can, however, be assessed by considering analytical uses found for beams of large diameter, bearing in mind the particular advantages that accrue from being able to irradiate a small volume of sample. Since development of the microprobe has been centred on nuclear laboratories, applications of primary interest have not unnaturally been those of concern to such establishments. Thus, metallurgical specimens have been most frequently investigated, but a sufficiently wide range of other matrices have now been examined to provide some idea of the overall potential of the method.

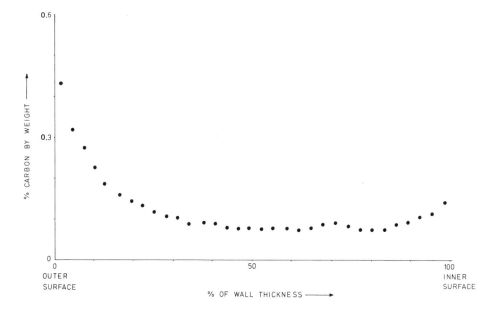

FIG.3. *Typical carbon profile obtained with the nuclear microprobe.*

(a) **Metallurgical analysis**

Investigation into distribution of light elements in metallurgical samples can be important, not only to obtain information about effects at grain boundaries, but also to identify the behaviour of metals when subject to particular processes. Considerable effort has been devoted to generating carbon [32] profiles to monitor the pickup of the elements during high-temperature experiments. Microprobe results have been shown to agree well with those obtained by micromachining the sample surface and analysing the separated fragments. Although machining did not offer a spatial resolution which could match the microprobe, results were adequate to show that the results were similar. An important characteristic of the nuclear microprobe is the ability to distinguish between elements present on and beneath the surface. This permits surface contamination to be identified which is particularly valuable during carbon analyses when carbon may be deposited from the machine vacuum system [33]. The sensitivity of the determination can thus be improved [34]. An example of a carbon-diffusion profile, typical of the many generated, is given in Fig.3. Correlation of the position of several light elements can sometimes be achieved by displaying counts selected from appropriate regions in the collected spectrum, and X-rays emitted from metals have been compared with carbon, oxygen, nitrogen and boron distributions [35]. The concentration of nitrogen has been compared with that of titanium distributions in partially nitrided, 20 chromium-25 nickel titanium-stabilized steel; nitrogen was determined by counting protons from the (d,p) reaction and Kα X-rays provided a measure of titanium present [20]. Nuclear reactions provide information about isotopic composition and, in certain cases, isotopic analysis can be carried out during scanning. As a specific example, the isotopic composition of boron has been determined by the (α,p) reaction, since protons from both the ^{10}B + α and ^{11}B + α reactions can be distinguished separately [36]. The presence of oxygen also plays a significant part in many metallurgical processes and a number of papers have reported examination of the distribution of oxygen. However, determination of specific isotopes can be extended to tracer studies and the distribution of oxygen in zirconium welds was measured by determining the yield of reaction ^{18}O(p,α)^{15}N [37]. Neutron counting has been exploited to permit the diffusion of beryllium in copper to be followed [20].

(b) **Minerals and allied materials**

The low conductivity of mineral samples has not precluded the use of the nuclear microprobe for their examination and both nuclear reactions and X-ray emission have provided appropriate elemental information. Gamma radiation from inelastic scattering processes has permitted the simultaneous determination of elements such as aluminium, silicon and iron [29] while X-rays have been used to provide analytical information from lunar samples, monazite crystals, mica foils and meteorites [38]. Observation of the characteristic X-rays emitted from slag during proton irradiation has permitted distributions of silicon, iron, zinc, copper and lead [39] to be identified. The determination of profiles of oxygen and silicon across silicon nitride has also proved possible [40].

(c) **Biological samples**

The medical applications of the nuclear microprobe are of considerable potential but have not so far been extensively investigated. However, the value of charged-particle excited X-rays for the examination of biological samples has already been demonstrated with large-diameter beams and there is no reason to assume that a reduction in beam diameter is likely to have any major effect, except that of reducing the beam current incident on the samples. In many cases, the X-ray yield is sufficiently high to mean that this is not an undue restriction upon possible use. Specimens examined with macrobeams include blood, serum, liver and autopsy samples. The application of the microprobe to biological analysis has been demonstrated [41]. Of particular interest is the possibility of applying the microprobe to samples that have been placed in air [42]. This has been acnieved by mounting the sample on a thin window through which the beam is extracted from the accelerator. Samples which cannot be examined in vacuo can be investigated by this technique.

REFERENCES

[1] ANDERSEN, C.A., Ed., Microprobe Analysis, John Wiley & Sons, New York (1973).
[2] HEVESY, G., LEVI, H., Dan. Vidensk. Selsk., Mat.-Fys. Medd. **14** 3 (1936).
[3] SEABORG, G.T., LIVINGGOOD, J.J., J. Am. Chem. Soc. **60** (1938) 1784.
[4] PIERCE, T.B., PECK, P.F., HENRY, W.M., Nature (London) **204** 4958 (1964) 571.
[5] PIERCE, T.B., PECK, P.F., HENRY, W.M., Analyst **90** 1071 (1965) 339.
[6] PIERCE, T.B., PECK, P.F., Proc. SAC Conference, Nottingham, 1965 (SHALLIS, P.W., Ed.), Heffer and Sons Ltd., Cambridge (1965) 159.
[7] DZEMARD'YAN, Y.A., MIKHAILOV, G.I., TARCHIK, L.P.S., Ind. Lab. **37** (1971) 708.
[8] SIPPEL, R.F., GLOVER, E.D., Nucl. Instrum. Methods **9** (1960) 37.
[9] MACEY, D.J., GILBOY, W.B., Nucl. Instrum. Methods **92** (1971) 501.
[10] RICCI, E., Private communication.
[11] ZELENSKI, V.F., KHAR'KOV, O.N., KULAKOV, V.S., SKAKUN, N.A., Prot. Met. **6** (1970) 235.
[12] MOLLER, E., STARFELT, N., TIEBELAGET, A.K., Atomenergi, Sweden, Report 237 (1966).
[13] PIERCE, T.B., Proc. 2nd Conf. Practical Aspects of Activation Analysis with Charged Particles, Liège, 1967 (EBERT, H.G., Ed.), Report EUR 3896, d-f-e, Brussels (1968) 389.
[14] DECONNINCK, G., DEMORTIER, G., Nuclear Techniques in the Basic Metal Industries (Proc. Symp. Helsinki, 1972), IAEA, Vienna (1973) 573.
[15] PIERCE, T.B., PECK, P.F., CUFF, D.R.A., Anal. Chim. Acta **39** (1967) 433.
[16] CHEMIN, J.F., ROTURIER, J., SABOYA, B., PETIT, G.Y., J. Radioanal. Chem. **12** (1972) 221.
[17] AMSEL, G., NADAI, J.P., D'ARDEMARE, E., DAVID, D., GIRARD, E., MOULIN, J., Nucl. Instrum. Methods **92** (1971) 481.
[18] ROBIN, R., COOPER, A.R., HEUER, A.H., J. Appl. Phys. **44** (1973) 3770.
[19] PRETORIUS, R., COETZEE, P., J. Radioanal. Chem. **12** (1972) 301.
[20] McMILLAN, J.W., HEARST, P.M., PUMMERY, F.C.W., HUDDLESTON, J., PIERCE, T.B., paper presented to 3rd Int. Conf. Ion Beam Analysis, Washington (1977).
[21] OLIVIER, C., PEISACH, M., J. Radioanal. Chem. **12** (1972) 313.
[22] WEBER, G., QUAGLIA, L., J. Radioanal. Chem. **12** (1972) 232.
[23] CUYPERS, M., QUAGLIA, L., ROBAYE, G., DUMONT, P., BARRANDON, J.N., Proc. 2nd Conf. Practical Aspects of Activation Analysis with Charged Particles, Liège, 1967 (EBERT, H.G., Ed.), Report EUR 3896, d-f-e, Brussels (1968) 371.

[24] TUROS, A., WIELUNSKI, L., BARCZ, A., Nucl. Instrum. Methods 111 (1973) 605.
[25] LIGHTOWLERS, E.C., NORTH, J.C., JORDAN, A.S., DERICK, L., MERZ, J., J. Appl. Phys. 44 (1973) 4758.
[26] WOLICKI, E.A., KNUDSON, A.R., Int. J. Appl. Radiat. Isot. 18 (1967) 429.
[27] PIERCE, T.B., PECK, P.F., CUFF, D.R.A., Analyst 97 171 (1972) 1152.
[28] Nucl. Instrum. Methods 142 (1977) 1/2.
[29] PIERCE, T.B., PECK, P.F., CUFF, D.R.A., Nucl. Instrum. Methods 67 (1969) 1.
[30] DYMNIKOV, A.D., FISHKOVA, T., YAVOR, S., Sov. Phys. − Tech. Phys. 10 (1965) 340.
[31] COOKSON, J.A., PILLING, F.D., United Kingdom Atomic Energy Report, AERE R6300 (1970).
[32] PIERCE, T.B., McMILLAN, J.W., PECK, P.F., JONES, I.G., Nucl. Instrum. Methods 118 (1974) 115.
[33] McMILLAN, J.W., PIERCE, T.B., Proc. Int. Conf. Karlsruhe, Ion Beam Surface Layer Analysis (KAPPLER, F., Ed.), Plenum Press, New York (1976).
[34] McMILLAN, J.W., PUMMERY, F.C.W., United Kingdom Atomic Energy Report, AERE R8457 (1976).
[35] PIERCE, T.B., HUDDLESTON, J., United Kingdom Atomic Energy Report, AERE R8514 (1977).
[36] OLIVER, C., McMILLAN, J.W., PIERCE, T.B., J. Radioanal. Chem. 31 (1976) 515.
[37] PRICE, P.B., BIRD, J.R., Nucl. Instrum. Methods 277 (1969) 69.
[38] NOBILING, R., TRAXEL, K., BOSCH, F., CIVELEKOGLU, Y., MARTIN, B., POVIY, B., SCHWALM, D., Nucl. Instrum. Methods 142 (1977) 49.
[39] COOKSON, J.A., FERGUSON, A.T.G., PILLING, F.D., J. Radioanal. Chem. 39 (1972) 12.
[40] COOKSON, J., PILLING, F.D., Thin Solid Films 19 (1973) 381.
[41] COOKSON, J.A., LEGGE, G.J.N., Proc. Anal. Div. Chem. Soc. 225 (1975) 12.
[42] COOKSON, J., Phys. Med. Biol. 21 (1976) 965.

DETECTION OF CHARACTERISTIC X-RAYS
Methods and applications

V. VALKOVIĆ
Rice University,
Houston, Texas,
United States of America
and
Institut "Ruđer Bosković",
Zagreb,
Yugoslavia

Abstract

DETECTION OF CHARACTERISTIC X-RAYS: METHODS AND APPLICATIONS.
Fundamental processes in production, absorption and detection of X-rays are discussed. The application of X-ray emission spectroscopy to the analysis of elements is presented. Sample excitation by X-ray tube, radioactive source and charged-particle beams from accelerators are described. Several experimental arrangements are discussed, and results on sensitivity, accuracy and precision are reviewed. The detection of X-rays characteristic for an element enables trace amounts of several elements to be detected simultaneously. Many applications in material sciences, industry, geology, archeology and forensic science are described. The most interesting studies done by X-ray emission spectroscopy involve the complex movement of elements in nature and their effects on health and diseases. There are numerous interesting applications in medicine, biology and environmental sciences. Of special interest are the biologically essential trace elements in fossil fuels and agriculture.

1. INTRODUCTION

It usually takes a decade, or even several decades, from a discovery to its practical applications. This was not the case with X-rays. They were widely applied in medical and industrial radiography within a year of their discovery (Röntgen, 1895). The milestones in the development of X-ray spectrometry are:

1895	W.C. Röntgen	discovery of X-rays
1896	A.W. Wright	first photographic paper röntgenogram
1896	J. Perrin	X-ray intensity measurement with an air ionization chamber
1909-1911	C.G. Barkla	discovery of absorption edges and emission line series
1912	M. von Laue, W. Friedrich, and E.P. Knipping	diffraction of X-rays by crystals
1912	J. Chadwick	detection of characteristic X-rays induced by α-particles
1913	W.L. Bragg, and W.H. Bragg	Bragg X-ray spectrometer

1913	H.G.J. Moseley	established the relationship between wavelength of X-ray lines and atomic number
1913	N. Bohr	model of the atom
1913	W.D. Coolidge	hot-filament, high-vacuum X-ray tube
1913-1923	W.D. Coolidge	measurement of X-ray spectra of the chemical elements
1922	A. Hadding	application of X-ray spectra to chemical analysis of minerals
1923	G. von Hevesy	quantitative analysis by secondary excitation of X-ray spectra
1928	R. Glocker, and H. Schreiber	applications of X-ray fluorescence spectrometry
1948	H. Friedman, and L.S. Birks	prototype of the first commercial X-ray emission spectrometer
1960-1970		development of high resolution Si(Li) detectors

Development of high resolution Si(Li) detectors in 1960's rejuvenated the interest in the detection of characteristic X-rays and possible analytical applications. The basic limitation of the energy aside from the preamplifier noise limitations, is the statistical fluctuation in the number of ion pairs created for a given photon energy. Geometrical effects have an important influence on the energy resolution. However, the first major step in improving energy resolution came with the development of low noise field-effect transistors (FET). Low-temperature operation of these devices was required in order to reduce the noise. Another significant improvement was the introduction of the light-emitting diode (LED) to provide low-frequency feedback to the preamplifier. In order to prevent contamination of the detector surface from condensation of impurities, the detector is protected by a part sealed by a beryllium window. The absorption in this thin beryllium foil reduces the detector efficiency for lower X-ray energies. Detector size and geometry will determine efficiency for higher X-ray energies. Efficiency for low X-ray energies can be further reduced by inserting an absorber between the radiator and the detector.

2. FUNDAMENTALS

X-rays are produced as a result of the removal of at least one inner-shell electron. Several processes might result in the removal of an electron from the inner shell in the atom. In principle one might distinguish between two different groups of processes. The first group consists of the scattering processes in which the incoming charged particle (electron, proton, α-particle, heavy ion) hits the electron in the K-shell and transfers part of its kinetic energy to the electron. As a result the vacancy is created which can be filled with the electron from one of the outer shells. Obviously the vacancy can be produced also in the process where X-ray or γ-ray transfer their energy to the K-shell electron. This transfer of energy can be partial or complete. In order to realize these processes charged particles (electron, proton, α-particles, heavy ions) produced by a variety of accelerators can be used.

The second group of processes involves the interaction of atomic nucleus with its electron cloud and as a result of this interaction a vacancy is created in the K-shell. Radioactive isotopes that decay by γ-emission may undergo internal conversion. The γ-photon is absorbed within the atom of its origin giving its energy to an orbital electron. Radioactive isotopes that decay by β-emission may also undergo internal conversion. The electron from the nucleus may transfer its energy to the orbital electron and create the vacancy in the K-shell. The third process which produces vacancies in the inner shells is orbital electron capture.

Once the atom is excited its de-excitation can occur not only by emission of electromagnetic radiation but also by some other processes. The fluorescence yield of an atomic shell is the probability that a vacancy in that shell will be filled through a radioative transition. The competing processes are: emission of Auger electrons and Coster-Kronig transitions, which are transitions between the subshells of the same principal quantum number.

When a target is bombarded by protons or heavier ions, electrons are ejected from the atomic shell of the target atoms, while the incoming charged particle loses its energy in passing through the target. In 1912 Chadwick [1] first observed and identified the characteristic X-rays of several elements exposed to α-particles from radioactive sources. Early theoretical and experimental work on charged-particle-induced X-ray emission has been summarized in the article by Merzbacher and Lewis [2]. The modern theoretical concepts and available experimental data on energy spectra of electrons ejected in ion-atom collisions are summarized by Ogurtsov [3]. Theoretical calculations are usually performed using the Born approximation for inelastic collisions.

Another theory which many authors used is classical binary-encounter approximation as developed by Gryzinski [4]. In the binary-encounter approximation an interaction is considered between two classical particles: the incident ion (or electron) and the atomic electron. The atomic features of the latter are taken into account by introducing a proper velocity distribution.

As pointed out by Garcia [5], the binary-encounter model offers a scaling law for the cross sections including only electron binding energy and particle energy. A universal curve is obtained if $u_i^2 \sigma$ (i = K or L) is plotted versus $E/\lambda u_i$, where u_i is the electron binding energy, E the proton energy, σ the ionization cross section and λ the proton-electron mass ratio.

Yields of X-rays produced by charged particles have been measured for a variety of charged particles at different bombarding energies. The early work has been summarized by Merzbacher and Lewis [2], while the partial list of recent work is presented by Duggan et al.[6]. Cross sections for K-shell ionization by p, d, ^{3}He and ^{4}He have been tabulated by Rutledge and Watson [7].

Protons are the most often used charged particles for the excitation of characteristic X-rays. Even so, the data on proton-induced X-ray emission cross sections are still scarce. Compilation of measured cross sections for proton energies most often used in X-ray production is presented by Valković [8].

Johansson and Johansson [9] have fitted all the available data on proton X-ray production with a fifth degree polynomial

$$\ln(\sigma u_i^2) = \sum_{n=0}^{\infty} b_n X^n$$

where $X = \ln(10^{-3} E/\lambda u_i)$. u_i is the ionization energy in eV and E is the proton energy in eV. The unit of the ionization cross section is $10^{-14} cm^2$. The obtained values for coefficients b_i for K X-rays are: $b_0 = 2.0471$, $b_1 = -0.65906$, $b_2 = -0.47448$, $b_3 = 0.99190$, $b_4 = 0.46063$, and $b_5 = 0.60853$.

3. TRACE ELEMENT ANALYSIS: EXPERIMENTAL SETUP

In recent years the detection of characteristic X-rays for trace element analysis has increased significantly, primarily because high resolution Si(Li) detectors have made possible energy dispersive analysis. X-rays from X-ray tubes and radioactive sources have been used to excite characteristic X-rays in different samples. Because of large cross sections for X-ray production, charged particles can also be used to produce characteristic X-rays from elements present in only trace amounts in target materials. Fig. 1 indicates the principle of X-ray emission spectroscopy.

A radioisotope X-ray fluorescence analyser consists of the following basic components (see Fig. 2): (i) a scaled radioactive source; (ii) a detection system which selects the energies of the excited characteristic X-rays and measures their intensities; (iii) an electronic and read-out system whose output may be related to the list of elements present in the sample and their concentrations.

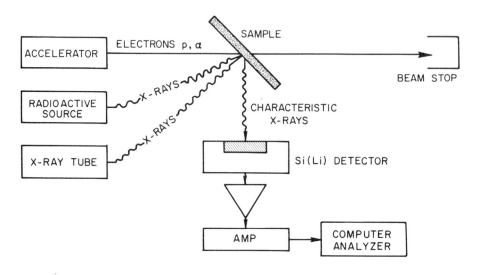

FIG.1. The principle of X-ray emission spectroscopy.

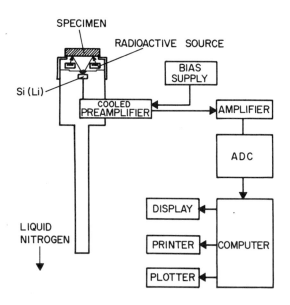

FIG.2. Schematic representation of a radioisotope X-ray fluorescence analyser.

Very useful is a review article by Rhodes [10]. He reported a compact X-ray energy spectrometer consisting of a radioisotope X-ray source, an Si(Li) or Ge(Li) detector and a small computer. The efficient excitation of K X-rays from Na to U and L X-rays from Ag to U was achieved using different interchangeable sources. This apparatus was reported to resolve K X-rays from adjacent elements down to Na. Depending on the source and on the part of the spectrum examined, characteristic X-rays from up to 15 elements can be simultaneously measured for either qualitative or quantitative analysis. A small computer was used to store spectra and to perform simple data processing.

There are several commercial systems on the market using radioactive source to excite characteristic X-rays in the sample. The most often used radioisotopes are: ^{55}Fe ($\tau_{\frac{1}{2}}$ = 2.7 years, E = 5.9 keV) and ^{109}Cd ($\tau_{\frac{1}{2}}$ = 1.3 years, E = 22.2 keV and 88.0 keV).

Sample excitation by X-rays produced by an X-ray tube has been reported by many authors. By exposing the sample to bremsstrahlung radiation from an X-ray tube it is possible to excite simultaneously all the elements present in the target. Electrons emitted from the heated filament are accelerated towards the target by a potential difference of up to 50 kV, and when they strike the target they generate X-rays characteristic of the elements present in it as well as bremsstrahlung. If one is interested in exciting a particular element in the target, the tube voltage should be adjusted so that the maximum of this energy distribution is just above the energy of the K absorption edge of that element, which gives the greatest excitation efficiency. In order to improve sensitivity (by reducing background), a monoenergetic X-ray beam should be used to excite the target. In principle there are two ways of obtaining monoenergetic radiation from an X-ray tube. One is to pass the radiation from the tube through suitable filters. Another way of obtaining monoenergetic X-rays is to use rays from an X-ray tube to excite a target of some pure metal and then make the secondary radiation from this excite the target to be examined.

There are several commercial systems available on the market using a variety of X-ray tubes. One should remember that the excitation by radioactive sources and X-ray tube requires a rather thick target and that matrix effects can be significant. Excitation by using radioactive sources or X-ray tubes is, however, completely nondestructive. As a result, it can be applied to a variety of problems where the sample is not supposed to be damaged.

Charged-particle-induced X-ray emission has been used as an analytical tool with a variety of machines accelerating different ions. Although it is difficult to compare results from different laboratories, this has often been done. These comparisons are subject to many subjective estimates. When different ions are compared, their energies have to be equivalent (MeV/amu). Comparisons of different modes of excitation have been recently summarized by Johansson and Johansson [9]. Excitations by protons, α-particles, heavy ions, electrons, and X-rays were discussed. Their discussion favors proton induced X-ray emission.

When a choice of beam is made, the scattering chamber should be constructed. Modification of a scattering chamber used in nuclear reaction studies is a possibility. Besides X-ray detector it is possible to mount charged particle detectors in such a chamber. This then allows simultaneous measurement of backscattered protons and beam intensity and target thickness monitoring. In some instances the use of scattering chamber may be very difficult (wet samples, for example). In cases like this the proton beam can be passed through thin foils onto samples in air.

Several scattering chambers constructed for the analytical work using charged-particle-induced X-ray emission have been reported in the literature. Let us mention the scattering chamber described by Valković et al. [11]. The target chamber was constructed with optimum efficiency for the collection of produced X-rays. Care was taken to prevent the Si(Li) detector from seeing any high-Z materials which could produce fluorescence by beam scattering from the last beam slit or Compton scattered X-rays. In the chamber the target frames ride up and down on an aluminum target ladder which is milled off center so that the target faces are in the exact center of the chamber. In order to eli-

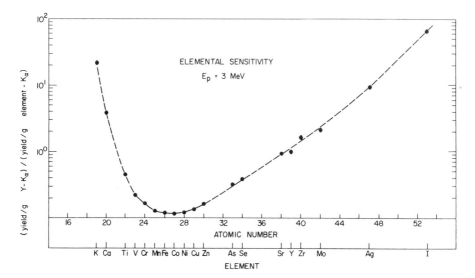

FIG.3. *The relative elemental efficiency of the system at $E_p = 3$ MeV shown as the ratio of the number of K X-rays of yttrium (dopant) and of the given element for the same number of atoms (From Ref.[11]).*

minate X-rays from elements below those of potassium and secondary electron-induced bremsstrahlung, a 0.1 cm polystyrene absorber was inserted between the target and the detector. A proton energy of 3 MeV was chosen to maximize sensitivity for elements in the region of Fe and Zn (essential trace elements) (see Fig. 3). While increasing the energy has the effect of increasing the characteristic X-ray production of all elements, the bremsstrahlung background is increasing at a faster rate. Thus, sensitivity is lost in the Fe to Zn region.

When the proton beam strikes the target in the scattering chamber a number of X-rays are produced. Some of them will hit the X-ray detector which will generate a pulse through the preamplifier. In most cases the preamplifiers with pulsed optical feedback are used. Because of the extremely long fall time of Si(Li) pulses, high-energy tailing of X-ray peaks due to pile-up can be a serious problem in trace element analysis. In order to avoid pile-up at higher counting rate a pulse pile-up rejector should be used to inhibit X-ray amplifier whenever piled up pulses are received. Pulses from the amplifier can then be fed into multichannel ADC which is interfaced to a computer. Pulse pile-up problems can be avoided by keeping the counting rate low. Each characteristic X-ray line will result in a peak in the spectrum. Once the data are stored in the computer or analyzer the reduction involves calculations of peak locations, amplitudes and subtraction of background counts.

Analysis of charged-particle-induced X-ray emission (using the accelerators) is greatly simplified because of the absence of matrix effects. In order to obtain absolute abundances of different elements in the target, the target material is usually doped with the known concentration of an element which one does not expect to find in the target. What is needed is only standards containing known ratios of the doping agent to the different elements which might be present in the target. Such a relative measurement does not require the knowledge of detector efficiency as a function of X-ray energy.

Target preparation is different for different types of specimen excitations. In the case of excitation by X-ray tube or radioactive sources, rather thick targets are needed. Loose powders or liquids are usually analyzed using a container of some kind. Filter papers are becoming increasingly important in

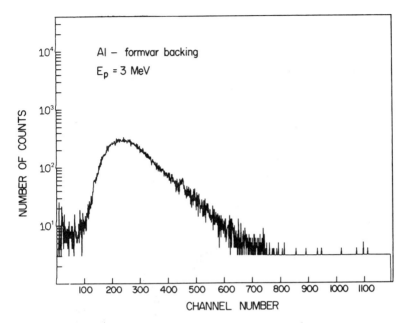

FIG.4. *An X-ray spectrum obtained by bombardment of aluminum formvar backing with 3-MeV protons (from Ref. [11]).*

the study of air pollutants, and also filters or membranes are used to concentrate trace elements from liquid. Interferences due to internal absorption and enhancement of the fluorescent radiation and the changes in emitted intensity due to specimen heterogeneity and finite particle size are different for different specimens. Minimization of errors due to these effects is a central problem in excitation by X-ray tube or by radioactive sources.

Charged-particle-induced X-ray emission has the advantage with respect to others that the targets used are much thinner and matrix effects can be neglected. The targets are usually prepared as deposits on some kind of backings. Different backings were investigated with the aim of finding one which would produce the least amount of background radiation and which would support the beam well. Any material to be deposited should be first reduced to a solution or a suspension of microscopic particles in an inert solvent. Pure water is the best solvent for most materials.

A very useful technique is the one involving the preparation of Al + formvar backings [11]. By reinforcing aluminum foils with formvar one obtains backings which may be used in charged-particle-induced X-ray spectrometry. The formvar provides mechanical strength while the liquid or suspension is drying, while the aluminum provides resistance on the beam of charged particles. Such a backing is free of any interference lines (see Fig. 4). Targets of water, blood serum and many solutions are very easily prepared by simply allowing a few drops to dry on such aluminum formvar backing. Another backing that can be used for powder samples is scotch tape (3M company, St. Paul, Minnesota). The main contaminant in this backing is bromine (see Fig. 5).

Two factors determine the choice of the particle bombarding energy: the X-ray production cross section dependence and the bremsstrahlung produced by secondary electrons. Small cross section eliminates the use of low energies.

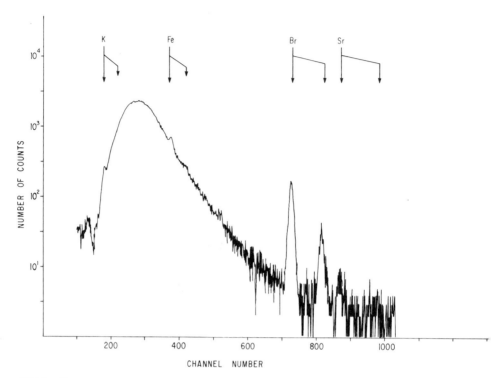

FIG.5. *X-ray spectrum resulting from 3-MeV proton bombardment of scotch tape (3M Company).*

On the other hand, bremsstrahlung of secondary electrons limits the use of higher energies. When all the parameters of the system have been chosen (including beam intensity and beam energy) the reproducibility of the method for many equivalent targets under routine analysis conditions should be studied. This should be performed for every type of target.

There are several good review papers published on the subject of X-ray emission spectroscopy [9, 12, 13]. The subject of particle-induced X-ray emission spectroscopy has been discussed at the International Conference, Lund, Sweden, 1976. The papers presented there will be published as a separate issue of Nuclear Instruments and Methods (1977). X-ray emission spectroscopy is a powerful analytical method; using it a variety of fundamental problems involving the movements of elements may be studied.

4. ELEMENTS IN THE NATURE

The problem of determining the "Universal abundances" of elements has been discussed by many authors. Beyond the Earth, the Moon and the meteorites, the abundance of at least some elements can be obtained for the solar system, for the Sun and other stars, for gaseous nebulae, including some of external galaxies, for interstellar space and for cosmic rays. Probably the best known compilation of abundances is that of Suess and Urey [14]. The predominance of hydrogen and helium in the Universe lead to the theories of development of the Universe and to the formulation of theories about element formation. It is

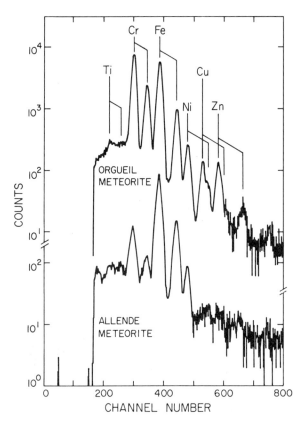

FIG.6. *X-ray spectra obtained by the proton irradiation of meteoritic dust. Top: Orgueil meteorite. Bottom: Allende meteorite.*

also now believed that the natural abundances of most elements have been established in the few final seconds of a star's lifetime as it disrupts explosively.

The "peninsula" of the known nuclei in the N-Z plane terminates because of nuclear fission. As one moves along the peninsula towards heavier nuclei, the disruptive Coulomb forces increase faster than the cohesive nuclear forces.

The "peninsula" of the known nuclei in the N-Z plane terminates because of nuclear fission. As one moves along the peninsula towards heavier nuclei, the disruptive Coulomb forces increase faster than the cohesive nuclear forces. "Island" of stability has been predicted on the basis of shell model calculations around $Z = 114$ and $N = 184$. Several experimental attempts have been made to detect the spontaneous fission of superheavy elements that might still be stored in natural materials. Evidence for superheavy nuclei has been reported by Gentry et al. [15] in an experiment where small inclusions of monazite in mica from Madagascar have been bombarded by a proton beam with subsequent detection of produced X-rays. Efforts in several other laboratories have not been able to reproduce their result, and the observed peak has been associated with the $(p,n\gamma)$ reaction.

The elemental composition of meteorites has been studied for some time. Of interest are isotopic ratios as well as elemental abundances. Since often only small amounts of material are available, proton-induced X-ray emission

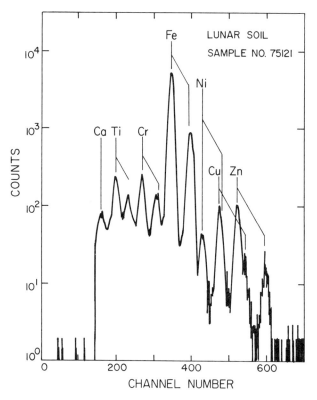

FIG.7. *X-ray spectrum obtained by proton irradiation of the lunar soil sample.*

is a valuable tool in determining the relative elemental abundances. Fig. 6 shows the X-ray spectra obtained by proton irradiation of meteoritic dust. Peaks corresponding to light elements are supressed by an absorber in front of the Si(Li) detector.

Trace elements in various rocks can give valuable clues as to their mode of formation. For example, the distribution of trace elements and their relative proportions in igneous, metamorphic and sedimentary rocks has been useful in establishing the genetic relationship between various rocks and between rocks and ore deposits.

The first direct chemical analysis of the Moon was performed by α-particle back-scattering [16]. Heavier elements can be easily measured by the X-ray emission spectroscopy. Fig. 7 shows an X-ray spectrum obtained by proton irradiation of the lunar soil sample.

In the natural system (rock-soil-aqueous solutions-organisms) soils are an exceptionally important link. Trace element composition of soils can be easily determined by X-ray emission spectroscopy. When protons are used for excitation, only small amounts of material are needed for target preparation. Fig. 8 shows X-ray spectra obtained from the proton irradiation of soil samples.

5. TRACE ELEMENTS IN LIVING MATTER

One of the most important characteristics of living cells is their ability to take up elements from a solution against the concentration gradient. This is most obvious for marine micro-organisms which obtain their nutrients

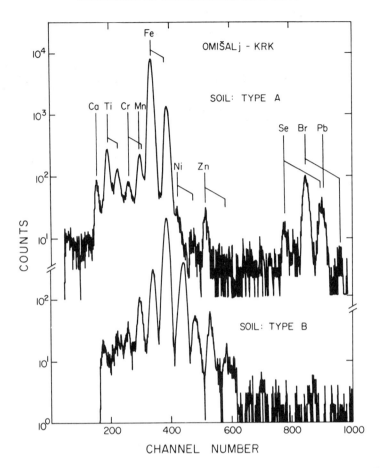

FIG.8. *X-ray spectra obtained by the proton irradiation of two types of soil samples.*

directly from the sea water. The concentration factor is then defined as the ratio of element concentration in the organism and the element concentration in sea water. The organisms concentrate all elements present in their environment. However, all of these elements are not essential for life. Criteria which an element must satisfy to be essential for life are well established [17].

The relation of the uptake of essential elements to yield or growth may be considered as a definition of essentiality. For an essential trace element there is a rather narrow range of adequacy of element concentration in the organisms. Smaller concentrations result in different abnormalities induced by deficiencies which are accompanied by pertinent specific biochemical changes. Higher concentrations result in toxicity.

The bulk of living matter consists of eleven elements which have very low atomic weights (H, C, N, O, Na, Mg, P, S, Cl, K, and Ca). These elements have been known to be essential for life for a long time because their presence is easy to detect. The problem of essential trace elements is much more difficult.

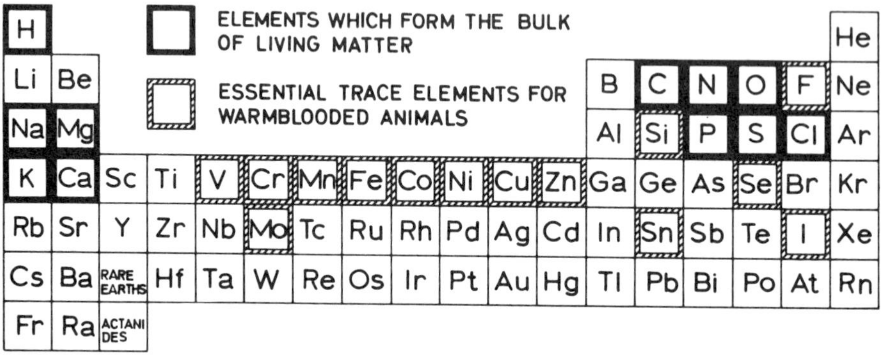

```
| H  |    |    |    |    |    |    |    |    |    |    |    |    |    |    |    |    | He |
| Li | Be |    |    |    |    |    |    |    |    |    |    | B  | C  | N  | O  | F  | Ne |
| Na | Mg |    |    |    |    |    |    |    |    |    |    | Al | Si | P  | S  | Cl | Ar |
| K  | Ca | Sc | Ti | V  | Cr | Mn | Fe | Co | Ni | Cu | Zn | Ga | Ge | As | Se | Br | Kr |
| Rb | Sr | Y  | Zr | Nb | Mo | Tc | Ru | Rh | Pd | Ag | Cd | In | Sn | Sb | Te | I  | Xe |
| Cs | Ba | RARE EARTHS | Hf | Ta | W  | Re | Os | Ir | Pt | Au | Hg | Tl | Pb | Bi | Po | At | Rn |
| Fr | Ra | ACTANIDES |    |    |    |    |    |    |    |    |    |    |    |    |    |    |    |
```

☐ ELEMENTS WHICH FORM THE BULK OF LIVING MATTER

▨ ESSENTIAL TRACE ELEMENTS FOR WARMBLOODED ANIMALS

FIG.9. Essential trace elements for warm-blooded animals.

So far, the elements F, Si, V, Cr, Mn, Fe, Co, Ni, Cu, Zn, Se, Mo, Sn, and I have been recognized as the essential trace elements for warm-blooded animals (see Fig. 9). The great majority of the essential trace elements serve as key components of the enzyme system or of proteins with vital functions. If the metal atom is removed, the protein usually loses its capacity to function as an enzyme. Of fourteen elements recognized until now as essential trace elements for warm-blooded animals, only fluorine and silicon are below 20 in atomic number. It is interesting to point out that none of the 30 elements beyond iodine have ever been shown to be of any physiological significance. The distribution of essential trace elements within the periodic table of elements might have some significance with respect to the development of life in prebiotic time.

5.1 Trace elements in microorganisms

Concentration factors for some elements in microorganisms are known [18]. Microorganisms require small quantities of Mn, Fe, Zn, Co, Cu, and Mo for their growth. Some diseases caused by microorganisms result in modified trace-element levels. Microorganisms must compete with host metal-binding agents for these essential elements. It is well established that in the distant past massive faunal extinctions have occurred. There is mounting evidence that a correlation exists between major faunal extinctions and geomagnetic polarity reversals. The validity of this correlation in recent geological time seems to have been well established by studies of fossil species of single-celled marine micro-organisms [19, 20].

Several mechanisms linking changes in the geomagnetic field with effects on living organisms have been proposed. Most of them are based on the assumption that during polarity reversals the dipole component of the geomagnetic field probably weakens or disappears for periods of a few thousands of years [21] allowing a much greater flux of both solar protons and galactic cosmic rays to bombard the surface of the Earth. The effect of increased exposure is, however, now believed to be probably insignificant. Other mechanisms include climate change [22] (which also may result from intense solar proton bombardment during polarity reversals), and the effects of a large reduction in the content of ozone in the atmosphere, which would increase exposure to ultraviolet radiation [23]. Direct magnetic field effects on growth were proposed by Hays [20] and Crain [24]. A direct cause-and-effect link between magnetic field reversals and species extinctions could explain how marine organisms easily could have been affected. There is also the possibility that extinctions discussed here may have resulted from the simultaneous effect of several factors.

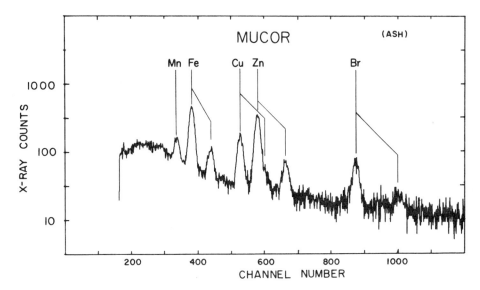

FIG.10. X-ray spectrum obtained from the bombardment of a mold sample on a scotch tape backing with a 3-MeV proton beam.

In order to try to explain the geomagnetic effects on living organisms the concentration factor dependence on magnetic field intensity has been recently proposed [25]. Magnetic field dependence of concentration factor can bring living organisms into the range of deficiency or toxicity without changing trace element availability in the environment. Experiments aiming to establish functional relationship between concentration factor and magnetic field intensity for some elements and some living organisms were reported [26]. The measurements involved the growth of single cell organisms in defined media and controlled external magnetic field intensity. Trace element analysis of harvest is performed by proton-induced X-ray emission spectroscopy.

The organism used is a respiratory deficient mutant of M. bacilliformis which in addition has lost the ability to grow as mycelium. Instead, it exists as spherical cells which reproduce only by budding at the expense of alcoholic fermentation [27]. A defined growth medium was used. It contains two vitamins, nine amino acids, glucose as an energy source, K salts, and the salts of five trace elements: Mg, Zn, Fe, Mn, and Cu.

Growth took place at room temperature in round-bottom flasks containing 14 ml of medium. Each flask was inoculated so as to have 10^3 cells/ml. Six flasks were placed in solenoids whose magnetic fields were extremely uniform over the active region of growth and six flasks were used as a control group. Growth (increase in cell number) was monitored daily by measurements of the turbidity of the cell suspensions. After ten days, when the cell number was maximum, the cells were harvested by filtration.

Fig. 10 shows a typical X-ray spectrum from a mold sample (Mucor). The microorganisms were harvested by filtration through filter paper which was then exposed to the beam of 3 MeV protons at 150 nA for 500 seconds. The observed spectra indicate that the following elemental concentration ratios can be measured for both the mold and the growth medium: Mn/Zn, Fe/Zn, Ni/Zn, and Cu/Zn. The results obtained so far are inconclusive about the existence of possible magnetic field dependence of concentration factor.

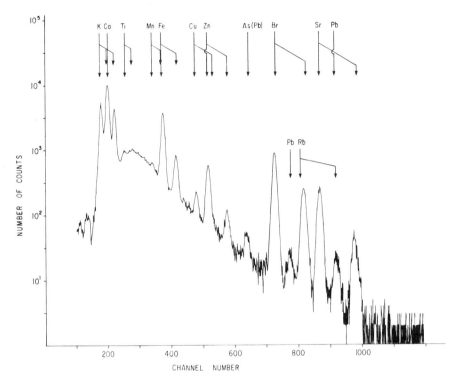

FIG.11. Characteristic X-ray spectrum from the tree target (ash on scotch tape). Beam intensity 10–20 nA, proton energy $E_p = 3\,MeV$.

5.2 Trace elements in plants

   The elemental composition of the plants reflects to some extent the composition of the soil or other growing medium. The relative ratios of the elements in a plant are not necessarily the same as those in the soil or even in the nutrient solution. In other words, plants are to some degree selective in their absorption of elements. The uptake of the elements from the soil is determined by many factors including: the element abundance, the form of element, the pH of the soil, the physical condition of the soil and the genetic constitution of the plant species. For an essential element there is a rather narrow range of the adequacy of element concentration in organisms. Smaller concentrations result in different abnormalities induced by deficiencies which are accompanied by pertinent specific biochemical changes. In plants it is possible to have, under severe deficiency conditions, a decrease in the concentration of an element which results in a small increase in the growth. This phenomenon is known as the Steenbjerg effect [28]. The elements shown to be essential for all the plants include B, Ca, Cl, Co, Cu, Fe, K, Mg, Mn, Mo, Zn, and possibly I, Na, Se, Si, and V.
   Data on the uptake of the elements from soils by higher plants are very scarce although it is known that certain species have the ability to accumulate uniquely high concentrations of a particular element.
   Tree-rings have been studied for some time. It seems that the trees store a record of atmospheric temperature in their rings. This is of interest to the

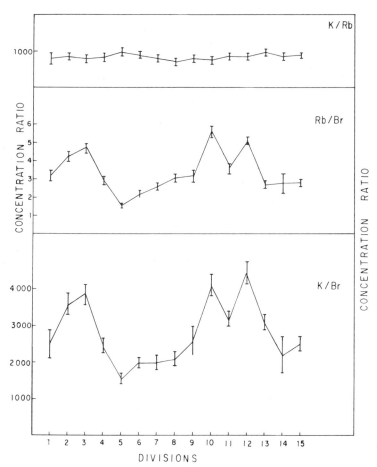

FIG.12. Variations in the Rb/Br and K/Br concentration ratios. K/Rb concentration ratios across the tree rings show no variations.

historical meteorology since in some areas the tree-ring records extend back for up to eight millenia. The science of dendroclimatology is based on the observation that the narrow rings represent a cold (or dry) year. Since the water conduction remains limited to the outmost annual ring tree rings offer also the possibility to study the historical trends in the movements of the elements. Elemental concentrations in the tree rings should reflect the availability of the elements in question. One possible source of the additional element loading is the atmospheric discharge.

In the recent measurements [29] samples from an approximately 500 year old tree were analyzed for trace elements using proton-induced X-ray emission spectroscopy. A linear cut across the cross section of the trunk was devided in 15 divisions; the division No. 1 being the closest to the centre of the tree. Each sample was then cut in small pieces and transferred into a separate crucible and ashed at 450°C. The obtained ash was smeared on the scotch tape and exposed to a 10-20 nA beam of 3 MeV protons. Fig. 11 shows the resulting X-ray spectrum. The peaks associated with K, Ca, Ti, Mn, Fe, Cu, Zn, As, Pb, Br, Rb, and Sr could be identified in all spectra. Fig. 12 shows the variations in the Rb/Br and K/Br concentration ratios. The top of the figure shows the variations in the K/Rb concentration ratios across the tree rings from the centre of the tree outwards. Within the error bars, the ratio K/Rb is 1000 across the tree. This fact should be explained by keeping in mind that Rb is a member of the series NH_4 - K - Rb - Cs, and that the members of this series are more similar in their chemical and physical properties than are the members of any other group, with the exception of the halogens. The radius of the Rb ion is 1.48 A (only about 10% larger than the potassium ion radius); as a result, the Rb ion can be accommodated into the same structures as the K ion.

The Pb/Zn concentration ratio shows two peaks corresponding to the divisions 5 and 8 (one division covers approximately 30 years). A recent increase in the Pb/Zn ratio (divisions 14 and 15) is probably due to the atmosphere particulate loading resulting from the industrial activities and automobile traffic. If the two peaks in the Pb/Zn concentration ratio distribution are due to the Pb increase, data would suggest past particulate loading of the atmosphere due to the natural events (volcano activities and similar). The alternate approach to Pb/Zn distribution data is to assume that the observed peak structure is due to Zn depletion. In this case Zn depletion is strongly correlated with the peak in K/Ca distribution.

Plants provide the main source of minerals to animals and to most human beings. Several interesting problems deserve further study:

(1) Developing a method of identification of the plant source (soil type, location) from the knowledge of plants' trace element composition. There is a possibility of using such a methodology in legal enforcement (drugs, for example).

(2) Determining the concentration factors for toxic elements (Hg, Cd, Pb) for plants in food chains. The dependence of concentration factors on physical and chemical properties of the soil should be determined.

(3) Search for plants with large concentration factors for some elements of interest. This may lead to plants which could be used in geoprospecting, in the "harvesting of elements" using plants as a collector of elements.

5.3 Trace elements in human blood

The serum concentrations of numerous trace elements have in recent years become an area of great interest in biochemical sciences [30]. Blood samples contain the trace quantities of Cu, Fe, Al, Ba, Mn, Ni, Cs, Sn, Sr, Cr, Zn, Pb, Mo, Cd, and others. However, only copper, iron, strontium and zinc appear in 100% of the specimens investigated. These elements attracted interest among many research groups, since it appears that the concentrations hold much promise as a clinical test in several pathological conditions. Several methods have been developed to determine the concentration levels of these elements. X-ray emission spectroscopy offers a very interesting approach to these problems because of its ability to detect simultaneously several trace elements. In addition, charged particle-induced X-ray emission has the advantage that only very small quantities of serum are needed in target preparation; good results can be obtained with a drop of blood serum on an aluminum formvar backing. Fig. 13 shows a typical spectrum of proton-induced X-rays from the blood target. The intensities of light elements P, S, Cl, Ca, and K are artificially suppressed.

In a comparative study of X-ray fluorescence, X-ray excitation with protons and atomic absorptions [31] it was found that all three techniques were routinely capable of measuring elemental concentrations of blood serum at ppm levels with relative standard deviations less than 10 percent. As practical analytical tools, each possess certain inherent advantages that must be weighed in deciding which technique is the most appropriate for a given study.

Clinical studies of trace element concentrations in Hodgkin's disease, non-Hodgkin's lymphoma, and acute leukemia and other diseases should be done. The specific aims of such investigations should be: (i) to determine the clinical usefulness of multiple trace element concentration measurements for the definitive relationships to disease activity and efficiency of therapy; (ii) to explore further the preliminary observations that changes in trace element levels precede other symptoms of disease activity; (iii) to determine trace element concentrations in normal versus diseased tissues.

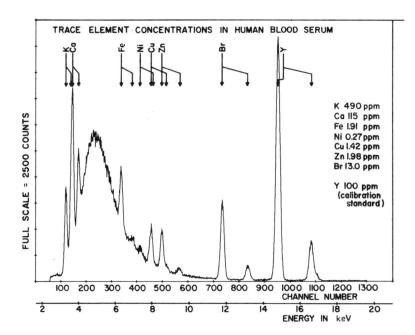

FIG.13. *A characteristic X-ray spectrum from a blood serum target prepared as a dried droplet on an aluminum formvar backing and bombarded with 3-MeV protons (from Ref.[11]).*

Several serum trace element concentration levels are also modified as a result of infectious diseases. There are several examples of infectious diseases that occur because of iron imbalance: patients with hyperferremias caused by such diseases as sickle cell anemia, malaria, hepatitis or relapsing fever frequently suffer from a variety of secondary bacterial infections. Serum zinc level decreases also upon invasion of the host by micro-organisms. Contrary to iron and zinc, serum copper levels rise in acute and chronic infections. The levels return to normal in severely ill patients as well as in cases in which the infection disappears.

5.4 Trace elements in tissues

Trace element concentrations in different organs have been studied for some time. The accumulation of some metals in some organs in malignant diseases has attracted special attention. This fact coupled with the observed geographical variations in death rates from cancer suggested that environmental factors may play an important role. Tumors depend on essential trace elements as does healthy tissue. By changing the trace element concentration, by substituting one element for another, it is a possibility that manipulations can be developed that are detrimental to the tumor without endangering the host.

Let us mention some examples of trace element-malignant disease relationships. For example, selenium is concentrated by rapidly growing tumors, possibly in an effort by the body to inhibit the unchecked growth. This fact allows radioactive selenium to be used to diagnose brain and bone tumors because of its concentration action in these tissues [32]. Human cancer death rates appear to be considerably higher in areas with low selenium concentrations [33].

The reported data demonstrate that cities with high selenium levels have fewer cancer death, on an age-adjusted basis, than cities with low selenium levels. Selenium has also been reported to inhibit carcinogenesis in animals.

Distinct differences were found between normal bone and human osteogenic sarcoma in regard to eight trace metals [34]. In osteogenic sarcoma Fe, Zn, Cu, and Mn were increased and Ca, Mg, Co, and Ni were decreased. There were also differences in the relationship between the metals when taken in pairs and in terms of the overall picture of metals in bone.

X-ray emission spectroscopy, with its capacity to detect several elements simultaneously in very low concentrations, offers a very convenient tool for the study of trace elements and their relationship to malignant diseases. Preliminary results [12] indicated differences in trace metal concentrations in the spleen of the normal C3H mouse and the spleen of the C3H mouse with radiation-induced fibrosarcoma. In a recent study [35] the contents of zinc, iron, antimony, chromium, cobalt, and scandium in DNA were investigated during the growth of transplanted sarcoma M-1. It was found that the concentrations of most elements varied over a wide range during the process of the disease.

Since tumors need essential trace elements for their growth as does the healthy tissue, time variations of trace element concentrations should be studied. The simultaneous measurement of the concentration of essential trace elements in different organs, blood serum, and hair as a function of time interval elapsed from the beginning of the tumor growth will provide the information on kinetics of trace elements. This information then can be used to understand the processes involved in tumor growth, the cross correlation of element concentrations and eventually to use measured trace element levels as the signature of tumor activity. The importance of such work is in the establishment of trace element levels and their interrelationships during tumor growth. This might help the development of : (i) new diagnostic tools for early detection of tumor growth; (ii) establish tumor dependence on trace elements which may lead to the affection of tumor by the change of trace element concentration.

5.5 Trace elements in hair

Hair is a unique biologic material which reflects the biomedical and environmental history of the subject. Since it is convenient to handle and sample, and since it has relatively high concentrations of metals, trace-element analysis of human hair has been applied widely. It has been shown [36] that the diagnosis of chronic plumbism in children can be confirmed by hair lead. Poisoning by other heavy metals and environmental pollution should also be accompanied by elevated levels in the growing hair. Head hair analysis appears to offer a unique approach to the investigation of human trace element nutrition and metabolism. There is a strong possibility that the trace element content of hair correlates with body stores, especially of bone. Analyses of feces and urine are of limited value as indicators of stores, and blood has restricted use because the hemostatic mechanisms operate to keep many of the components of blood constant. Human head hair is a recording filament which can reflect metabolic changes of many elements over a long period of time, and thus it reflects past nutritional events. The idea of hair analysis is very inviting, because hair is easily sampled, shipped, and analyzed.

The concentration of trace elements in human hair may vary for most of the elements from individual to individual, but some correlation due to sex, age, and color has been found. Assuming that the growth of hair is approximately constant and continuous during the lifetime of the subject, the concentrations of trace elements should differ with increasing distance from the scalp. The concentrations of most trace elements should increase with increasing distance from the scalp if the exposure of hair to the elements of the environment is constant. Therefore, the influence of the environment on a population group can be indicated by trace-element analysis. Renshaw et al. [37] have studied the concentration of lead along a single hair and found a nearly linear increase in concentration with distance from the scalp.

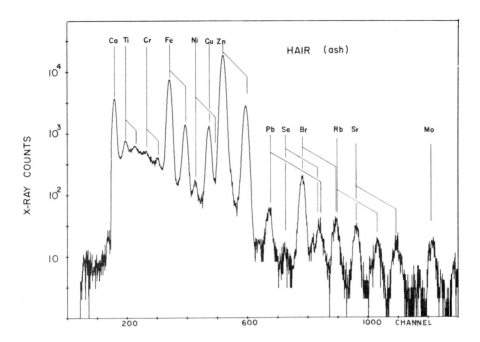

FIG.14. *X-ray spectrum from the proton bombardment of hair target (ash).*

Because of its growth, hair reflects previous elemental concentrations in serum and body (history of previous biochemical and medical events in man), as well as previous environmental effects. Several measurements of trace element distributions along the hair length have been reported [38, 39, 40, 41]. As an example, Fig. 14 shows the X-ray spectrum from the proton bombardment of hair target. Peaks corresponding to Ca, Ti, Cr, Fe, Ni, Cu, Zn, Pb, Se, Br, Rb, Sr, and Mo are present in the spectrum. The investigators agree that the variations are characteristic of the subject and that Zn variations along hair are negligible. This latter fact may be also due to the relation of Zn with the production of melanin pigment in hair. Assuming constant distribution of Zn along the hair, only the ratios of elemental concentrations to Zn concentration need to be measured to obtain elemental concentrations along hair. Such relative measurements can be easily performed on single hairs using proton-induced X-ray emission spectroscopy. It has been shown [42] that the elements whose concentrations increase monotonically along the hair can be identified as pollutants in the area. For widespread pollutants (such as lead) even the medium value of a group of subjects can be used as a measure of the exposure. Actually only elemental concentration ratios (relative to Zn) need to be measured. Fig. 15 shows the concentration ratios As/Zn, Pb/Zn, and Ni/Zn for the hair of one subject as described in ref. [42]. It is important to note that the shape of the dependence on the distance from the scalp is similar for all three ratios. This would indicate that elements As, Pb, and Ni are contaminants in the environment of the subject. Note that the Pb concentrations are an order of magnitude higher than those of Ni and As.

Examining blood and urine provides immense insight into human diseases. It is natural that one would like to add to these examinations a routine analysis of hair. The difficulties associated with such an idea are due to the fact that, as yet, normal and accepted standards have not been established. A definition of "normal" hair for different population groups is different for

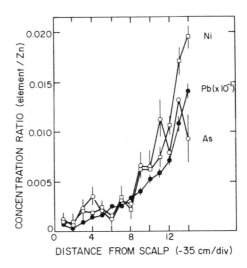

FIG.15. Concentration ratios: As/Zn, Pb/Zn and Ni/Zn as functions of distance from the scalp. The Pb/Zn ratio has been divided by a factor of 10. The errors shown are the statistical ones.

various geographical areas because of the strong environmental influence. X-ray emission spectroscopy was found to be a very convenient method in the determination of these normal values [43].

Kopito and Schwachman [45] have suggested that alterations in hair elemental composition may serve as diagnostic aids in cystic fibrosis, celiac disease, phenylketonuria, intoxication with heavy metals, severe nutritional deficiencies, and geophagia. Hair analysis offers possibilities in determining the interdependencies of the trace elements, and the possible disease identification by trace element pattern. Such an approach has been reported by Larsen et al. [44] in arthritic patients. A similar approach can be undertaken in a number of other diseases. For example, it is known that trace element levels in hair during pregnancy deviate significantly from normal values.

The effects of ionizing radiation on hair have been studied to some extent. In man, scalp hair is the most sensitive hair to irradiation. First, only temporary depilation occurs; with the exposure to a higher dose, permanent depilation results. The pigment anomalies have been observed in the hair growth after the temporary depilation: the hair is usually darker. In an effort to evaluate the radiation effects the trace element analysis of mouse hair has been performed [46]. Hair from mice irradiated with 600 rad (γ-rays) and hair from control group have been analyzed. Fig. 16 indicates some of the changes observed.

6. CHARACTERISTIC X-RAY SPECTRA FROM FOSSIL FUEL SAMPLES

In addition to hydrocarbons, compounds of sulphur, oxygen, and nitrogen and traces of metals are present in crude oils. All these substances can cause problems affecting refining processes and the quality of products. For example: nickel adversely affects catalysts used in refinery processes to increase gasoline yield from heavy gas oil. Vanadium has adverse effects when present in too great a concentration; sulphur in crude oil can form corrosive compounds that may damage the vessels of a refinery. The difference in vanadium and nickel content of various crude oils provides a way to identify the source of

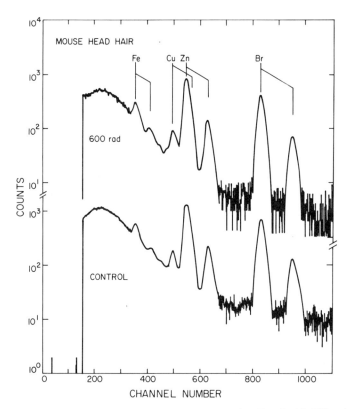

FIG.16. *X-ray spectra from mouse head hair. Top: mouse irradiated with 600 rad (Cu/Zn = 0.061; Fe/Zn = 0.115). Bottom: control (Cu/Zn = 0.069; Fe/Zn = 0.132).*

a crude oil. Crude oil "fingerprints" are used already several times to identify the source of oil contamination. However, the impact of burning fossil fuels on the atmosphere is probably the most important aspect of trace elements present in oil and coal.

For most of metals in crude oils some preconcentration procedure is required to concentrate these metals; crude-oil samples are prepared for analysis and reduced to ash by one of these two methods: (i) dry ashing; and (ii) wet oxidation followed by ashing. The samples to be analyzed usually contain materials other than oil, such as brine and sand, which have to be removed before the oil is analyzed. A dry ashing method is described by Horr et al. [47]. (This is a slightly modified ASTM-46 procedure.) Morgan and Turner [48] have showed by radioactive tracer technique that no significant losses in organic ash occur if the ashing is carried out below 550°. The losses of metals in the dry ash method have been found not appreciable for crude oils and residual stocks; however, in charge stocks and overhead fractions obtained by vacuum distillation losses of metals may be considerable. Loss of some metals during the ignition has been recognized [49]. To eliminate such losses a number of investigators have utilized methods of fixing the metals bre pre-sulfating the oil before igniting it.

Wet oxidation of oil is similar to wet ashing procedure of biologic and other organic material. A technique frequently used is to char the sample by

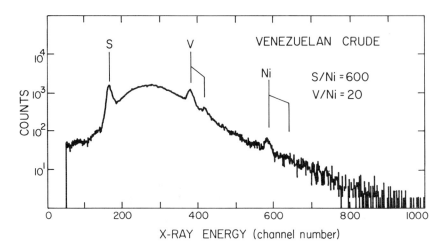

FIG.17. X-ray spectrum obtained by the bombardment of the target made by depositing a few drops of Venezualan crude oil on filter paper.

heating it with sulfuric acid and then adding concentrated nitric acid in one or two ml increments; alternatively or additionally, strong hydrogen peroxide is added dropwise directly into the charred digestion mixture. In general, the characteristics of the wet-oxidation techniques are that relatively large acid-to-sample ratios are needed and limited amounts of sample can be decomposed in a reasonable length of time. The two methods of ashing show generally close agreement for the metals contained in oil.

Several other methods of preconcentration or separations have also been applied. The element of interest may sometimes be extracted either after or before decomposition. It is often possible to find a suitable extractant for the determination of some specific element or compound.

High-speed burning is also applied particularly for the determination of sulphur and hydrogens. Rapid burning is done in an oxyhydrogen flame. Combustion in oxygen has numerous advantages. In the oxygen bomb method organic matter can be decomposed without introducing large amounts of other reagents or risking loss of volatile components.

All three modes of sample excitation have been used for the trace element analysis of oils. For example, Rhodes [50] has described a system for the analysis of metals which arrive, as the result of wear, in jet-engine oil. The radioactive sources used were ^{238}Pu and ^{55}Fe. Targets were prepared as thin film deposited on filter-paper discs. Standards in the form of solutions were also deposited on filter paper.

There are several commercial tube excited X-ray fluorescence analyzers on the market. Many laboratories are using X-ray fluorescence for their routine analysis. For example, Marangoni et al. [51, 52] measured the determination of wear metals in lubricating oils from aircraft motors. Cr, Fe, Ni, Sn, and Pb were determined with a detection limit better than 1 ppm and good reproducibility at the 5 ppm level. Lutrario et al. [53] achieved limits of detection of 0.3 ppm for iron, 0.2 ppm for nickel and chromium in lubricants by optimizing instrumental parameters. In another report Lutrario et al. [54] reported a 0.5 ppm detection limit for Cu in lubricants using a Cr target tube operated at 20-60 kV and 20-60 mA, and a plastic sample holder with an aluminized polyester film window. A procedure for the determination of vanadium in the concentration range of 0.05-0.5% in heavy distillate fuels by the X-ray fluorescence after concentration by distillation was reported by Boyle [55].

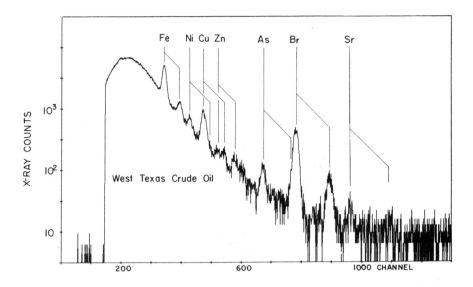

FIG.18. West Texas crude oil: X-ray spectrum from ash target.

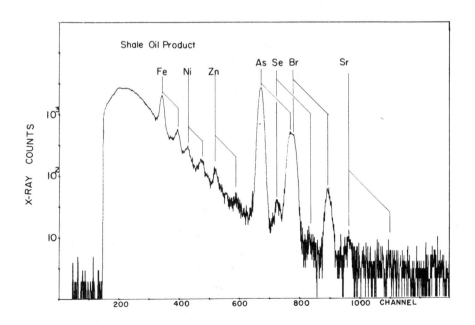

FIG.19. Shale oil product: X-ray spectrum from ash target.

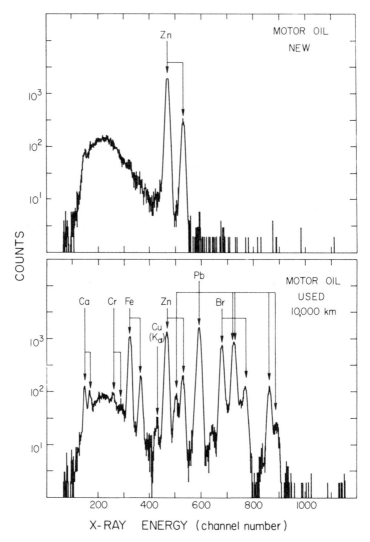

FIG.20. X-ray spectra obtained by bombarding new (top) and used lubricating oil (bottom) targets with 3-MeV protons.

Charged particle induced X-ray spectroscopy has also been used in some laboratories. Some elements are present in oil in high enough concentrations that they can be measured without preconcentration. For example, Fig. 17 shows an X-ray spectrum obtained by the bombardment of the target made by depositing a few drops of Venezuelan crude oil on filter paper. 3 MeV protons, I = 10 nA, were used. The spectrum is dominated by bremsstrahlung radiation due to the target backing. However, peaks associated with S, V, and Ni can be easily identified. Concentration ratios, S/Ni = 600 and V/Ni = 20, are easily determined with a measurement lasting only a few minutes. In order to be able to measure the other elements present in crude oil, dry ashing of oil was performed. Targets were then prepared by smearing ash on the scotch tape backing.

West Texas crude oil X-ray spectrum is shown in Fig. 18, while the shale oil product X-ray spectrum is shown in Fig. 19. In both cases, Br is due to the contamination in the scotch tape.

Determination of wear metals in lubricating oil from aircraft and other motors has been studied by many investigators. Detection limits better than 1 ppm are reported for many metals including chromium, iron, nickel, tin, lead, and copper. Fig. 20 shows X-ray spectra obtained by the bombardment of new and used lubricating oil targets by 3 MeV protons. Targets were prepared by putting a few drops of oil on the filter paper. New motor oil reveals only peaks associated with zinc (which is a known additive to oil). The used motor oil from the author's car (10 000 km) shows a variety of peaks; some of them can be associated with the additives of gasoline.

Concerns with air pollution and its effects on human health and disease have resulted in demands for fuels with lower and lower sulphur content. Measurements of sulphur content accurately and quickly are vital to refinery operations, to minimize quality give-away and delays in product testing. Routine laboratory tests have been developed which require only a few minutes for the analysis and are accurate to 0.01-0.10%. However, on-line measurement of S is a better approach, especially where direct product dispatch is required.

Gamage and Topham [56] have described such an on-line system for the determination of S which was based on X-ray fluorescence technique. They have used a miniature air-cooled X-ray tube which was run at 27 kV and 200 µA. The apparatus consisted of flow cell, measurement head containing the X-ray tube with collimator, flow cell window and leakage detection unit, power supplies. The X-ray beam strikes the middle of the flow cell and the sample flows continuously (20 lit/min). The window used was 8 µm Kapton film supported by photoetched Ni grid. X-rays were detected by gas proportional counter. Authors concluded that in such geometry there are no effects from 100 ppm Ni and 200 ppm Na. Vanadium interferes as a result of the resolution of gas filled proportional counter: 200 ppm V gave effect of 100 ppm S. A system for continuously monitoring low lead levels in gasoline has been developed by Exxon Research and Engineering and Princeton Gamma Tech. The system measures lead concentration in the range of 0.003 to 0.03 kg·m^{-3} every five minutes by X-ray fluorescence as the gasoline passes through a pipeline. The continuous monitoring gives a more accurate average lead content for gasoline than does conventional sampling from a storage tank. In addition, the system can be used to pinpoint sources of lead contamination in unleaded gasoline.

X-ray fluorescence analysis proved to be rapid, and reasonably accurate for determining the concentration of about 20 elements in whole coal. Although major elements in coal, carbon, hydrogen, oxygen and nitrogen cannot be analyzed by X-ray fluorescence, most other elements at levels greater than a few ppm are readily determined. Results of analysis of whole coal samples by X-ray fluorescence agreed well with values determined by several other independent methods [57]. Trace elements determined by the X-ray fluorescence method are limited to those occurring in whole coals at a few ppm at least. The list of these elements can be a long one.

Because of the speed and simplicity of the method, X-ray fluorescence is highly adaptable to large-scale surveys of coal resources. As an example, Fig. 21 shows the X-ray spectrum obtained by the bombardment of a coal target by 3 MeV protons. The target was prepared by affixing some coal dust to scotch tape. The spectrum shown has been collected in a few minutes of irradiation with the beam intensity of 10 nA.

7. ENVIRONMENTAL POLLUTION

The distribution of elements in Nature, their concentration and their movements can be seriously affected by the activities of Man. Environmental pollution results from Man's introduction of different substances into the marine, atmospheric or soil environment. Sometimes this can result in deleterious effects such as harm to living resources and hazards to human health.

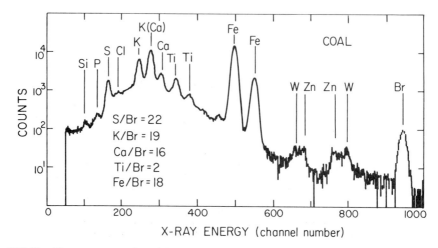

FIG.21. *X-ray spectrum obtained by the bombardment of a coal target by 3-MeV protons.*

7.1 Air pollution

Many effects of air pollution on plant and animal life have been reported. Because most metals can be extremely toxic to plants even in low concentration, airborne pollution may be causing significant changes in the genetic structure of the plant population exposed to them. There is evidence that a great increase in particular loading of the atmosphere has occurred during the past few years. It has been suggested that this increase is due to human activities, and that it is the probable cause of the observed decline in the average temperature of the planet. Particles in the size range of 0.1-1.0 μm have the greatest influence on the absorption and scattering of solar radiation in the atmosphere. Therefore concentrations of these particles in the atmosphere are of special importance to the planetary heat budget.

Transportation sources account for over 60% by mass of the total pollution. Exhausts from automobile engines produce most of the atmospheric pollution. The exhausts include smoke, carbon monoxide, oxides of nitrogen, hydrocarbons, lead, and other things. These pollutants, once in the atmosphere, participate in different photochemical reactions. The elemental composition of atmospheric aerosols has been determined in several areas around the world. Obviously the composition is influenced mainly by the local source distribution and meteorological conditions. Several institutions are actively monitoring air pollution using X-ray emission spectroscopy. The samples are collected on a fiberglass filter paper or on a cellulose membrane filter. The cellulose membrane filter is much more convenient for X-ray emission analysis because of the lower level of contamination. For the summary of recent work see references [9] and [30]. Lead is present in rather large amounts in the motor-engine exhausts. Altogether, about 50% of the lead in gasoline is emitted into the atmosphere as a fine sub-micron particulate aerosol. There is a clear evidence of lead pollution caused by the motor vehicles close to the busy roads. In addition, more and more evidence is being accumulated about the world-wide spread of pollutants originated in the urban areas. It has been shown that the atmospheric lead, mainly from the combustion of the leaded gasolines in the internal combustion engines, has increased on at least a hemispheric scale [58]. This has suggested its use as a tracer for the large-scale atmospheric movement of anthropogenic particulate material. As an illustration, Fig. 22 shows an X-ray spectrum from the aerosol sample collected on the filter paper. The sample was

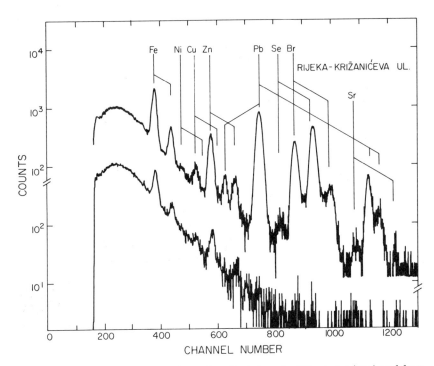

FIG.22. *X-ray spectra from an aerosol sample collected on the filter paper (top), and from the filter paper (bottom).*

bombarded with 3 MeV protons, I = 10 nA, for a few minutes. Cahill and collaborators [59, 60] at the Crocker Nuclear Laboratory (University of California at Davis) have developed an extensive network of aerosol monitoring stations through California and are using their fully automated particle-induced X-ray emission (cyclotron: α-particles).

Most studies of suspended particulate pollutants in urban air are limited to estimating the quantity of total particulate matter (μg of particulate matter per m^3 of air) and its elemental composition. In order to assess the inhalation health hazard, a knowledge of particle-size distribution is needed. Particle size is one of the very important factors in determining the degree of respiratory penetration, the extent of visibility reduction, the nature of particle-particle interaction and the mechanisms of a wide range of atmospheric phenomena.

7.2 X-ray emission spectroscopy of water

Elemental composition of the water is of great importance to life. Water has to provide the essential elements for the animals living in its environment and through the food chain to the man. Drinking water is a significant source of essential elements to man and animals living on the soil. In addition, it has to be free of elements which can have deteriorating effects on health.

Let us mention a few examples of trace-element concentrations in water and their relation to human health. Many investigations have found a correlation between cardiovascular deaths and water composition. Total death rates were reported [61, 62] to be inversely correlated to hardness of water (total calcium and magnesium concentrations). This association may be due to the fact

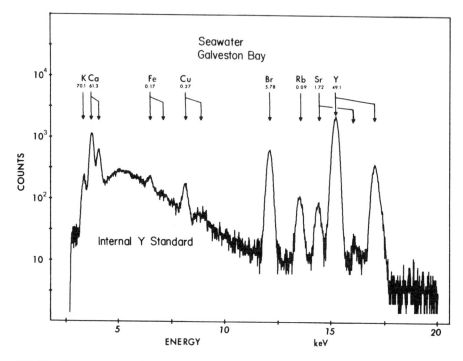

FIG.23. X-ray spectrum of a seawater sample bombarded with 3-MeV protons. Yttrium was used as the dopant.

that soft waters lack the beneficial elements present (Cr, Cn, F, Mn, Si, V, and Zn seem to exert beneficial effects) in hard waters, or extract harmful elements (Cd, Co, Pb) from pipes. Certain associations were found also between the chemical composition of rocks and soils and cardiovascular death rates.

There are other numerous examples indicating the importance of the knowledge of trace element composition of water. As a result many techniques for the analysis of water samples have been developed. The applications of X-ray emission spectroscopy to multi-elemental analysis of different water samples will be discussed here.

Water targets are very easily prepared by simply allowing a few drops to dry on an aluminum formvar backing [11]. Such a simple target preparation allows the detection of elements present in the seawater with concentrations > 0.1 µg/ml. With proton-induced X-ray emission spectroscopy, concentrations of K, Ca, Br, and Sr can be determined in clean seawater. However, in many cases the concentration of metals can be significantly higher. For example, Fig. 23 (from ref. [63]) shows the X-ray spectrum of a seawater sample bombarded with 3 MeV protons at 200 nA for 600 s. Yttrium was used as a dopant. Iron and copper concentrations are measured to be Fe: 0.17 µg/ml; Cu: 0.27 µg/ml (normal values are Fe: 0.0034 µg/ml; Cu: 0.003 µg/ml).

The generation of water samples which are truly representative of any aquatic environment is a problem concerning itself with sampling methodology. Once a sample has been taken it should have no possibility of transporting trace metals either to or from the sampling container walls or suspended material within the aqueous soultion. Care taken at the sampling stage is wasted if the representative nature of the sample is altered during handling and/or storage. The transfer of trace metals at low concentrations from solution to

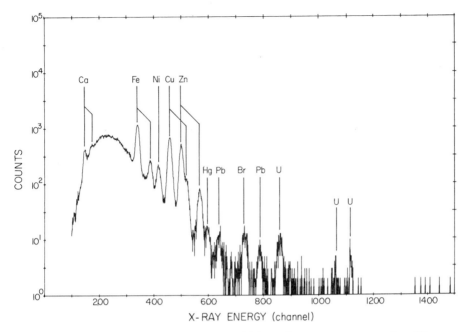

FIG.24. X-ray spectrum from seawater target; 0.1 liter of seawater was prepared on filter paper (see text).

container walls is a problem familiar to all water scientists. There are several means by which to deal with this problem. Naturally, the optimum method is to prepare and analyze the sample as soon as possible. When information is needed on elements which occur in small concentrations (< 0.1 ppm) some preconcentration procedure is required.

Trace elements are isolated under certain definite experimental conditions as complexes such as dithiozonates, quinolates and dithiocarbamates, which can either be precipitated in an aqueous medium, because of their poor solubility in water, or can be extracted by an organic solvent [64].

For the analysis of dissolved trace metals in water one often uses a method which involves formation of insoluble metal chelates via coordination with dithiocarbamate (ammonium pyrrolidine dithiocarbamate or diethyl dithiocarbamate), filtration through a membrane filter, and the analysis of the precipitate. The complexes formed are highly insoluble in water and can therefore be trapped on a membrane filter which serves as a target matrix. (The technique has been applied to analysis of water from various different types of material systems as well as to urine samples.) Simultaneous analysis of most of the transition elements is possible, but alkali and alkaline-earth metals are excluded. Following elements can be separated with dithiocarbamate: V, Cr, Mn, Fe, Co, Ni, Cu, Zn, Ga, As, Mo, Rh, Pd, Ag, Cd, In, Sn, Sb, Pt, Au, Hg, Tl, Pb, Bi, and U [64]. The dithiocarbamate precipitation technique of target preparation enjoys several advantages over procedures based upon evaporation, among them uniformity of distribution of elements on the target, simplicity which reduces both handling time and contamination, and high efficiency within a wide pH range.

Fig. 24 shows an X-ray spectrum from a seawater target prepared on millipore filter paper from 0.1 liter of seawater. The target was bombarded with only 1-2 nA beam of 3 MeV protons. This technique of selective complexation by dithiocarbamate chelates affords one a very nice way of analyzing seawater.

The chelates are very selective against the sodium, potassium, and other cations found in percent quantities in seawater. Hence the trace metals can be extracted, and their concentration measured even when present only in 10^{-3} ppm range.

8. CONCLUSIONS

Research interest in several disciplines has begun to focus on the complex problem of movements of elements in nature, and the relationship between the environment and living organisms. A number of fascinating problems deserve more attention. The detection of characteristic X-rays has led to new analytical techniques capable of simultaneous detection of a large number of elements. The excitation by the radioactive sources and X-ray tube allows that such work be done even in small laboratories and without extensive funding. A number of systems are already commercially available. Both "fundamental" and "applied" problems related to the trace element composition should be studied. In addition, the development of trace element analysis by charged-particle induced X-ray emission spectroscopy as an analytic technique suggests a nuclear accelerator laboratory as an important facility for the study of trace elements. Large cross sections for X-ray emission along with low background radiations allow the usage of thin sample targets in which matrix effects are negligible. At the same time, thin samples can be of considerable importance where only small quantities of the sample are available. Available computer facilities of a typical nuclear physics laboratory allow fast collection, processing, and analysis of data. With standardized target-making procedures, short data collection and analysis periods, and simultaneous sensitivity to a large number of elements, charged-particle-induced X-ray emission spectroscopy compares favorably with other trace element analysis techniques.

REFERENCES

[1] CHADWICK, J., Phil. Mag. 24 (1912) 594.
[2] MERZBACHER, E., LEWIS, H.W., Encyclopedia of Physics (S. FLÜGGE, Ed.), Springer-Verlag, Berlin (1958).
[3] OGURTSOV, G.N., Rev. Mod. Phys. 44 (1972) 1.
[4] GRYZINSKI, M., Phys. Rev. 138 (1965) A305.
[5] GARCIA, J.D., Phys. Rev. A1 (1970) 280.
[6] DUGGAN, J.L., BECK, W.L., ALBRECHT, L., MUNZ, L., SPAULDING, J.D., Adv. X-Ray Analysis 15 (1972) 407.
[7] RUTLEDGE, C.H., WATSON, R.L., Atomic Data and Nuclear Data Tables 12 (1973) 195.
[8] VALKOVIĆ, V., Nuclear Microanalysis, Garland Publishing Inc., New York (1977).
[9] JOHANSSON, S.A.E., JOHANSSON, T.B., Nucl. Instrum. Meth. 137 (1976) 473.
[10] RHODES, J.R., Energy Dispersion X-Ray Analysis (RUSS, J.C., Ed.), ASTM publication 485 (1971).
[11] VALKOVIĆ, V., et al., Nucl. Instrum. Meth. 114 (1974) 573.
[12] VALKOVIĆ, V., Contemp. Phys. 14 (1973) 415.
[13] FOLKMAN, F., J. Phys. E 8 (1975) 429.
[14] SUESS, H.E., UREY, H.C., Rev. Mod. Phys. 28 (1956) 53.
[15] GENTRY, R.V., CAHILL, T.A., FLETCHER, N.R., KAUFMANN, H.C., MEDSKER, L.R., NELSON, J.N., FLOCCHINI, R.G., Phys. Rev. Lett. 37 (1976) 11.
[16] TURKEVICH, A.L., FRANZGROTE, E.J. PATTERSON, J.H., Science 158 (1967) 635.
[17] UNDERWOOD, E.J., Trace Elements in Human and Animal Nutrition, 3rd ed., Academic Press (1971).
[18] BOWEN, H.J.M., Trace Elements in Biochemistry, Academic Press (1966).
[19] KEATING, B., HELSLEY, C.E., PESSAGNO, E.A., Geology 3 (1975) 73.
[20] HAYS, J.D, Bull. Geol. Soc. Am. 82 (1971) 2433.
[21] HARRISON, C.G.A., SOMAYAJULU, B.L.K., Nature 212 (1966) 1193.

[22] HARRISON, C.G.A., PROSPERO, J.M., Nature 250 (1974) 563.
[23] REID, G.C., ISAKSEN, I.S.A., HOLZER, T.E., CRUTZEN, P.J., Nature 259 (1976) 177.
[24] CRAIN, I.K., Bull. Geol. Soc. Am. 82 (1971) 2603.
[25] VALKOVIĆ, V., Origins of Life (1977), to be published.
[26] BIEGERT, E.K., CRAIG, M., VALKOVIĆ, V., STORCK, R., Proc. 4th Conf. on Applications of Small Accelerators, Denton, Texas, 1976.
[27] STORCK, R., MORRILL, R.C., Biochem. Genetics 5 (1971) 467.
[28] STEENBJERG, F., Plant Soil 3 (1951) 97.
[29] VALKOVIĆ, V., RENDIĆ, D., BIEGERT, E.K., ANDRADE, E., to be published.
[30] VALKOVIĆ, V., Trace Element Analysis, Taylor and Francis, London (1975).
[31] WHEELER, R.M., et al., Medical Physics 1 (1974) 68.
[32] CAVALIERI, R.R., SCOTT, K.G., J. Amer. Med. Assoc. 206 (1968) 591-595.
[33] ALLAWAY, W.H., KUBOTA, J., LOSSEE, F., ROTH, M., Arch. Environ. Health 16 (1968) 342.
[34] JANES, J.M., MCCALL, J.T., ELVEBACK, L.M., Mayo Clin. Proc. 47 (1972) 476-478.
[35] ANDROMIKASHVILI, E.L., et al., Cancer Res. 34 (1974) 271-274.
[36] KOPITO, L., BRILEY, A.M., SCHWACHMAN, H.J., JAMA 209 (1969) 243.
[37] RENSHAW, G.C., POUNDS, C.A., PEARSON, E.F., Nature 238 (1972) 162.
[38] VALKOVIĆ, V., et al., Nature 243 (1973) 543.
[39] RENSHAW, G.C., POUNDS, C.A., PEARSON, E.F., Nature 238 (1966) 59.
[40] EADS, E.A., LAMBDIN, C.E., Environ. Res., 6 (1973) 247.
[41] OBRUSNIK, I., et al., Forens. Sci. 17 (1972) 426.
[42] VALKOVIĆ, V., RENDIĆ, D., PHILLIPS, G.C., Environmental Science and Technology 9 (1975) 1150.
[43] RENDIĆ, D., et al., J. Invest. Dermatol. 66 (1976) 371.
[44] LARSEN, W.B., LORD, R.S., MILLER, J.J., J. Am. Osteopath. Assoc. 74 (1974) 131-136.
[45] KOPITO, L.E., SHWACHMAN, H., The First Human Hair Symposium (BROWN, A.C., Ed.), Medcom Press (1974).
[46] VALKOVIĆ, V., OTTE, V., ANDRADE, E., BIEGERT, E.K., to be published.
[47] HORR, C.A., MYERS, A.T., DUNTON, P.J., U.S. Geological Survey Bulletin 1100-A (Washington: US Government Printing Office) 1961.
[48] MORGAN, L.O., TURNER, S.E., Anal. Chem. 23 (1951) 978.
[49] MILNER, O.I., et al., Anal. Chem. 24 (1952) 1728.
[50] RHODES, J.R., et al., Environ. Sci. Tech. 6 (1972) 922.
[51] MARANGONI, O., LUTRARIO, P., TRONCA, A., Met. Ital. 64 (1972) 381.
[52] MARANGONI, O., et al., Met. Ital. 78 (1973) 6199.
[53] LUTRARIO, P., TRONCA, A., CAPRIOTTI, R., Rass. Chim. 24 (1972) 397.
[54] LUTRARIO P., et al., Rass. Chim. 78 (1973) 1617.
[55] BOYLE, J.F., Amer. Soc. Test. Mater., Spec. Tech. Pub. 531 (1973).
[56] GAMAGE, C.F., TOPHAM, W.H., Advances in X-Ray Analysis 17 (1974) 542.
[57] KUHN, J.K., HARFST, W.F., SHIMP, N.F., Trace Elements in Fuel, (BABU, S.P., Ed.), Advances in Chemistry Series 141 (1975).
[58] MUROZUMI, M., CHOW, T.J., PETERSON, C.C., Geochim. Cosmochim. Acta 33 (1969) 1247.
[59] CAHILL, T.A., University of California at Davis Report (1973), UCD-CNL-162.
[60] FLOCCHINI, R.G., et al., Advances in X-Ray Analysis 18 (1975) 579.
[61] SCHROEDER, H.A., J. Chronic. Dis. 12 (1960) 586; ibid 8 (1965) 647.
[62] MORRIS, J.N., CRAWFORD, M.D., HEADY, J.A., Lancet 1 (1961) 860.
[63] ALEXANDER, M.E., BIEGERT, E.K., JONES, J.K., THURSTON, R.S., VALKOVIĆ, V., WHEELER, R.M., WINGATE, C.A., ZABEL, T., Int. J. Appl. Radiat. Isotopes 25 (1974) 229.
[64] PINTA, M., Detection and Determination of Trace Elements (English translation), Jerusalem (1966).

NEUTRON ABSORPTION PHYSICS IN THE DEVELOPMENT AND PRACTICE OF ACTIVATION ANALYSIS

S.S. NARGOLWALLA
Nuclear Applications Research Laboratory,
SCINTREX Ltd,
Concord, Ontario,
Canada

Abstract

NEUTRON ABSORPTION PHYSICS IN THE DEVELOPMENT AND PRACTICE OF ACTIVATION ANALYSIS.

The review outlines the contribution of neutron absorption physics toward the development and practice of neutron activation analysis techniques. In so far as they concern this development, fundamental nuclear parameters of neutron energy, absorption cross-section and induced nuclear interactions are briefly discussed. The basic analytical tools, such as the neutron source, sample irradiation and radiation measurement instrumentation, are described. In this respect, a special emphasis is placed on the utilization of essentially monoenergetic fast neutrons obtained from neutron generators for analytical applications. The role of activation techniques, in basic cross-section studies and in analytical chemistry, is discussed with a view to establishing the technique as a fundamental tool in nuclear physics research and materials characterization, respectively. The importance of neutron activation analysis is demonstrated by a critical comparison of its capabilities with those offered by competing chemical and other instrumental analytical methods. This comparison, in effect, provides an insight into the many areas where the contribution of neutron absorption physics has served in the development of a highly sensitive tool for trace analysis. The impact of neutron activation analysis in biological and biomedical studies, material science and industrial applications, environmental and ecological assessments, geo- and cosmo-chemical investigations, mineral and energy resource evaluations, and in archaeological and forensic sciences, is evidenced in the above areas of application. Some comments regarding the application of neutron activation anlysis to future challenges in the realm of general materials characterization are offered.

1. INTRODUCTION

Neutrons interact with elemental nuclei over a wide range of incident neutron energies. Because of their lack of electrical charge, neutrons of very low energies will easily penetrate the electrical field surrounding nuclei and produce interactions. Neutron sources of sufficient intensities are available for the study of nuclear reactions over an energy range from about 10^{-4} to 10^8 eV. The nuclear reactor offers an opportunity to conduct such kinematic reaction studies through an energy range of 10^{-4} to 10^7 eV. The reaction types induced depend upon the incident neutron energies utilized. It is therefore convenient to classify incident neutrons in terms of their energy groups. These

81

classifications can be broadly related to the possible types of neutron induced reactions, and to some extent to the various types of neutron sources and experimental methods applied to neutron absorption studies.

The most commonly used classifications are: thermal neutrons - the most probable velocity of the distribution at 293°K is 2200 m.s^{-1}, corresponding to an energy of about 0.025 eV; intermediate-energy neutrons - this range extends from thermal to about 10 keV and exhibits sharp resonances for neutron absorption for most elemental nuclei; fast neutrons - this energy group is defined to include neutron energies from 10 keV to approximately 20 MeV. The upper limit also coincides with the capability of useful neutron sources; and relativistic neutrons - this range essentially includes neutrons above 20 MeV and exhibit relativistic effects.

In considering the probability of neutron absorption interactions of analytical importance due regard must be given to the energy classifications outlined above. The interaction probability or cross section (σ) combined with the neutron intensity or flux (ϕ) form the foundations upon which the technique of neutron activation analysis is based. Therefore, all factors affecting the energy of the incident neutrons will control the probability of neutron absorption, and the nature and extent of the probable activation processes. The importance of this fact is brought out in this review in terms of the excitation functions for neutron absorption for each classification of neutron energy. Selected references, containing cross section tabulations, cross section vs. energy dependence and other relevant data, as annotated in this review, provide the working tools for the development and practice of neutron activation analysis.

The importance of activation analysis, in general, was recognized almost immediately upon the discovery of artificial radioactivity. However, its genesis is linked with the classic studies of Hevesy and Levi in 1936, [1] and Seaborg and Livingood in 1938 [2] . During the first ten years of its development, less than two dozen published works could be found in open literature. Nevertheless, the exponential growth during the past two decades has resulted in the inclusion of almost 10,000 publications in the current acquisition files. This fantastic growth can be attributed, to a large measure, to the ever increasing acceptance of the basic technology as a tool to probe the composition of matter. In acknowledgement of the technique's broad capabilities, a considerable effort has been expended toward the development of neutron sources, radiochemical analytical techniques, sampling and nuclear instrumentation design. Today, neutron activation analysis rightfully belongs in the rapidly expanding analytical arsenal of measurement techniques. Rather than present itself as a competitive analytical tool, its usefulness is more evidenced in its role as a complementary or comparative technique directed towards the pursuit of achieving better accuracy in the overall chemical measurement process.

With the availability of portable medium intensity neutron sources such as radioisotopic, fission and from small accelerators, considerable emphasis is currently being placed to establish neutron activation as a basic technique for scientific and technological applications. As such, its use in the field of biological and biomedical studies, environmental and ecological assessments, material science and industrial applications, geo-and cosmo-chemical analyses, evaluation of mineral and energy resources, and in archaeological and forensic sciences, can be evidenced through a review of the vast literature in existence.

In this review an attempt is made to trace the development of neutron activation analysis from fundamental nuclear concepts. Certain basic parameters associated with neutron absorption physics and governing the development of activation techniques are considered. These are the incident neutron energy, the probability or cross section of the absorption process and the energy/abundance characteristics of the interaction decay products. A special discussion of the analytical application of fast neutrons in research, and in some of the important areas described above, is presented. The contribution of neutron absorption physics is indirectly assessed by a presentation of a comparison of neutron activation analysis with other non-nuclear techniques for both trace-and macrochemical characterization of matter. Although not a panacea for all analytical problems, neutron activation analysis will continue to offer a viable means for the study of tomorrow's challenges associated with the characterization of materials.

2. NEUTRON ABSORPTION PHYSICS

2.1 General Considerations

In contrast to the generally predictable behaviour of gamma-ray attenuation, the physical processes governing neutron absorption are quite complex. This fact has lead to a general diversion from the pure analytical approach to the use of empirical methods for the determination of neutron absorption cross sections.
Upon interaction with atomic nuclei, neutrons can be absorbed and/or scattered with differing probabilities. The scattering events, be they elastic or inelastic, isotropic or assymmetric, also result in degradation of the incident neutron energy. Therefore the relative proportion of the absorbed to the scattered component of the incident neutron intensity in its passage through a given target medium is varying. The problem is further compounded by the consideration of the cross section versus energy behaviour of neutrons. This dependence is often an erratic function punctuated by sharp excursions or resonance absorption peaks. In view of these difficulties, neutron absorption kinematics are better evaluated and understood by monitoring the radioactivity of expected decay products formed by single or multiple interactions. Such methods are dependent upon the availability of source neutrons possessing well defined energies. From an examination of literature, it is evident that studies related to the utilization of neutrons with energies less than 1 keV and greater than 100 keV have been abundantly documented. Cross section values in the above mentioned range are sufficiently reliable for the purpose of calculating interaction yields. Unfortunately, in the neutron energy range of 1 - 100 keV, the cross section information is far from satisfactory; principally due to the experimental difficulty of making accurate cross section measurements.
In view of the energy dependence of neutron absorption cross sections, a rather arbitrary classification of neutron energies is found to be useful and is tabulated (Table I). From the neutron activation analysis viewpoint, only those classifications defined as thermal, resonance and fast neutrons need be considered. The following discussion of absorption cross sections is therefore restricted to the above mentioned analytically useful neutron energies.

2.2 Neutron absorption cross sections

The probability of a given neutron absorption process independent of all other events can be denoted in terms of a cross section for that specific interaction. In its simplest form, the concept of a nuclear cross section may be visualized as the cross-sectional area presented by a target nucleus to an incident neutron. The cross section versus energy dependence, i.e. the excitation function, exhibits differing distributions depending upon the neutron energy group being considered. The absorption cross section for the three earlier mentioned neutron energy groups are briefly discussed in the following subsections.

2.2.1 Thermal neutron cross sections

A study of the excitation functions in the thermal region show irregularities with an almost complete absence of sharp resonance peaks, due to the very limited energy range typified by the thermal energy region. However, as the neutron wavelength (~ 1.8 Å) is comparable to the interatomic distance (~ 1 Å), neutron wave scattering from nuclei occur. Consequently, complex diffraction effects are evidenced and result in sharp changes in cross section values for the elements. Perhaps the most useful bibliography documenting sources for microscopic neutron cross section data are the CINDA [3] and CINDU

TABLE I. CLASSIFICATION OF NEUTRON ENERGIES

Term	Description	Neutron energy range
Cold	Neutrons in equilibrium with a cold medium	Energy at 25°K corresponds to 0.005 eV
Thermal	Neutrons in equilibrium with a medium at 293°K	Most probable energy at 293°K is 0.0253 eV. The Maxwellian distribution at 293°K extends to about 0.1 eV
Epithermal	Neutron energies in excess of thermal	0.2 - 1.0 eV
Resonance	Neutrons of energies corresponding to resonances in ^{238}U, In, Au etc.	1 - 300 eV
Intermediate	-	300 eV - 10 keV
Fast	Neutron energies capable of inducing threshold reactions	0.5 -20 MeV
Ultrafast	Relativistic neutrons	Greater than 20 MeV

[4] series. Compilation of early cross section values are published by the Brookhaven National Laboratory [5] and in a more recent update [6] issued by the same laboratory. The latter publication includes data on thermal neutron cross sections, resonance parameters and an extensive bibliography.

The great importance of neutron absorption systematics in the thermal energy range is best exemplified by the large number of possible interactions with atomic nuclei. Therefore cross section studies for the entire Maxwellian distribution of thermal neutrons have been carefully carried out using neutron velocity selectors. Such experimental results have been documented in user oriented tabulations. In general, the accuracies assigned to such data in recent compilations are about an order of magnitude better than those for data documented in the late 1950's. Certain absorption cross sections are currently known to a very high degree of certainty, e.g. the thermal activation cross section for ^{197}Au is given [6] as 98.8± 0.3 barns, while cross sections in the rare earth region are known only to a relative accuracy of about ±20%. The sometimes large errors in cross section values arise from numerous sources. By far the most significant error is introduced when converting the measured counts of a decay product into an absolute disintegration rate. Another significant source of error results from the difficulty of measuring the effective neutron flux of a specific energy impinging upon a sample of the element under consideration. Perturbation of the neutron energy and intensity by the sample itself can introduce significant error in the cross section determination. Recently, vast efforts have been expended in standardization of cross section measurement techniques and neutron flux evaluations. A direct result of such efforts is reflected in the good accuracy limits assigned to values recently documented [6].

2.2.2 Resonance neutron cross section

Except for some interactions with certain elements, e.g. $^3\text{He}(n,p)^1\text{H}$; $^6\text{Li}(n,\alpha)^3\text{H}$; $^{10}\text{B}(n,\alpha)^7\text{Li}$; $^{14}\text{N}(n,p)^{12}\text{C}$; $^{17}\text{O}(n,\alpha)^{14}\text{C}$; $^{235}\text{U}(n,\text{fission})$; thermal neutron reactions are all of the (n,γ) type. Resonance neutrons also induce (n,γ) reactions. In this neutron energy range, and for isotopes with atomic numbers usually in excess of 30, certain elements exhibit sharp, high intensity cross section peaks. In principle, therefore, by the use of substantially monochromatic beams from neutron spectrometers, a high degree of selective neutron absorption is possible. The utilization of the resonance neutron absorption phenomenon for analytical purposes is however limited to a neutron energy upper limit of about 20 eV because the resonance neutron intensity in a reactor neutron spectrum falls off as $\frac{1}{E^2}$. In the compendium [6], radiative neutron capture resonance integral cross sections are tabulated. In general, the tabulation contains results of computations based upon the Breit-Wigner shape for resonance cross sections occurring close to the cadmium cutoff energy of about 0.5 eV. For resonances sufficiently removed in energy from the above cutoff, the sum of single-level Breit-Wigner contributions are used to develop a single expression for the calculation of the resonance integral. In compiling this tabulation, the authors have taken great care in assessing the discrepancies between computed and measured values.

The resonance absorption effect is particularly useful in increasing the selectivity for element determination, since in most cases the resonance peaks do not overlap. In the rare earth region and in the region of heavy elements, resonance integral cross sections can vary from about 300 - 30,000 barns. From a knowledge of the total peak resonance integrals and the neutron flux as a function of energy, the expected induced radioactivity for a given resonance (n,γ) reaction can be calculated. The emphasis on the rare earth isotopes is obvious, when consideration is given to the fact that from the analytical viewpoint, determination of individual rare earths is difficult. Therefore, by utilizing the concept of resonance neutron absorption, it is possible to illustrate [7] that substantial enhancement in the induced radioactivity of a given rare earth element present in a matrix of another can be achieved. Such enhancements make it possible to apply the resonance neutron activation technique for the solution of some difficult analytical problems.

2.2.3 Fast neutron cross section

The discussion of fast neutron cross sections is intentionally addressed to select fast energy groups which are readily available to the activation analyst. The principal sources of such energy groups are the fission neutron spectrum from reactors or from spontaneous fission isotopes, and discrete monoenergetic groups available for low-voltage positive ion accelerators such as neutron generators and small cyclotrons, and from certain nuclear reactions.

2.2.3.1 Fission neutron cross section

Neutrons produced by fission of ^{235}U can be described by a semi-empirical expression as proposed by Leachmann [8].

$$f(E) = 0.7725 E^{1/2} \exp(-0.775 E) \quad \ldots \ldots \quad 1$$

where $f(E)$ is the fission neutron flux at energy, E, in the units of $n.cm^{-2}s^{-1}$. This expression is however only good for the fission neutron spectrum <9 MeV. The utilization of the reactor fast neutron spectrum poses considerable problems in analysis due to the interfering reactions induced by the preponderant thermal neutron component. Published work [9] describing the applicability of such a generalized neutron group is worthy of study. Excitation functions relevant to reactor fast neutrons can be developed by a study of the induced threshold type interactions. For this purpose a knowledge or a measurement of the average neutron absorption cross section for the fission flux is necessary. To facilitate such studies, concepts such as "effective threshold energy" and "effective absorption cross sections" have been introduced. Examples of some experimental and calculated excitation functions for fission neutron absorption

are given by Liskien and Paulsen [10]. Jung et al. [11] describe simple
experimental procedures for the determination of effective threshold energies
and cross sections. Cross section information for reactor fast neutrons is
also available [12].

2.2.3.2 Accelerator produced neutron cross section

In the following discussion, the information is heavily drawn from some
basic dissertations [13-15], sections of which elaborate on neutron absorption
cross sections for neutrons of discrete monoenergetic groups. Specifically,
the neutron energies of interest are the 2.5 MeV and 14 MeV obtained by accelerating a beam of deuterons on to a deuterium and tritium targets, respectively.
These low-voltage accelerators, or neutron generators, have been successfully
employed for the analysis of light elements with a high degree of sensitivity
and specificity. The emphasis here will be placed on the utilization of 14-MeV
neutrons, although for certain specific applications, 2.5-MeV neutrons have
also been used for analysis. The principal neutron absorption processes of
interest using 14-MeV neutrons are:

(n,γ): $^A Z + ^1 n \longrightarrow ^{(A+1)} Z + \gamma$

$(n,n'\gamma)$: $^A Z + ^1 n \longrightarrow ^A Z + ^1 n + \gamma$

(n,p): $^A Z + ^1 n \longrightarrow ^A (Z-1) + p$

(n,α): $^A Z + ^1 n \longrightarrow ^{(A-3)} (Z-2) + \alpha$

$(n,2n)$: $^A Z + ^1 n \longrightarrow ^{(A-1)} Z + 2n$

In order to provide some insight into the probabilities of occurrence
of these reactions, a qualitative description of the interaction energetics
involved, is presented.

For incident neutron energies much less than the nuclear binding energy,
the reaction promoted will strongly depend on the magnitude of the binding
energy. For instance, for target mass numbers A<20, the neutron binding energy
exhibits large and periodic fluctuations from nucleus to nucleus. For mass
numbers between 20 and 130, the neutron binding energy shows, on the average,
a very slow increase from about 8 to 8.5 MeV, followed by a slow decrease to
about 7.5 MeV for the heaviest elements. However, even this slow trend is
interrupted by a few instances of anomalous variations. Examples of such
anomalous behaviour are evidenced by nuclei with odd mass and neutron numbers
whose binding energies are about 1 to 2 MeV greater than those for even nuclei
adjacent to them.

The (n,γ) reaction is in all cases exoergic, and the absorption cross
section in almost all instances decreases with increasing neutron energy. In
contrast to the large variations, often by as much as 10^{10} barns in the isotopic
absorption cross sections for thermal neutron absorption, 14-MeV (n,γ) absorption
probabilities are of the order of a few millibarns and are usually neglected.
In general, for neutron energies from 1 to 15 MeV, the (n,γ) cross section
increases as the third power of the mass number A up to a mass number $A \approx 100$
when a saturation value is reached. For 2.5-MeV neutrons, certain nuclei exhibit relatively high (n,γ) absorption probabilities which can be useful for
certain analytical applications. An assessment of documented [14] 14-MeV (n,γ)
cross sections indicate rather large errors associated with the cross section
values; in some instances of the order of ±30%. These errors, to a large measure,
result from chemical separation processes employed for the separation of product
nuclei, and from difficulties in making absolute counting measurements. In
general however, the (n,γ) cross sections are of the order of 5 mb. If the
cross section is plotted against the neutron number as shown in Fig. 1, it can
be readily seen that evidence of lower values in the region of neutron magic
numbers is indeed slight. For 2.5-MeV neutrons, cross section values as high
as 100 mb have been measured and documented. An excellent compilation by
Csikai et al. [13] is summarized in Fig. 2. This figure illustrates the (n,γ)
cross section distributions for 1-, 3- and 14-MeV neutrons as a function of the
target neutron number. Cross section maxima of ∼100 mb, ∼30 mb and ∼15 mb for
1-, 3- and 14-MeV neutrons, respectively, can be observed.

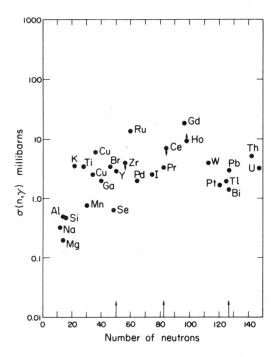

FIG.1. *Radiative capture cross sections for 14.5-MeV neutrons.*

The neutron scattering reactions leading to isomeric states and designated as (n,n'γ) are slightly endoergic. Only limited cross section information is currently available. Among these reactions, only a few produce metastable states with half-lives sufficiently long to be useful for analytical purposes. For neutron energies above 10 MeV, the (n,2n) reaction offers strong competition to the (n,n'γ) interaction. From a limited listing [14] of some useful (n,n'γ) reactions, it can be observed that for 14-MeV neutrons, the cross section values lie in the range from about 100 to 600 mb. However, the errors are of the order of ±10 to ±20%. Most of the (n,n'γ) reactions are also induced by 2.5-MeV neutrons due to the existence of low energy levels of target nuclei.

Insofar as the threshold reactions are concerned, the cross section values for (n,p), (n,α) and (n,2n) threshold reactions range from about 5 to 500 mb. Some examples of excitation functions are illustrated in Figs. 3-5. Research studies have indicated that systematic trends observed in the (n,p), (n,α) and (n,2n) cross sections bear great significance, especially for the estimation of unknown cross section values. In this review the measurement techniques employed for the determination of the above mentioned cross sections are not discussed. Suffice it to note that a host of techniques have been utilized and described in the literature [9,16,17]. A series of symposia [18-21] covering a broad range of neutron cross section studies and appropriate technologies are worthy of study. The subject matter includes both the measurement of cross sections and their critical evaluations, and application of neutron absorption processes for the study of nuclear decay schemes. An examination of such collected works leaves the user or activation analyst with the distinct impression that, in general, poor communication exists between the pure physicist who is most qualified and experienced in cross section measurement technology, and the analytical nuclear chemist who invariably demands

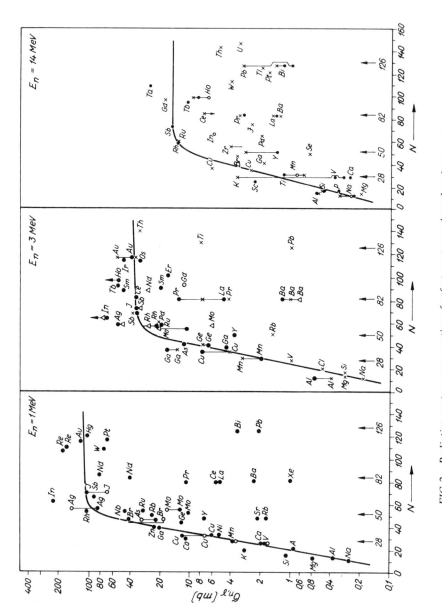

FIG.2. *Radiative capture cross sections for fast neutrons plotted against target neutron number at 1, 3 and 14 MeV.*

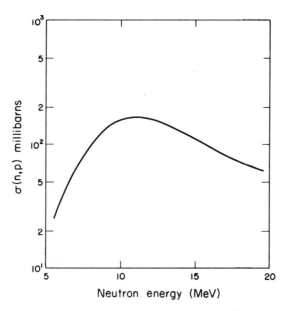

FIG.3. Excitation function for (n,p) reaction: $^{60}Ni(n,p)^{60}Co$.

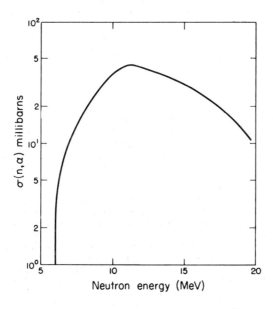

FIG.4. Excitation function for (n,α) reaction: $^{63}Cu(n,\alpha)^{60}Co$.

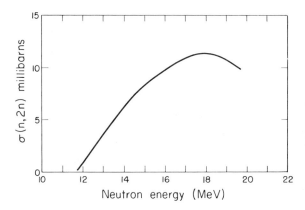

FIG.5. Excitation function for (n,2n) reaction: $^{14}N(n,2n)^{13}N$.

highly accurate nuclear data. Cox [19] in his classic summary of fast neutron cross sections emphasizes the fact that the user is primarily interested in obtaining reliable cross section information for his research, while the cross section measurer is essentially interested in physics. Expanding on this observation, Cox reiterates that experimental physicists would do the data user and data evaluator a great deal of service by including all pertinent details of the experimental procedures so that intercomparisons between data from different sources could be objectively analyzed.

3. NEUTRON SOURCES

Neutron sources used in activation analysis can be classified into four main categories: (1) nuclear reactors, (2) radioisotopic, (3) fission, and (4) accelerator neutron sources. Depending upon the type and irradiation assemblies used, fast, resonance or thermal neutrons can be produced. The choice of an appropriate source depends principally on the application being pursued. Criteria governing neutron source selection are: (1) cost, (2) physical dimensions and shielding requirements, (3) source handling facilities, (4) source strength, and (5) available space.

3.1 Nuclear reactor

It is well known that when a thermal neutron is absorbed in a fissile nucleus such as ^{235}U, the binding energy added to the target nucleus is sufficient to separate the nucleus into two fission fragments e.g. $^{235}U + ^{1}n \longrightarrow ^{148}La + ^{88}Br$. Since 1942, when the first graphite-uranium pile was constructed, numerous reactors have been built using a variety of geometrical configurations, fuel elements and moderators. The reactor, therefore, is classified in terms of the fuel and moderator used, and its specific function.

From earlier discussions it can be gathered that (n,γ) interaction cross sections are much larger than those for threshold reactions with a given target isotope. Since the former interaction is quite prolific, it is obvious that any reactor providing access to a high thermal neutron flux can be used for activation analysis. To minimize the production of interfering threshold interactions, a highly pure thermal neutron spectrum is desirable. Reactors used for radioisotope production are generally equipped with a "thermal column" constructed from graphite blocks and substantially removed in distance from the core itself. If however, the activation analyst is interested in using threshold reactions as means for material characterization, then a pure fission neutron spectrum is desirable. Often, this condition is reasonably satisfied by carrying out sample irradiations inside a hollow fuel element where moderation of fission neutrons is at a minimum. Special reactor assemblies such as those described by Lukens et al.[22] , and Yule and Guinn [23] can be pulsed by rapid removal and reinsertion of the reactor control rod. Neutron fluxes of the order of 10^{16} n. $cm^{-2}s^{-1}$ can be obtained during the "pulsed" time interval of a few milliseconds. This procedure results in a significant enhancement of the induced activity of short-lived nuclides.

In view of the operational aspects of nuclear reactors, samples for neutron irradiation have to be transported to and from the irradiation site by remote methods. The sample transfer container, or "rabbit", is carefully selected to be essentially insensitive to neutron absorption and radiation damage. The container is transported by pneumatic or hydraulic power inside a "rabbit" tube whose terminal is located in some desired neutron environment inside the reactor containment vessel. In order to ensure that the sample inside the "rabbit" is exposed to a reasonably homogeneous neutron flux, a judicious choice of the terminal emplacement is necessary. The sample must, at the same time, have the benefit of a high neutron flux for maximum sensitivity of analysis. Sample transfer times of the order of 1 to 3 seconds are typical. After irradiation, the "rabbit" is exited into a small shielded enclosure for "cooling" off prior to handling. In the design of such sample transfer facilities due attention is given to the effect of in-core gamma heating and radiolysis of liquid samples. Perhaps the most difficult irradiation parameter to evaluate is the expected temperature rise in the sample. The sample temperature is dependent upon such conditions as perturbation of neutron flux by the sample, attenuation of gamma-rays and their buildup in the materials surrounding the sample, and the sample matrix itself. Details, essential to the user, have been summarized [24,25] .

3.2   Radioisotopic source

Certain naturally occurring and artificially produced low-mass elements have very low neutron binding energies and suitable absorption cross sections for neutron production. The most commonly used radioisotopic neutron sources are of the (α,n) and (γ,n) type. The source consists of an alpha-or gamma-emitting radioisotope intimately mixed with a target element capable of interaction. The target element used in (α,n) sources is beryllium. Principal α-emitters used include ^{226}Ra, ^{210}Po, ^{227}Ac, ^{238}Pu, ^{239}Pu, ^{241}Am and ^{242}Cm.

The (γ,n) sources utilize such high-energy γ-emitters as ^{124}Sb, ^{88}Y, ^{24}Na and ^{140}La in conjunction with such targets as beryllium and deuterium oxide. Some radioisotopic neutron sources are fabricated to permit the separation of the radioisotope and the target, thus offering a means for turning off the neutron supply. From a detailed survey of (α,n) and (γ,n) radioisotopic neutron sources, relevant data useful to the user is tabulated Table II . Some of the obvious advantages in the utilization of such sources are; mechanical stability, compact size and cost. The neutron output, though decaying with the pertinent half-life, is predictable from day to day; and in most cases a constant over reasonable irradiation periods. The major disadvantage is, of course, the relatively low-neutron output when compared to the nuclear reactor capability. Furthermore, difficulties are envisaged due to the inability for turning off neutron production without some sacrifice of the optimum source strength possible for a given source design. Typical neutron energy distributions available from such sources are shown in Figs. 6-9. The reader is also directed to a comprehensive characterization of radioisotopic neutron sources as described by Ansell and Hall [26] .

3.3   Isotopic spontaneous fission source

Some transuranic elements disintegrate to a significant degree by spontaneous fission and are therefore potential sources of energetic neutrons. In Table III is presented a listing of spontaneous fission sources and their emission characteristics. Most of the sources listed are produced in minute quantities from the irradiation of fuel elements in extremely high-flux reactors

TABLE II. ALPHA-NEUTRON AND GAMMA-NEUTRON RADIOISOTOPE SOURCES

Type	Source	Main Reaction	Q (MeV)	Ave Neutron Energy (MeV)	Neutron Yield per sec per curie (nps/curie)	Half-life of Alpha emitting nuclide
(α,n)	^{242}Cm-Be	^{9}Be(α,n)^{12}C	5.65	4 MeV	4×10^6	163 d
	^{244}Cm-Be	-	5.65	4 MeV	3×10^6	18.1 y
	^{241}Am-Be mixture	^{9}Be(α,n)^{12}C	5.65	4 MeV	2.2×10^6	433 y
	^{226}Ra-Be mixture	^{9}Be(α,n)^{12}C	5.65	3.6 MeV	1.5×10^7	1 620 y
	^{210}Po-Be mixture	^{9}Be(α,n)^{12}C	5.65	4.3 MeV	2.5×10^6	138 d
	^{239}Pu-Be mixture	^{9}Be(α,n)^{12}C	5.65	4.5 MeV	2×10^6	24 360 y
	^{227}Ac-Be	^{9}Be(α,n)^{12}C	5.65	4 MeV	2.4×10^7	21.8 y
	^{228}Th-Be	-	5.65	4 MeV	2.8×10^7	1.9 y
	^{238}Pu-Be	-	5.65	4 MeV	2.8×10^6	89 y
(γ,n)*	^{24}Na-Be	^{9}Be(γ,n)^{8}Be	-1.67	0.2	1.4×10^5	15 h
	^{24}Na-D$_2$O	^{2}H(γ,n)^{1}H	-2.23	0.8	2.9×10^5	15 h
	^{88}Y-Be	^{9}Be(γ,n)^{8}Be	-1.67	0.16	1×10^5	108 d
	^{88}Y-D$_2$O	^{2}H(γ,n)^{1}H	-2.23	0.3	3×10^3	108 d
	^{124}Sb-Be	^{9}Be(γ,n)^{8}Be	-1.67	0.02	1.9×10^5	60 d
	^{140}La-Be	^{9}Be(γ,n)^{8}Be	-2.23	0.6	2×10^3	40.2 h
	^{140}La-D$_2$O	^{2}H(γ,n)^{1}H	-2.23	0.15	7×10^3	40.2 h

* Neutron yields are those for a 1 g. beryllium or heavy water target at 1 cm from the gamma source of one curie

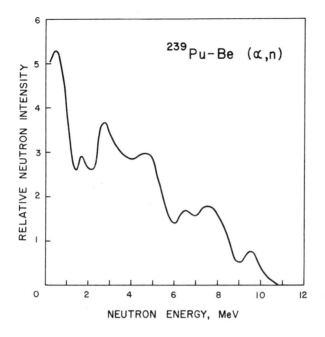

FIG.6. *Neutron energy distribution of ^{239}Pu-Be (α,n) source.*

for an extended period of time. Successive neutron absorption events produce beta-unstable products. After several beta decays, the unstable nuclide can decay by spontaneous fission and alpha decay, to generate high fluxes of neutrons whose energy distributions are comparable to that obtained from the thermal neutron fission of ^{235}U. Among these sources, ^{252}Cf is, by far, enjoying wide-spread utilization for analytical applications. It is of rugged construction, small in size, provides a stable flux ($\sim 2.3 \times 10^{12}$ n.s^{-1}. g^{-1}) for a reasonably long period of time (half-life 2.65 years). The source is generally encapsulated in stainless-steel, platinum-rhodium alloy, zircaloy or certain types of ceramics. Currently the source is manufactured and marketed by the Energy Research Development Administration (ERDA) of the United States at a cost of about $10 per microgram plus encapsulation costs. For a 100 microgram source, the encapsulation costs are of the order of $1000-$2000. If a high source strength is needed without adding to the ^{252}Cf weight, a uranium blanket [27] is used to boost the neutron output. Boosting factors approaching a factor of $\sim$40 have been achieved. Concise information regarding the purchase, encapsulation and research applications relevant to this source can be found in periodicals released by ERDA.

3.4 Accelerator neutron source

During the last twenty years accelerator produced neutrons have played a significant role both in terms of experimental neutron physics and neutron activation analysis. The theoretical capabilities [28] of such devices suggests the utilization of a wide variety of neutron producing interactions. However the choice is often dictated by the cost of irradiations. These accelerators can be used to induce useful neutron producing interactions as summarized in Table IV. The neutron yields as a function of incident particle bombarding energy for some of the reactions listed in Table IV are shown in Figs. 10 and 11. Neutron yield information given herein has been abstracted from a guide

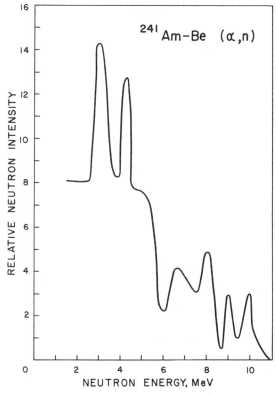

FIG.7. *Neutron energy distribution of* ^{241}*Am-Be* (α,n) *source.*

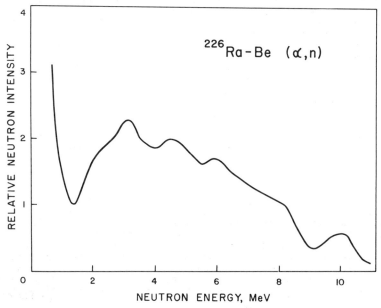

FIG.8. *Neutron energy distribution of* ^{226}*Ra-Be* (α,n) *source.*

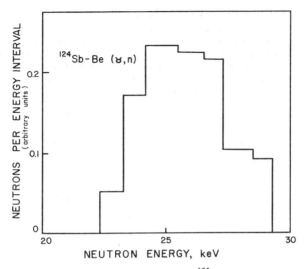

FIG.9. Neutron energy distribution of ^{124}Sb-Be (γ,n) source.

TABLE III. SPONTANEOUS FISSION NEUTRON SOURCES

Nuclide	Alpha Decay T^{1}_{2}(years)	Fission T^{1}_{2} (years)	Fissions per $10^6 \alpha$'s	Neutrons per Fission
^{238}U	4.5×10^9	6.5×10^{15}	5.6×10^{-1}	-
^{236}Pu	2.85	3.5×10^9	8.1×10^{-4}	1.89 ± 0.20
^{238}Pu	86.4	4.9×10^{10}	2.3×10^{-3}	2.04 ± 0.13
^{240}Pu	6580	1.3×10^{11}	5.4×10^{-2}	2.09 ± 0.11
^{242}Pu	3.8×10^5	7.1×10^{10}	5.3	2.32 ± 0.16
^{244}Pu	7.6×10^7	2.5×10^{10}	3.0×10^{-3}	-
^{242}Cm	0.45	7.2×10^6	6.2×10^{-2}	2.33 ± 0.11
^{244}Cm	17.6	1.3×10^7	1.3	2.61 ± 0.13
^{252}Cf	2.6	85	3.3×10^4	3.51 ± 0.16
^{254}Cf	Long	0.17	Large	-

TABLE IV. NEUTRON PRODUCTION BY ACCELERATORS

Reaction	Threshold energy of incident radiation (MeV)	Q value (MeV)	Emitted Neutron Energy at Threshold Bombarding Energy (MeV)	Principal Accelerator* Used
$^2H(d,n)^3He$	-	+ 3.266	2.448	NG
$^3H(p,n)^3He$	1.019	- 0.764	0.0639	VG, C
$^3H(d,n)^4He$	-	+17.586	14.064	NG
$^9Be(\alpha,n)^{12}C$	-	+ 5.708	5.266	VG, C, NG
$^{12}C(d,n)^{13}N$	0.328	- 0.281	0.0034	VG, C
$^{13}C(\alpha,n)^{16}O$	-	+ 2.201	2.07	VG, C
$^7Li(p,n)^7Be$	1.882	- 1.646	0.0299	VG, C
$^9Be(d,n)^{10}B$	-	+ 3.79	1-6	VG, C
$^7Li(d,n)^8Be$	-	+15.0	-	VG, C
$^9Be(\gamma,n)^8Be$	1.66	-	Polyenergetic	LA
$^2H(\gamma,n)^1H$	2.2	-	Polyenergetic	LA

*NG = Neutron generator; VG = Van der Graaff; C = Cyclotron
LA = Linear accelerator

[29] on neutron production using small accelerators and a comprehensive treatise [14] on the subject of neutron generators. Since this review emphasizes the utilization of small low-voltage accelerators or neutron generators, the following descriptions pertain to this particular type of accelerator only. Readers interested in the larger and more expensive particle accelerators are referred to references [29] and [30].

Small neutron generators have been widely used for activation analysis. The commercial instruments are essentially scaled-down charged-particle accelerators producing moderate fluxes of 14- and 3-Mev neutrons by exoergic reactions such as $^3H(d,n)^4He$ and $^2H(d,n)^3He$, respectively. Since all positive-ion accelerators are basically similar in design, they can be represented by the diagram shown in Fig. 12. Principally a neutron generator is composed of an ion source, an accelerating tube and a suitable target. Positive ions generated in the ion source are accelerated through a potential difference of about 60-200 kV in the accelerator tube which is maintained continuously at low pressures of about 10^{-4} to 10^{-6} mm Hg. Upon striking the target, the accelerated ions enter into interactions to generate neutrons which are essentially isotropic in emission.

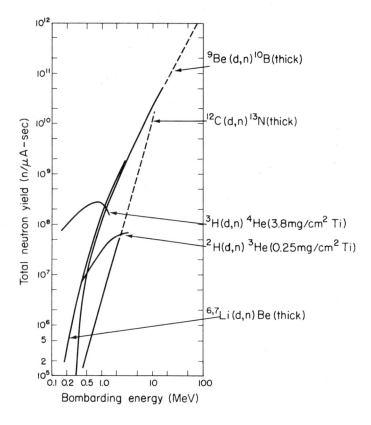

FIG.10. Neutron yields (d,n) reactions.

Average fluxes of about 5×10^{10} n.cm^{-2}s^{-1} (14 MeV) and 5×10^{8} n.cm^{-2}s^{-1} (3 MeV) at about 1 cm from the target, are available for activation analysis. The high energy neutrons can be moderated in a paraffin block placed adjacent and around the target to yield small fluxes of thermal neutrons. The application and mechanical aspects of such low-voltage accelerators are comprehensively documented [13-15]. Use of small sealed-tube type of high energy neutron generators has been demonstrated by their application in the geological field. Such tubes are capable of pulsed operations for the purpose of performing special analyses for uranium in mineral deposits. An example of a sealed-tube generator is shown in Fig. 13.

The application of neutron generators offers the following advantages: (1) the neutron beam can be turned off by simply de-exciting the acceleration high voltage, and (2) threshold reactions often provide a unique solution to certain nuclear interference problems, not possible with thermal neutron activation. On the other hand, the low-voltage accelerators present some problems in terms of the difficulty of accurate flux determination of the primary neutron emission from the target, and mechanical difficulties associated with the limited life of targets.

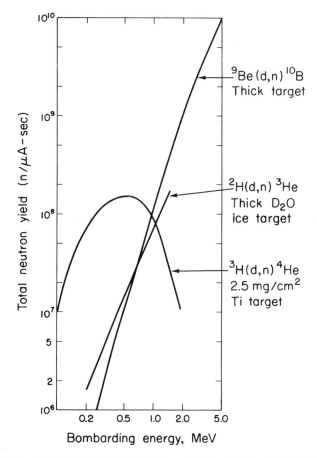

FIG.11. Experimental neutron yields under normal laboratory conditions.

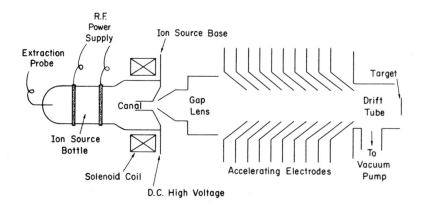

FIG.12. Schematic of a neutron generator.

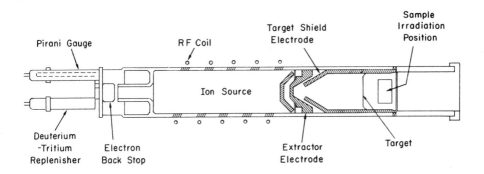

FIG.13. *Sealed-tube neutron generator.*

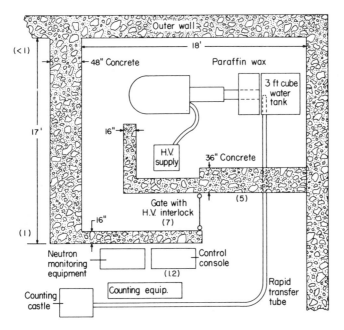

FIG.14. *Schematic of a neutron generator activation analysis facility.*

When using neutron generators, two paramount problems must be recognized if reproducible results of analysis are to be expected. Firstly, the inhomogeneity of the accelerated particle beam and the distribution of target element nuclei result in the production of a neutron flux which is far from homogeneous. Secondly, the degradation of the neutron flux and its normal gradient through a sample introduce a corresponding gradient in the induced radioactivity. As a general solution to the two problems, great efforts [31-39] have been made to design irradiation facilities which permit the sample to be subjected to a homogeneous exposure to neutrons.

Because of the limited target life it is obvious that neutron generators are most suited for short irradiations, therefore their application is directed to the analysis of short-lived decay products. This fact automatically implies the inclusion of a very fast automatic and programmable sample transfer system as an integral part of an activation facility utilizing a neutron generator. A typical facility is schematically shown in Fig. 14. The large amount of concrete shielding necessary for radiation safety may be observed.

In order to correct for differences in the neutron flux impinging upon a sample from irradiation to irradiation, neutron flux monitoring is essential. Many methods have been used and the reader is referred to detailed descriptions given in a treatise by Nargolwalla and Przybylowicz [14], and a dissertation by Iddings [40].

4. NEUTRON ACTIVATION ANALYSIS

As mentioned in the Introduction, neutron activation analysis has gained wide acceptance in recent years as a very sensitive analytical tool for the characterization of materials. In the early days, radioisotopic neutron sources were used; these, however, were low in neutron output and consequently of limited use. With the availability of nuclear reactors and the development of small, relatively inexpensive neutron generators, the growth of activation analysis was considerably accelerated. Today the practitioner has available several monographs [13-15, 41-54] to provide adequate background in all aspects of this technique. The popularity of the method has resulted in the establishment of a series of conferences [55-59] named "Modern Trends in Activation Analysis" to provide a common ground for the exchange of ideas and comparison of developments. Comprehensive bibliographies [14, 44, 60-65], including almost 10,000 studies of a very broad range of applications, provide ready references for the uninitiated and also serve as a general compendium of information for the veteran.

4.1 Principle of the technique

The fundamentals of neutron activation analysis have been adequately described in texts [14, 41-47, 50] and general reviews [13, 15, 48, 49, 51-54]. In general, the method involves the irradiation of a sample with a flux of neutrons. Upon neutron absorption, the irradiated nuclei are converted into various radionuclides or stable isotopes. The radionuclides decay with the emission of characteristic radiation at a rate that is well defined and through a specific decay scheme [66-70] unique to the particular decaying nuclide. The decay characteristics and the decay rate permit the unambiguous identification of the radionuclides in the irradiated sample.

The radioactivity generated from the irradiation of a target nucleus for a specific nuclear reaction can be expressed as a function of the operational parameters of the activation technique in the following manner:

$$A = DCN\sigma\phi \left[1-\exp\left(\frac{-0.693 t_i}{T_{1/2}}\right)\right]\left[\exp\left(\frac{-0.693 t_d}{T_{1/2}}\right)\right] \quad \ldots\ldots\ldots(2)$$

where,

A = the measured count rate at a time t_d, counts per second.
D = the branching ratio for the emission being measured.
C = the detector efficiency of the counter.
N = the number of target nuclei exposed to the incident flux.
σ = the neutron absorption cross section of the particular reaction with the target isotope at a given incident neutron energy, cm^2
φ = the neutron flux of constant intensity and energy, $n.cm^{-2} s^{-1}$
t_i = irradiation time
t_d = decay time from the termination of the irradiation to the start of counting.
$T_{1/2}$ = the half-life of the product nucleus being measured.
The units of time should of course be the same throughout Eq. 2.

In Eq. 2, factors such as ϕ, C and σ are not generally evaluated because their determination is both difficult and inaccurate. To improve accuracy the more practical comparative approach is used. This method involves the simultaneous irradiation of a standard and the sample. Thus, the amount of a given element in an unknown sample is given by,

$$\text{Element (in unknown)} = \text{Element (in standard)} \times \frac{A_u}{A_s} \quad \ldots\ldots(3)$$

where,

A_u = Measured count rate of the element in the unknown sample at time t_d, counts per second.

A_s = Measured count rate of the element in the known sample at time t_d, counts per second.

Although almost all analyses are performed using the comparative approach, Eq.2, or some modification of it, is very useful in establishing optimum conditions for a given analysis, and for the compilation of sensitivity tabulations [53, 72-76]. In regard to reported detection limits or sensitivities given in the tabulations, the reader is cautioned to use these figures of merit only as a very approximate indication of expected sensitivities. Uncontrolled factors, such as the energy distribution of the incident neutron flux and the user's choice of irradiation, decay and counting conditions preclude a ready comparison of expected radioactivities with those published in the tabulations.

Activation analysis differs from all other non-nuclear chemical analysis methods in that it is based on a nuclear interaction rather than on an interaction with the orbital electrons surrounding a nucleus. This immediately implies the independence of the activation technique from any chemical or physical state of the target element. The technique is sensitive only to the nuclear properties such as neutron absorption cross sections, half-lives of the product isotopes and their energy emission decay scheme, and is unrelated to the chemical properties of the target elements. Therefore, elements that are chemically similar, such as rare earths and the transition element groups, and thus difficult to differentiate by normal chemical methods, can often be discriminated by activation analysis.

To fully appreciate the significance of neutron activation as a technique for compositional analysis, one must compare its operational parameters and capabilities with those of other techniques designed to perform similar functions. A chart showing such a comparison is given in Table V. Note that the neutron activation technique has been subdivided according to the type of neutron source used. The sensitivity aspects of the listed techniques have been deliberately excluded as much controversy can arise from the selection of basic analytical parameters used in the calculation of detectable limits. Suffice it to mention that all of the techniques listed, with the exception of 14-MeV neutron activation and x-ray fluorescence, are capable of reaching detection limits in the sub-ppm range. Since in this review, a special consideration is given to the utilization of 14-MeV neutrons from small accelerators, it is necessary to include Table VI which gives the detection limits for pure elements under acceptable laboratory conditions. For the interpretation of the detectable limits, certain analytical conditions need to be qualified. A 14-MeV neutron flux of 10^9 n.cm^{-2} s^{-1} impinging upon a 10 gram sample of the pure element is assumed. The counting arrangement of a 3-in right circular cylindrical sodium iodide (NaI(Tℓ)) scintillation detector and a 2π counting geometry, is considered. The detection limit is calculated by: totalizing counts in the gamma-ray photopeak region; computing the background counts under the photopeak by a single base-line subtraction technique; computing the square root of this background; and expressing the limit of detectability as the amount of target which will generate counts in the photopeak which are 6 times the square root of the background counts.

TABLE V. COMPARISON OF TECHNIQUES FOR COMPOSITIONAL ANALYSIS

Technique	Qualitative analysis		Quantitative analysis			Specificity	Time of analysis (Long: > 30 min / Short: < 30 min)	Multicomponent analysis capability	Instrumentation cost (High > $ 30 000 / Low < $ 15 000)
	Property measured	Evaluation	Property measured	Evaluation					
14-MeV neutron activation	Gamma energy and decay rate of product nucleus	Excellent	Gamma energy and amount of induced activity	Excellent in the trace-element region		Excellent	Short	Good to excellent	Medium to high
Classical chemical (gravimetric, volumetric combustion etc.)			Volume or weight	Excellent		—	Long	—	Low
Mass spectrometry	Mass/charge of ionized particles	Excellent	No. of particles specific mass/charge	Excellent (gases, liquids) Good (solids)		Excellent	Long (solids) Short (gas and liquids)	Excellent	Medium (gases and liquids) High (solids)
X-ray spectroscopy	Wavelength of emitted X-ray	Excellent	Intensity of X-ray emission at specific wavelength	Good in semi-micro concentrations		Excellent	Short	Good	Medium
Atomic absorption spectrophotometry	Not applicable for qualitative survey		Absorption of specific line emission	Excellent		Excellent	Short	Poor	Low
Infrared absorption	Wavelength distribution	Good	Absorption at specific wavelength	Good		Fair	Short	Fair	Medium
Ultraviolet absorption	Wavelength distribution	Poor	Absorption at specific wavelength	Excellent		Fair	Short	Fair	Low-medium

NOTE: Only 14-MeV neutron activation and X-ray spectroscopy offer non-destruction analyses; however, the latter technique provides analytical information about the compositional nature of the sample surface.

TABLE VI. DETECTION LIMITS FOR 14-MeV NEUTRON ACTIVATION ANALYSIS

Note: s = seconds; m = minutes; h = hours; d = days.

Element	Nuclear reaction product	Half-life	Gamma-ray energy analyzed (MeV)	T_{act}	T_{dec}	T_{cnt}	Detection limit (mg)	Relative standard deviation
Aluminium	^{27}Mg	9.5 m	0.84	5 m	1 m	5 m	0.3	± 6%
Antimony	^{120}Sb	15.9 m	0.51	5 m	1 m	5 m	0.4	± 6%
Arsenic	^{75}Ge^m	49 s	0.14	3 m	15 s	3 m	10	± 5%
Barium	^{137}Ba^m	2.6 m	0.662	5 m	1 m	5 m	0.2	± 4%
Boron	^{11}Be	13.6 s	2.12	50 s	15 s	50 s	0.01	± 10%
Bromine	^{78}Br	6.5 m	0.51	5 m	1 m	5 m	0.1	± 6%
Cadmium	^{111}Cd^m	49 m	0.25	5 m	1 m	5 m	0.8	± 6%
Cerium	^{139}Ce^m	55 s	0.74	3 m	1 m	3 m	0.3	± 6%
Cesium	^{132}Cs	6.58 d	0.67	5 m	1 m	5 m	20	± 7%
Chlorine	^{34}Cl^m	32 m	0.51	5 m	1 m	5 m	45	± 6%
Chromium	^{52}V	3.77 m	1.43	5 m	1 m	5 m	0.45	± 6%
Cobalt	^{56}Mn	2.58 h	0.845	5 m	1 m	5 m	10	± 7%
Copper	^{62}Cu	9.9 m	0.51	5 m	1 m	5 m	0.1	± 6%
Dysprosium	^{165}Dy^m	75.4 s	0.108	4 m	80 s	4 m	0.5	± 8%
Fluorine	^{18}F	110 m	0.51	5 m	1 m	5 m	2	± 5%

TABLE VI. (Cont.)

Element	Nuclear reaction product	Half-life	Gamma-ray energy analyzed (MeV)	T_{act}	T_{dec}	T_{cnt}	Detection limit (mg)	Relative standard deviation
Erbium	$^{167}Er^m$	2.5 s	0.208	8 s	3 s	8 s	8	±13%
Europium	^{150}Pm	2.7 h	0.33	5 m	1 m	5 m	40	± 7%
Gadolinium	^{161}Gd	3.7 m	0.36	5 m	1 m	5 m	30	± 6%
Gallium	^{68}Ga	68 m	0.51	5 m	1 m	5 m	0.4	± 6%
Germanium	^{74}Ga	8 m	0.60	5 m	1 m	5 m	12	± 6%
Gold	$^{197}Au^m$	7.2 s	0.279	25 s	8 s	25 s	1.2	±12%
Hafnium	$^{179}Hf^m$	19 s	0.217	60 s	20 s	60 s	2	± 6%
Indium	$^{116}In^m$	54 m	1.27	5 m	1 m	5 m	0.8	± 8%
Iron	^{56}Mn	2.58 h	0.845	5 m	1 m	5 m	3	± 7%
Iridium	$^{191}Ir^m$	4.9 s	0.129	15 s	5 s	15 s	40	±12%
Lead	$^{203}Pb^m$	6.1 s	0.83	20 s	6 s	20 s	45	±16%
Magnesium	^{24}Na	15 h	1.37	5 m	1 m	5 m	8	± 7%
Manganese	^{52}V	3.77 m	1.43	5 m	1 m	5 m	1	±18%
Mercury	$^{199}Hg^m$	44 m	0.37	5 m	1 m	5 m	8	± 8%
Molybdenum	^{91}Mo	15.5 m	0.51	5 m	1 m	5 m	2	± 6%
Neodymium	$^{139}Ce^m$	55 s	0.76	—	—	—	—	—
	$^{141}Nd^m$	64 s	0.74	3 m	1 m	3 m	0.7	± 7%

TABLE VI. (Cont.)

Element	Nuclear reaction product	Half-life	Gamma-ray energy analyzed (MeV)	T_{act}	T_{dec}	T_{ent}	Detection limit (mg)	Relative standard deviation
Niobium	$^{90}Y^m$	3.2 h	0.20	5 m	1 m	5 m	40	± 7%
Nickel	^{57}Ni	36 h	0.51	5 m	1 m	5 m	1.0	± 6%
Nitrogen	^{13}N	9.96 m	0.51	5 m	1 m	5 m	1.0	± 6%
Oxygen	^{16}N	7.2 s	6.13	25 s	8 s	25 s	0.01	±10%
Palladium	$^{109}Pd^m$	4.8 m	0.18	5 m	1 m	5 m	0.8	± 5%
	^{107}Ru	4.2 m	0.19					
Praseodymium	^{140}Pr	3.4 m	0.51	5 m	1 m	5 m	0.1	± 6%
Phosphorus	^{28}Al	2.3 m	1.78	5 m	1 m	5 m	0.1	±10%
Potassium	^{38}K	7.7 m	0.51	5 m	1 m	5 m	7.0	± 6%
Rubidium	$^{86}Rb^m$	1 m	0.56	5 m	1 m	5 m	0.5	± 7%
Ruthenium	^{101}Tc	14 m	0.31 + 0.34	5 m	1 m	5 m	3	± 8%
	^{104}Tc	18 m						
	^{95}Ru	99 m						
Samarium	$^{141}Nd^m + {}^{143}Sm^m$	64 s + 1 m	0.75	5 m	1 m	5 m	5	± 7%
Selenium	$^{79}Se^m + {}^{81}Se^m$	3.9 m + 57 m	0.10	5 m	1 m	5 m	0.5	± 7%
Silicon	^{28}Al	2.30 m	1.78	5 m	1 m	5 m	0.2	± 6%

TABLE VI (cont.)

Element	Nuclear reaction product	Half-life	Gamma-ray energy analyzed (MeV)	T_{act}	T_{dec}	T_{cnt}	Detection limit (mg)	Relative standard deviation
Silver	^{106}Ag	24 m	0.51	5 m	1 m	5 m	0.2	± 6%
Sodium	^{23}Ne	38 s	0.44	2 m	40 s	2 m	1	±12%
Strontium	^{87}Sr^m	2.8 h	0.39	5 m	1 m	5 m	2	± 6%
Tin	^{123}Sn	40 m	0.153	5 m	1 m	5 m	1.5	± 7%
Titanium	^{49}Se	57.5 m	1.76	5 m	1 m	5 m	15	±10%
Vanadium	^{51}Ti	5.8 m	0.32	5 m	1 m	5 m	0.3	± 6%
Yttrium	^{89}Y^m	16 s	0.91	50 s	20 s	50 s	0.6	± 6%
Zinc	^{63}Zn	38 m	0.51	5 m	1 m	5 m	2.0	± 6%
Zirconium	^{89}Zr^m	4.2 m	0.59	5 m	1 m	5 m	0.5	± 6%

4.2 Analytical considerations

Upon neutron absorption most of the isotopes in a sample become radioactive. The gamma rays emitted during subsequent decays are therefore representative of all decay species generated in the sample. The method of measurement should therefore be able to detect and discriminate between the radioisotopes. Two principal approaches can be followed, and are briefly described below.

4.2.1 Radiochemical separation

The traditional method of detecting a particular radioelement is radiochemical separation. In almost all of these procedures, the element of interest is isolated by chemical means and its radioactivity measured. The chemical separation process is however time consuming except when only one or two elements are analyzed. The method is also destructive and the sample is consumed in the procedure. However, the exceptional discrimination factors possible from separation procedures permit extremely high sensitivities to be obtained. Considerable efforts have been expended in the development and practice of radiochemistry. Fast and automatic radiochemical separation procedures and apparatus are currently used in conjunction with activation analysis techniques, and are well documented in texts [44, 46].

4.2.2 Instrumental gamma-ray spectrometry

The counting of radionuclides is invariably done by gamma-ray spectrometry in a non-destructive manner. The gamma rays are measured by placing the irradiated sample over a sodium iodide (NaI(Tℓ)) scintillation detector or a solid state detector of lithium-drifted germanium or intrinsic germanium. The germanium system provides excellent resolution but with a considerably reduced efficiency when compared to an equivalent size sodium iodide (NaI(Tℓ)) scintillator. The output from the detector system is fed into pulse processing and sorting instrumentation called multichannel analyzers which have the capability of presenting graphical, oscilloscopic and printed gamma-ray spectrometric data. Modern gamma-ray spectrometry systems are equipped to perform with a considerable degree of automation. The instrumentation can be computer coupled with a high degree of flexibility in terms of software usage. Such systems have the capacity of raw data storage, location of specific gamma-ray photopeaks, identification of these peaks by comparison with library spectra, introduction of relevant analytical parameters such as neutron flux normalization factors, detector efficiencies etc., and finally to perform required computations and produce analytical answers. Nuclear instrumentation requirements for the activation analyst are given in recommended [77-80] texts and monographs. Gamma-ray spectra catalogues useful in neutron activation analysis with neutron generators [76, 81] and reactor neutrons [82, 83] are mandatory additions to the practitioner's reference files.

It is opportune at this juncture to emphasize some of the advantages and demerits of neutron activation analysis. The judgements to follow are primarily related to the use of neutron generators for activation analysis; however, some of the statements are applicable to the field as a whole. The choice of any analytical method is governed by four principal factors: selectivity, accuracy, precision, and sensitivity. The selection will depend on the stress given each of these factors based on the objectives of the analysis and the nature of the sample matrix. Furthermore, speed and economic considerations sometimes control the final choice of a method. Analytical chemists are continually seeking to develop new methods which will complement existing methods of analysis and thereby provide a full spectrum of techniques and approaches for the characterization of all types of materials. The role of experimental nuclear physics has played a significant part in the development of the neutron generator and made it possible for most analytical laboratories to set up its own activation analysis facility. From the earlier mentioned cross section discussion it is obvious that the neutron generator offers some unique approaches to the analysis of certain elements. In particular, the light elements such as N, O, F, Al and Si can be rapidly determined by 14-MeV neutron activation

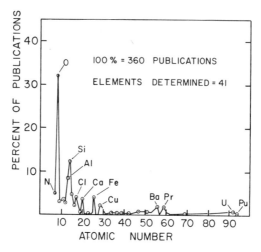

FIG.15. Frequency plot of element determinations in the literature.

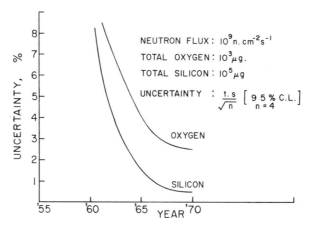

FIG.16. Improvement with time in calculated uncertainties of reported analyses for oxygen and silicon by 14-MeV neutron activation.

techniques. A recognition of this fact has resulted in concentrated efforts for the analysis of elements in the low mass number region. A survey of literature carried out in 1970 [51] revealed a distribution of studies as shown in Fig. 15. This survey indicated that fifty-odd elements had been successfully determined in the concentration range between 10 ppm to 50%. Almost 50% of the analytical effort was however devoted to the determination of oxygen and silicon. The potential for automation, especially in industrial process control, has made this technique quite attractive. The technique is also highly selective, owing to several parameters that are under control of the analyst. For instance, nuclear properties such as half-life, reaction cross section, and decay scheme permit the analyst to differentiate between elements in a matrix optimally through a judicious selection of irradiation, decay and counting times. Through a choice of neutron targets and moderating media, neutron energies of thermal, 3-, or 14-MeV can be obtained. Workers interested in multiple neutron energy utilization are referred to the development of Cosgrove [84] who has described a neutron producing system involving a "turret" type target holder which permits a rapid interchange of neutron producing targets.

Perhaps the most significant advances made with this technique have been in improved accuracy and precision. In spite of the anisotropic and time-varying nature of the neutron output from neutron generators, extensive research and development has found great success in combating this inherent source of analytical imprecision. A graphic representation of this improvement is illustrated in Fig. 16. The selection of the elements was based on their frequency of determination, and their respective quantities are based on the average quantities determined in the studies examined. To a large measure, the improvements indicated are a direct result of the recognition of possible sources of error followed by in-depth studies of systematic error corrections.

The existence and evaluation of systematic error have been thoroughly discussed in a comprehensive treatise [14], and will not be elaborated on herein. However it is instructive to list some of the principal sources of systematic error: 1) nuclear constants, 2) interfering nuclear reactions, 3) nuclear reaction recoil effect, 4) sample blank, 5) neutron flux normalization, 6) sample attenuation, and 7) instrumental. The reader is specifically referred to published works [13, 31, 85-90] dealing specifically with error analysis.

There are, of course, certain limitations to the technique when it is compared to others in the analytical "arsenal". The principal drawback appears to be the lack of sufficient flux to enable analyses in the low ppm range. From an operational standpoint, the limited target life for neutron production is perhaps the greatest disadvantage. However, considerable research continues in this area of target development. Hopefully, both of the above mentioned disadvantages will be either eliminated or reduced substantially in the near future. Cost of a neutron generator facility can be comparable to equivalent capability non-nuclear instrumentation. But this cost is by no means inconsequential. The total cost of a system including measurement equipment and biological shielding can be as high as $50,000-$75,000. However, generator facilities for so called "one element" analyses, can be assembled for half the cost.

In summary it can be mentioned that many elements can be determined at the milligram level rapidly and nondestructively. Neutron generators are commercially available and have placed this technique within reach of every well-equipped analytical laboratory. Although limited in sensitivity and neutron-target life, continuing research will undoubtedly result in significant improvements in the future.

5. APPLICATION OF NEUTRON ACTIVATION ANALYSIS

In turning to the scope and applications of neutron activation analysis it is necessary to assert that all applications in a sense are related to the investigation of the nature of the environment and man's interaction with it. Although this statement can be applied to many other branches of knowledge, the principal areas of application have been classified under nomenclatures readily understood in the language of modern day technology. The application of reactor and accelerator produced neutrons in these areas will be discussed with a view to establishing this technique among others performing equivalent tasks. In doing so, the role of neutron absorption physics in the development and practice of activation analysis is inherently emphasized.

The selection of particular studies pertaining to each area of application from the thousands described in literature is obviously beyond the scope of this review. The reader is urged to consult extensive annotated bibliographies [14,44,52,54,60-64] and conference reports [55-59], previously referenced, for such reported applications. In this review the stress is applied to the various analytical activities which can be classified under an area of interest. As such an attempt is made to itemize the most significant areas of application in the following subsections.

5.1 Biological and biomedical

Analytical problems related to the biological environment are also experienced in the cosmic and geological surroundings. This is to be expected since both the biosphere and the geosphere represent essentially the same raw materials i.e. living matter is nothing else but a rearrangement of atoms constituting the geosphere. Consequently, the wide variations in elemental compositions is evidenced in both of the above environments. However, in the biological world, selective concentration of elements such as carbon, calcium, oxygen and sulphur at the expense of other elements such as helium and silicon, occurs. The importance of activation analysis for biological studies does not lie completely in the determination of macro constituents, but in the examination of elements in the ppm and ppb range for which little if any data exists. In any given situation, the activation analyst must decide upon a specific technique which will allow the best possible discrimination for the elements of interest. Through a choice of bombarding particles and the utilization of such nuclear physics parameters as thresholds and resonances, it is entirely possible to apply this technique to almost all elements. The application of activation analysis can be discussed in terms of various aspects of the biological and biomedical fields. For the study of the intact human body, the in vivo method of analysis offers an interesting aspect of future research. The analyst, of course, will be sensitive to the fact that a typical human specimen cannot be exposed to high radiation dosages without severe damage. Therefore in this specific area, the analyst must exercise great care and be highly skilled. One of the main areas, i.e. the estimation of total body sodium can be studied by the irradiation of the subject with 14-MeV neutrons. Although fast neutrons are highly penetrating, the absorbed radiation dose would be excessive compared to an equivalent flux of thermal neutrons. Unfortunately thermal neutrons are effectively absorbed in the human tissues and homogeneous irradiations are therefore difficult to achieve. After irradiation with 14-MeV neutrons, the subject is transferred to a whole body counter and the most conspicuous radioactivities ^{24}Na and ^{38}Cl are measured. Cyclotron-produced high energy neutrons can also be used to achieve the same purpose. Suggestions by Palmer et al. [91] indicate that the elements sodium, calcium and chlorine can be determined by thermal neutron activation; nitrogen and phosphorus by fast neutrons using the reactions ^{14}N(n,2n)^{13}N and ^{31}P(n,α)^{28}Al; hydrogen by a measurement of the 2.23 MeV capture gamma rays from the ^{1}H(n,γ)^{2}H interaction. Except for the element carbon which requires the use of photon activation, the technique of neutron absorption can be effectively used for the in vivo analyses of biologically important elements. Perhaps the greatest difficulty in the application of neutron activation is in the availability of a suitable standard. A reasonable accuracy is gained by utilizing a plastic phantom filled with an appropriate solution of sodium; however for elements such as calcium which is heterogeneously distributed in the human body, the above technique of standardization does not work very well.

One of the best examples of in vivo analysis of limited areas of the human body is the determination of iodine in the thyroid. The general method includes the injection of the long-lived isotope of iodine (^{129}I, $T_{\frac{1}{2}}$ = 1.6x10^7 years). Before injection this isotope is made slightly radioactive by the ^{129}I(n,γ)^{130}I interaction. The isotope ^{130}I has a convenient half-life of about 12 hours. The gamma emission from ^{130}I is measured by external counting to estimate the amount in the thyroid gland. Once deposited, the ^{129}I is used as a flux monitor for the subsequent in vivo neutron irradiation. The indigenous thyroidal ^{127}I is converted to the 25 minute decay isotope ^{128}I. Since the element iodine is concentrated in the thyroid gland while interfering elements such as sodium and chlorine are rapidly removed by circulation, the determination of iodine concentration is considerably facilitated.

Although in vivo neutron activation analysis has made some strides in the measurement of the macro elements in the human body, its application for trace element determination in the human system is faced with severe difficulties. For the nutrition of man only a limited number of trace elemts are considered to be essential. These include zinc, copper, iodine manganese, molybdenum, cobalt and selenium. Although almost seventy additional elements are found in the human system, their effect on the biochemical processes is not yet understood with any degree of confidence. Progress in this regard is however being made and the function of such elements as chromium, vanadium and zirconium is being studied.

The biological significance of trace elements in various tissues has provided the activation analyst with a means for careful sampling and subsequent trace analysis. With the threat of radiation exposure removed, novel lines of investigation have been carried out. For example, it was found that the accumulation of such elements as tin and cadmium bear some correlation with the advancing age of the sample donor. However the ingestion of these elements from the environment as a function of time has not been discounted as a possible source of accumulation. The increased concentration of copper and chromium in newborne is another example on which hypotheses can be construed.

In general activation analysis has served in the discovery of trace elements which require highly sensitive analytical tools. The significance of discrete trace elements in terms of the biochemical processes is difficult to assess, mainly because of the rather wide concentration limits present in the human system. Statistical approaches have been followed to establish the demarcation limits below which elemental deficiencies can be contemplated. The biological and biomedical fields offer challenges to the analyst unsurpassed by any other field of investigation. The relevance of analytical observations, be they from neutron activation or from other analytical methods, will not be obvious until the biological and physiological significance of trace element distributions are fully understood.

5.2   Material science and industrial applications

In the area of materials science neutron activation has enjoyed its greatest success. In all industrial manufacturing operations materials quality is a basic specification, and most of the world's analytical chemists are engaged in industry with the almost sole objective of materials characterization. Because of the keen competition offered by other non-nuclear techniques, neutron activation analysis has had to bear its sharpest criticisms and in-depth assessments. Nevertheless the nuclear technique has been exploited to advance the state of the art of the semiconductor industry in the examination of ultra pure materials. To this end, the activation technique has been pushed to the very limit of its sensitivity. In industry, activation analysis has been applied in the analytical laboratory, and for process control on an on-line basis. In many cases this approach has rendered great economic benefits and has been considered to be most expedient. The non-destructive aspect of neutron activation has often been the determining factor for its selection. A prime example illustrating all of the above advantages is evidenced in the analysis of oxygen in metals, particularly in steel. The popularity enjoyed by neutron generators was due, to a large measure, in its ability to determine ppm amounts of oxygen in metals on a rapid and non-destructive basis. Each analysis could be performed at a cost of a few cents and with a high through-put. The 14-MeV neutron interaction $^{16}O(n,p)^{16}N$ results in the emission of very high energy (6.1, 7.1 MeV) gamma rays, so that the measurement process is essentially interference-free. Much of the research done to improve this technique was initiated by this great need in the metals industry. Today, the 14-MeV neutron activation technique for oxygen in metals is firmly established and offers advantages in cost, rapidity and reliability.

The estimation of nitrogen in foodstuffs for the purpose of determining the protein content is another example of 14-MeV neutron activation analysis. Despite some difficulties from matrix interferences, the $^{14}N(n,2n)^{13}N$ is utilized for protein estimation of grain on a routine basis.

Standard reference materials, so critical for instrumentation and calibration control, have also been subjected to neutron activation analysis. In general such materials must be analyzed by two or more unrelated methods of analysis prior to certification. Neutron activation analysis has made significant contribution to analytical data issued by standards disbursing agencies such as the National Bureau of Standards of the United States. Since the ultimate goal of an analytical measurement is to obtain a result of known accuracy, the importance of standard materials in the materials science cannot be overly emphasized. In the area of standard materials for industry, the neutron generator has made significant contribution. The example of oxygen in various types of iron and steels has already been cited. The determination of the metal components such as cobalt, silver, magnesium and silicon in organometallic compounds used to determine engine wear is another example where 14-MeV neutron activation analysis offers significant advantages.

On-line activation analyses such as: nitrogen in grain products, oxygen in coal, fluorite in fluorspar, sodium and phosphorous in detergents, silicon in iron ore or in coal, aluminum in coal, copper in copper ore and barium in barytes, are only a few examples in which the small accelerator generating high energy neutrons has been successfully exploited. In most of the above instances, the ingenuity of the activation analyst has resulted in accurate analysis by the elimination of interferences and sources of systematic error.

5.3 Geo-and cosmo-chemical analyses

The geochemist is primarily concerned with the formation, distribution, and chemical and physical interactions of minerals of rocks and ores. In order to test his theories and hypotheses for the above geochemical properties it is essential that compositional knowledge of the geological sample be known with good accuracy. Therefore analytical information for terrestrial rocks, minerals, ores, soils, water and the atmosphere for the geochemist; and the composition of meteorites, tektites and lunar material for the cosmochemist is of fundamental importance. In many instances the compositional knowledge is used to determine the age of the material.

In general the elemental abundances in the geo- and cosmo-sphere are grouped into three classifications: major (>1%), minor (0.01 - 1%) and trace (<0.01%) levels of concentration. In the major class, elements such as O, Mg, Al, Si, Ca and Fe are found. Elements H, C, P, S, Cl, V, Cr, Mn, Sr, Zr and Ba form the minor constituent group, while the noble gases, Li, Be, B, N, Sc and Cu, and all elements beyond Cu constitute the trace group of elements. Elements such as Na, K, Ti, F, Co, Ni can belong to one or more of the groups defined above.

A review of elemental abundances published by Suess and Urey in 1956 [92] indicated serious errors in the earlier reported abundance data for meteorites. Since a large amount of data were highly suspect an immediate need arose for an analytical method capable of analysis of small samples with high sensitivity. Neutron activation was therefore immediately introduced to accurately establish a new generation of abundance data. In fact activation analysis was one of the two principal methods used in the analysis of lunar materials retrieved by the Apollo 11 and 12 missions. The popularity of this method is shown by the existence of over 600 published works in the field of geo- and cosmo-chemistry.

The 14-MeV neutron activation method has been applied to the determination of Si and O on a non-destructive basis. As is well known, the classical methods for these elements are rather tedious. Efforts in these analyses were directed towards the achievement of high degrees of precision and accuracies. Using 2 - 3 gram samples of rock, average fractional deviation of ±0.54% for O and 0.44% for Si have been reported for U.S.G.S. standard rocks [93]. In these standard rocks, the oxygen content varies from about 41 - 49% and the silicon composition between 18 - 31%. The 14-MeV neutron activation method has also been applied to other elements such as Ti, Nb, Ce, Pr, Cu and Y in ores and minerals. The application of activation analysis to geochemistry and cosmochemical studies is described in several proceedings [55-59, 94].

Instrumental thermal neutron activation analysis has been widely used for multielement determinations. Combined with high resolution Ge(Li) detector systems, this method of analysis provides the geochemist with a very powerful tool to probe the composition of the geosphere. When combined with adequate radiochemical separation techniques, the general activation method has been applied to the analysis of 45-50 elements in standard rock samples. A few of the trace elements can be determined in the ppb range of concentrations.

Some of the areas in which neutron absorption processes have played a significant part include: meteoritic studies with 14-MeV neutron activation; geological age determinations by a measurement of K/Ar and I/Xe ratios; determination of the nickel content of tektites; and abundances of over 30 elements in lunar samples. Perhaps neutron activation analysis faces one of its greatest challenges in the areas of geo- and cosmo-chemistry.

5.4 Mineral and energy resources

The application of neutron activation analysis in the mineral and resource fields has been well documented in the bibliographies referenced previously. A selection of such applications is also given in international symposia [96,97] devoted to nuclear techniques and mineral resources. However, the majority of this research has been carried out in highly controlled laboratory environments. Despite the fact that many of these studies purport direct applications to the mineral and resource industry, very few have been tested under typical field conditions. Furthermore, only a handful are actually being utilized for mineral exploration and mine development. It is instructive to list the principal causes for the difficulties experienced by activation analysts in their attempts to "sell" their individual developments to the mineral resource industry:

1) The displacement of existing chemical techniques by neutron activation or by other nuclear techniques has been met with great reluctance by the industry.
2) A perusal of activation literature reveals numerous instances where the activation analyst has failed to recognize the specific needs of a particular exploration or prospecting technique. In all fairness the analyst has had to face variable requirements from different developers of the same mineral.
3) Neutron activation analysis time as visualized in the laboratory often takes much longer when applied to field conditions. In mining circles, the time factor is the major determinant as to whether or not a given method is utilized.
4) The activation analyst is often guilty of overlooking the fact that laboratory nuclear instrumentation may not function in the hostile environment synonymous of field conditions.
5) The high degree of sophistication built into neutron activation methods often acts as a deterrent for its field acceptance.

It is generally found that those nuclear methods being practised on a routine basis were developed with a full recognition of the above factors. Consequently, their application in the mineral resource industry has led to substantial economic benefits.

Although the petroleum industry has been using nuclear techniques for well logging of oil bearing formations since the 1930's, the mineral industry has been rather tardy in their acceptance of such techniques. In the literature, it is difficult to search for such applications for specific elements since the methods themselves are generally classified according to the nuclear principles used. However, some excellent reviews on the general application of nuclear excitation techniques are available [97-100].

Among nuclear techniques, the ones involving neutron absorption processes are considered to be the most promising. Studies by Eisler et al. [101], Hoyer and Locke [102], Landström et al. [103], Moxham et al. [104], and Nargolwalla et al. [105], are examples of the researchers recognizing the needs of industry. Neutron absorption methods used in the mineral resource industry have depended heavily on the (n,γ) reaction induced by radioisotopic, spontaneous fission and accelerator-produced neutron sources. The detection systems used include the sodium iodide scintillator and the Ge(Li) solid state assembly. To date there are several case histories developed from the use of the sodium iodide detector. In spite of the ingenious efforts [106, 107] to adapt the Ge(Li) system for downhole applications, an adequate production logging case history is yet to be reported.

Fast neutron-induced reactions have been extensively used for mineral exploration. With the rapid advancement in electronic technology, it is possible to construct small sealed-tube type accelerators which can be introduced into boreholes of only 3 inches in diameter. In one of the early applications of the neutron generator for surface analysis, it was found that 3-MeV neutrons were more practical than 14-MeV neutrons. Elements amenable to this type of application include: Na, F, O, Si, B, P, Ag, Br, Mg, Cl, Cu, Rh, S and Fe. Both Dibbs [108] and Eisler et al. [101] have studied the applicability of the neutron generator for the determination of copper. In general, the elements capable of being analyzed by fast neutron activation without significant interference from other elements are oxygen, silicon and aluminum.

The availability of ^{252}Cf has considerably accelerated the growth of applications utilizing thermal neutron activation techniques. The introduction of thermal neutron activation for both surface and downhole applications in the resource industries will go a long way towards a general acceptance of nuclear techniques by the mining community.

5.5 Archaeology

The non-destructive attribute of neutron activation analysis is perhaps best demonstrated in its application in the fields of art and archaeology. It is, of course, of paramount importance that any technique utilized must leave the object in an unaltered state. In many cases, the art object can be analyzed without sampling. It is sometimes possible to obtain minute samples from the surface of the art object for certification purposes. Neutron activation can then be applied to the specimen, and is one of the few methods with the sensitivity of determining a large number of trace elements. The trace element abundances are used to identify materially or chronologically the composition or age of the artifact, respectively. The origin of an object sometimes can be assessed by the identification of elements present in the ppb range. The composition of art objects can not only be determined by activation techniques, but their structure can also be examined. This examination is done by neutron activation autoradiography to reveal subsurface structure of oil paintings.

Fast neutrons have been used to identify gold and silver in ancient coins, silver in lead roof tiles of historic edifices and the like. The ability to determine low atomic weight elements has made the neutron generator a popular choice for investigating this social area. Studies of ancient pottery dominated the number of applications devoted to art and archaeology. Pottery symbolic of Grecian, Roman, Mesoamerican, British, Eastern Mediterranean and Western Asia cultures have been subjected to neutron activation analysis. Coins and metal objects have also been analyzed. Stain-glass windows and pigment from paintings have also been authenticated by using comparative methods.

Entire oil paintings have been irradiated by neutron fluxes of the order of 10^9 n.cm^{-2}s^{-1}. Following irradiation, a series of autoradiographs are obtained by placing x-ray films in contact with the painting. Experts examining such autoradiographs have been able to determine the distribution of individual pigments.

The field of archaeology offers the analyst an exciting area of research, and thereby demonstrates the important role of neutron absorption physics in the social sciences.

5.6 Forensic sciences

In forensic sciences the application of neutron activation analysis as an identification tool has been limited. Only in a few instances can it be said that the nuclear technique yields comparable information to those from "fingerprint" methods. In all others, it serves as a means to confirm or deny the possibility of common origin of two samples and generally as a characterization tool for materials evaluation.

This is one field of application where the activation analyst must exercise great caution in his expression of data and conclusions because of the obvious social and legal ramifications. To deduce comparisons or dissimilarities from a research study of a limited number of samples pertinent to a given type of evidence material is a dangerous path to follow; and has often led to discredit of the basic technique. At times the proponents of this technique are all too quick to make hurried judgements based on their analytical results.

Trace analysis has become an integral part of a modern day forensic laboratory. Neutron activation analysis must compete with others such as atomic absorption and x-ray fluorescence. The attraction for neutron activation analysis is due to its survey capability. Routine and simultaneous determinations of many elements on a non-destructive basis have attracted major efforts in forensic laboratories around the world. As a result, forensic activation analysis is now routinely used for the characterization of evidence materials in certain types of criminal cases. The forensic scientist must exercise good judgement in the adoption of the given technique or techniques depending upon the particular circumstance of the investigation. Speed of analysis is invariably a key factor. An excellent discussion on the merits and demerits of forensic activation analysis is given by Krishnan [109]. From the forensic standpoint, a brief discussion of some types of evidence materials to which neutron activation analysis has been applied follows.

A common evidence material, namely hair, has received the most attention. Thousands of trace element distributions in human hair have been obtained through neutron activation. From these results, concentration patterns and frequency distributions for up to thirty odd elements are available. Forensic activation has been particularly suited for these analyses, especially when the evidence material is very small and elemental sensitivities down to the ppm and ppb range are desired. However, uncontrolled parameters such as human diet, environment, growth phase and others which have a marked influence on elemental distribution, have precluded the utilization of such evidence for positive identification. In the absence of other types of evidence however, neutron activation analysis of hair does offer a means for possible identification.

The examination of metal fragments taken from the scene of some criminal actions can also be done by forensic activation analysis. The purpose of such examination may be to identify whether or not the fragment itself is that from a shattered bullet. In such circumstances it is only necessary to analyze for elements present in bullet alloys, and attempt to establish the common origin through trace element analysis. Therefore the concentration patterns of elements such as antimony, arsenic, copper and silver are studied.

Other examples of forensic activation analysis include the examination of glass chips, paint fragments, soil samples, certain types of fibres and toxic materials. The analysis of gunshot residue is one examination in which neutron activation has made a significant contribution. Systematic research has yielded valuable information on the effect of parameters such as washing of hands and elapsed time after the incident.

The field of forensic science offers the neutron activation analyst a wide field of practice. Evidence samples include biological, biochemical, inorganic and organic materials. In this area of application the activation analysis technique serves as a good complement to other non-nuclear methods of analysis.

5.7 Environmental and ecological

During the last decade or so communities throughout the world have become increasingly conscious of the ever increasing deterioration of the environment. Much of this deterioration has been attributed to industry. Consequently, governments have taken steps to formulate laws and regulations that are strictly administered and offenders are dealt with effectively. The application of analytical methods for pollution control and environmental research forms the basis of many such judgements. The quality of our water, bioenvironment and atmosphere must, at all cost, be protected. In recognition of this fact, the International Atomic Energy Agency convened their first conference [110] to deal with nuclear techniques in environmental pollution. Shortly afterwards the U.S. Energy Research and Development Administration sponsored a major symposium [111] which brought together a collection of environmental and ecological studies in which the application of neutron activation techniques was significantly evident.

The element mercury among all others has been the subject of much debate and analysis. Perhaps this has been in part due to its property of being methylated in nature, thereby being transformed from a less toxic form to one of higher long-term toxicity with genetic effects. In addition to mercury several other elements such as arsenic, cadmium, selenium, antimony, chromium and indium have been rigorously studied in a host of environmental matrices. The preponderance of such studies has precluded the selection of any one or more for inclusion in this review. Suffice it to state that this area of investigation poses considerable difficulties for the activation analyst; and therefore offers the greatest challenges. If it is at all possible to predict the future application of neutron absorption techniques for the solution of problems in the environmental sciences, it can be said that the activation analyst will play a very important role.

6. CONCLUSION - PRESENT AND FUTURE

The current status of neutron absorption techniques can be summarized as follows. In general neutron activation offers a rapid method for compositional analysis. The technique can be easily adapted for automatic operation and data processing, and can be developed into a reasonable survey method for multi-element analyses. It is generally recognized that no matter how sensitive the method might appear to be on the basis of theoretical yield calculations, the practical sensitivity limit, in general, is somewhere between 0.05 to 0.5 ppm. In this respect the method is quite comparable to some of the non-nuclear methods, such as emission and x-ray spectrometry. In order to obtain a significant improvement in sensitivity, it is necessary to resort to some form of discrimination by performing radiochemical separations. In many cases such a procedure can result in an improvement by an order of magnitude or more.

In considering neutron activation analysis among the host of other measurement techniques, it is possible to state that when combined with reliable and reproducible radiochemical methods, it can yield accurate results with high sensitivities. With proper sample handling in a "clean room" environment this method of analysis can conceivably be said to have no sample "blank" problems, and analyses down to the 1 to 10 ppb level appear feasible.

Insofar as the future is concerned, it can be emphasized that this technique must now attack the world of practical samples. In the past, the development of many a technique was based on the laboratory sample. The conclusions of such studies invariably included extrapolations illustrating ultra low sensitivities that would have been in the grasp of the analyst had the neutron flux available been a thousand times higher. It appears that, somewhere in such extrapolations, the investigator loses contact with the real world, i.e. the practical sample. The continuous battle with the practical sample and obtaining of results with meaningful error estimates is perhaps the greatest challenge this technique faces in the future. Success in such an area will no doubt lead to a redetermination of fundamental nuclear constants such as the cross section with an added degree of accuracy. The goal for higher sensitivities with minimum error is not only necessary but mandatory to achieve, if a better understanding of the environment is to be gained. Too often has this method been used as a last resort. Yet it offers some unique advantages to the analyst. The full exploitation of its attributes in the future will undoubtedly lead to a better understanding of the many problems inherent in materials characterization.

REFERENCES

[1] HEVESY, G., LEVI, H., The action of neutrons on the rare earth elements, Det. Kgl. Danske Videnskabernes Selskab. Mathematisk - Fysiske Meddelelser, 14(5),(1936), 3-34.

[2] SEABORG, G.T., LIVINGOOD, J.J., Artificial radioactivity as a test for minute traces of elements, J.Am.Chem.Soc., 60,(1938),1784-1786.

[3] CINDA 68, An index to the literature on neutron microscopic data, USAEC Division of Technical Extension, USSR Nuclear Data Information Centre, ENEA Neutron Data Compilation Centre, IAEA Nuclear Data Unit, (1967), (1968), (1969) (and Supplements).

[4] CINDU, IAEA Nuclear data unit, Vienna.

[5] BROOKHAVEN NATIONAL LABORATORY, "Neutron Cross Sections", Rep. BNL-325, 2nd ed., Suppl. 1 (1960) and Suppl. 2, five volumes (1964-1966).

[6] MUGHABGHAB, S.F., GARBER, D.I., "Neutron Cross Sections", Rep. BNL-325, 3rd ed., Vol 1, Brookhaven National Laboratory, (1973).

[7] DEVOE, J.R., ed., "Radiochemical Analysis", National Bureau of Standards,Technical Note 404, (1966), 47-51.

[8] LEACHMANN, R.B., Proc.Int.Conf. on Peaceful uses of Atomic Energy, Geneva, 2, (1956), 193.

[9] HUGHES, D.J.,"Pile Neutron Research",Addison-Wesley, Cambridge, Massachusetts, (1953).

[10] LISKIEN, H., PAULSEN, A., "Compilation of Cross-Sections for some Neutron Induced Threshold Reactions", EUR-119e, Euratom, Brussels, (1963).

[11] JUNG, R.C., EPSTEIN, H.M., CHASTAIN, J., "A simple experimental method for determining effective threshold energies and cross sections", Rep. BMI-1486, Batelle Memorial Institute, (1960).

[12] ROY, J.C., HAWTON, J.J., "Tables of estimated cross sections for (n,p), (n,α) and $(n,2n)$ reactions in a fission neutron spectrum", CRC-1003, AECL, (1960).

[13] CSIKAI, J., BUCKZKÓ, M., BÓDY, Z., DEMÉNY, A., "Nuclear data for neutron activation analysis", At.Energy Rev., Vol. VII(4), IAEA, Vienna, (1969).

[14] NARGOLWALLA, S.S., PRZYBYLOWICZ, E.P., "Activation Analysis with Neutron Generators",Vol.39, Chem. Analysis Series, John Wiley & Sons, New York, (1973).

[15] CSIKAI, J., "Use of small neutron generators in science and technology", At.Energy Rev., Vol.II(3),IAEA, Vienna, (1973).

[16] MARION, J.B., FOWLER, J.L., eds., "Fast Neutron Physics", Part II, Interscience Publishers Inc., New York, (1960).

[17] SEGRE, E., ed., "Experimental Nuclear Physics, Vols. I-III, Wiley, New York, (1953).

[18] Proceedings of Conf. on Neutron Cross Section Technology, USAEC, Rept. Conf-660303, (1966).

[19] Ibid.,National Bureau of Standards, Special Publication 299, Vols. I & II, (1968).

[20] Ibid.,USAEC, Rept. Conf-710302, Vols. I & II, (1971).

[21] Ibid.,National Bureau of Standards, Special Publication 425, Vols. I & II, (1975).

[22] LUKENS Jr., H.R., YULE, H.P., GUINN, V.P., Nucl. Instr. Methods, $\underline{33}$, 273(1965).

[23] YULE, H.P., GUINN, V.P., Proc. Int. Conf. on Radiochem. Method of Analysis. IAEA, Vienna, Vol. II (1964) 111.

[24] SCHREIBER, R.E., ALLIO, R.J., Nucleonics, $\underline{22}$(8), (1964) 120.

[25] TAYLOR, C., WEST, R., WHITING, M., Proc. Int. Conf. on Production and Uses of Short-lived Radioisotopes from Reactors, IAEA, Vienna, Vol. I, (1963) 67.

[26] ANSELL, K.H., HALL, E.G., "Recent developments in (α,n) sources", in Proc. of Conf. on Neutron Sources and Applications, American Nucl. Soc., National Topical Meeting, Augusta, Georgia, USAEC, Rep. CONF-710402, Vol. II, (1971) 90-99.

[27] CALIFORNIUM 252 PROGRESS, Californium-252 News, No. 17,(1974) 3-6.

[28] BURRILL, E.A., MACGREGOR, M.H., Nucleonics, $\underline{18}$ (12), (1960),64.

[29] BURRILL, E.A., "Neutron Production and Protection", High Voltage Engineering Corporation, Burlington, Massachusetts, (1963).

[30] FLEISCHER, A.A., "The Production of Fast Neutrons by Small Cyclotrons", The Cyclotron Corporation, TCC Rep. 2003, Berkeley, California, (1968).

[31] ANDERS, O.U., BRIDEN, D.W., Anal.Chem., $\underline{36}$, (1964) 287.

[32] MOTT, W.E., ORANGE, J.M., Anal. Chem., $\underline{7}$, (1965) 1338.

[33] WOOD, D.E., JESSEN, P.L., JONES, R.E., Pittsburgh Conf. on Analytical Chemistry and Applied Spectroscopy, Pittsburgh, Pennsylvania, (1966).

[34] DYER, F.F., BATE, L.C., STRAIN, J.E., Anal. Chem., $\underline{39}$ (1967), 1907.

[35] LUNDGREN, F.A., NARGOLWALLA, S.S., Anal. Chem., $\underline{40}$, (1968), 672.

[36] NARGOLWALLA, S.S., Ph.D. Thesis, Univ. of Toronto, Toronto, Canada, (1965).

[37]   WOOD, D.E., "Development of Fast Neutron Activation Analysis for Liquid Loop Systems", Kaman Nuclear Corporation, Colorado Springs, Colorado, Rep. KN-67-458(R), (1967).

[38]   AL-SHAHRISTANI, H., Ph.D. Thesis, Univ. of Toronto, Toronto, Canada, (1969).

[39]   PRIEST, H.F., BURNS, F.C., PRIEST, G.L., Anal. Chem., $\underline{42}$, (1970), 499.

[40]   IDDINGS, F.A., Anal. Chim. Acta, $\underline{31}$, (1964), 206.

[41]   KOCH, R.C., "Activation Analysis Handbook", Academic Press, New York, (1960).

[42]   LYON, Jr., W.C., ed., "Guide to Activation Analysis", Van Nostrand, Princeton, New Jersey, (1964).

[43]   LENIHAN, J.M.A., THOMSON, S.J., eds., "Activation Analysis, Principles and Applications", Academic Press, New York, (1965).

[44]   DE SOETE, D., GIJBELS, R., HOSTE, J., "Neutron Activation Analysis", Chemical Analysis Series, Vol. 34, John Wiley & Sons, New York (1972).

[45]   TAYLOR, D., "Neutron Irradiation and Activation Analysis", George Newnes, London, (1964).

[46]   BOWEN, H.J.M., GIBBONS, D., "Radioactivation Analysis, Oxford University Press, London, (1963).

[47]   HOSTE, J., OP DE BEECK, J., GIJBELS, R., ADAMS, F., VAN DEN WINKEL, P., DE SOETE, D., "Instrumental and Radiochemical Activation Analysis", CRC Press, Cleveland, Ohio, (1971).

[48]   LENIHAN, J.M.A., THOMSON, S.J., eds., "Advances in Activation Analysis", Vol. 1, Academic Press, New York, (1969).

[49]   LENIHAN, J.M.A., THOMSON, S.J., GUINN, V.P., eds., "Advances in Activation Analysis", Vol. 2, Academic Press, New York (1972).

[50]   KRUGER, P., "Principles of Activation Analysis", John Wiley & Sons, New York, (1971).

[51]   NARGOLWALLA, S.S., "Application of Neutron Generators to Activation Analysis", in Proc. of the Second Oak Ridge Conf. on the Use of Small Accelerators for Teaching & Research, USAEC, Rep.Conf- 700322, (1970), 185-204.

[52]   COLEMAN, R.F., PIERCE, T.B., "Activation Analysis, A Review", The Analyst, $\underline{92}$ (1090), (1967), 1-19.

[53]   DIBBS, H.P., "Activation Analysis with a Neutron Generator", Dept. of Mines and Technical Surveys, Mines Branch, Ottawa, Res. Rep. R155, (1965).

[54]   ADAMS, F., HOSTE, J., "Non-destructive Activation Analysis", At. Energy Rev., Vol. IV(2), IAEA, Vienna, (1966).

[55] Proc., 1961 International Conf. on "Modern Trends in Activation Analysis", College Station, Texas, (1961).

[56] Proc., 1965 International Conf. on "Modern Trends in Activation Analysis", College Station, Texas, (1965).

[57] Proc., 1968 International Conf. on "Modern Trends in Activation Analysis", National Bureau of Standards, Special Publication 312, Vol. I&II, (1969).

[58] Proc., 1972 International Conf. on "Modern Trends in Activation Analysis", Paris, J. Radioanalytical Chemistry, Vol. 18 (1&2), Vol. 19(1&2),(1974).

[59] Proc., 1976 International Conf. on "Modern Trends in Activation Analysis",Munich, Vol. I & II, Repro-Mayer-offset, Munich, Federal Republic of Germany, (1976).

[60] BOCK-WERTHMANN, W., AED Information Service, Series C: Bibliographies, Section 14: "Activation Analysis" AED-C-14-1, Hahn-Meitner-Institut für Kernforschung, Berlin, (1961).

[61] Ibid.,AED-C-14-02, (1963).

[62] Ibid., AED-C-14-03, (1964).

[63] LUTZ, G.J., BORENI, R.J., MADDOCK, R.S., WING, J., eds., "Activation Analysis: A Bibliography through 1971", National Bureau of Standards, Technical Note 467, (1971).

[64] LUTZ, G.J., ed., "14-MeV Neutron Generators in Activation Analysis: A Bibliography", National Bureau of Standards, Technical Note 533, (1970).

[65] VAN GRIEKEN, R., HOSTE, J., "Annotated Bibliography on 14 MeV Neutron Activation Analysis", Eurisotop, Office information booklet 65, Series: Bibliographies -8, (1972).

[66] LEDERER, C.M., HOLLANDER, J.M., PERLMAN, I., "Table of Isotopes", John Wiley & Sons, New York, (1967).

[67] ALIEV, A.I., DRYNKIN, V.I., LEIPUNSKAYA, D.I., KASATLIN, V.A., "Handbook of Nuclear Data for Neutron Activation Analysis", Translation from Russian, Israel Program for Scientific Translation, Keter Press, Jerusalem, (1970).

[68] PAGDEN, I.M.H., PEARSON, G.J., BEWERS, J.M., "An Isotope Catalogue for Instrumental Activation Analysis",I, J. Radioanalytical Chemistry, $\underline{8}$, (1971) 127-188.

[69] Ibid., II, J. Radioanalytical Chemistry, $\underline{8}$, (1971), 373-479.

[70] Ibid.,III, J. Radioanalytical Chemistry, $\underline{9}$, (1971), 101-189.

[71] SOYKA, W., ERDTMANN, G., "Die γ-Linien der Radionuklide", KFA Zentralinstitut für Analytische Chemie, Jul. 1003-AC, Julich, Bundesrepublik, Deutschland, Vols. 1, 2, 3, (1974).

[72] BAUMGARTNER, F., "Tables of Neutron Activation Constants", translated from Kerntechnik, $\underline{3}$, (1961) 356.

[73]   GIRARDI, F., GUZZI, G., PAULY, J., "Data Handbook for Sensitivity Calculations in Neutron Activation Analysis", Euratom Rep. 1898.e, (1965).

[74]   NARGOLWALLA, S.S., NIEWODNICZANSKI, J., SUDDUETH, J.E., "Experimental Sensitivities for 3-MeV Neutron Activation Analysis", J. Radioanalytical Chemistry, 5, (1970), 403.

[75]   KENNA, B.T., CONRAD, F.J., "Tabulation of Cross Sections, Q-values, and Sensitivities for Nuclear Reactions of Nuclides with 14 MeV Neutrons", Rep. SC-RR-66-229, Sandia Laboratory, Albuquerque, New Mexico, (1966).

[76]   CUYPERS, M., CUYPERS, J., "Gamma-Ray Spectra and Sensitivities for 14-MeV Neutron Activation Analysis", Texas A & M, College Station, Texas, (1966).

[77]   FRIEDLANDER, G., KENNEDY, J.W., MILLER, J.M., "Nuclear and Radiochemistry", John Wiley & Sons, New York, (1964).

[78]   KRUGERS, J., ed., "Instrumentation in Applied Nuclear Chemistry", Plenum Press, New York, (1973).

[79]   Proc. of Conf. on "Nuclear Electronics", Vols I - III, IAEA, Vienna, (1962).

[80]   BROWN, W.C., HIGINBOTHAM, W.A., MILLER, G.L., CHASE, R.L., eds., "Proc. of Conf. on Semiconductor Nuclear-particle Detectors and Circuits", National Academy of Sciences, Publ. 1593, Washington, D.C., (1969).

[81]   NARGOLWALLA, S.S., NIEWODNICZANSKI, J., SUDDUETH, J.E., "Gamma-ray Spectra and Experimental Sensitivities for 3-MeV Neutron Activation Analysis", National Bureau of Standards, unpublished internal report, (1969).

[82]   HEATH, R.L., "Scintillation Spectrometry Gamma-ray Spectrum Catalog", Vols. I, II, 2nd edn., USAEC, Rep. IDO-16880, (1964).

[83]   HEATH, R.L., "Gamma-ray Spectrum Catalog Ge(Li) and Si(Li) Spectrometry", ANCR-1000-2, UC-34C, 3rd edn., Aerojet Nuclear Co., Idaho Nuclear Engineering Lab., Idaho Falls, Idaho, (1975).

[84]   COSGROVE, J.F., "Routine determination of major components by activation analysis", in Proc. 1968 Int. Conf. Mod. Trends in Activation Analysis, National Bureau of Standards, Special Publ. 312, Vol. 1, (1969), 457.

[85]   CURRIE, L.A., Nucl. Instr. Methods, 100, (1972) 387.

[86]   RICCI, E., DYER, F.F., Nucleonics, 22 (6),(1964), 45.

[87]   MATHUR, S.C., OLDHAM, G., Nuclear Energy, (Sep-Oct 1967), 136-141.

[88]   NARGOLWALLA, S.S., CRAMBES, M.R., DEVOE, J.R., Anal. Chem., 40, (1968) 666.

[89]   GIJBELS, R., "Activation Analysis with Neutron Generators", in Instrumental and Radiochemical Activation Analysis, Adams, F., Op de Beeck, J., Van den Winkel, P., Gijbels, R., DeSoete, D., Hoste, J., eds., CRC Press, Cleveland, Ohio, (1971).

[90] WOOD, D.E., "Problems in Precision Activation Analysis with Fast Neutrons", in Proc. of Conf. on Small Accelerators for Teaching and Research, Oak Ridge Associated Universities, Oak Ridge, Tennessee, USAEC, Rep. Conf-680411, (1968).

[91] PALMER, H.E., NELP, W.B., MURANO, R., RICH, C., Phys. Med. Biol., 13, (1968), 269.

[92] SUESS, H.E., UREY, H.C., Rev. Mod. Phys., 28, (1956), 53.

[93] VINCENT, H.Z., VOLBORTH, A., Nucl. Appl., 3, (1967), 753.

[94] Proc. of the Nato Advanced Study Institute on "Activation Analysis in Geochemistry and Cosmochemistry", Steinnes, E., Brunfelt, A.O., eds., Kjeller, Norway, (1970).

[95] Proc. of a Symposium on "Nuclear Techniques and Mineral Resources", Buenos Aires, IAEA, Vienna, (1969).

[96] Proc. of an International Symposium on "Nuclear Techniques in Exploration, Extraction and Processing of Mineral Resources", IAEA, Vienna, (1977), in press.

[97] BERZIN, A.K., BESPALOV, D.F., ZAPOROZHETS, V.M., KANTOR, S.A., LEIPUNSKAYA, D.I., SULIN, V.V., FELDMAN, I.I., SHIMELEVICH, YU.S., "Present State and Use of Basic Nuclear Geophysical Methods for Investigating Rocks and Ores", At. Energy Review, IAEA, Vienna, 4(2), (1966), 59-111.

[98] KEYS, W.S., BOULOGNE, A.R., "Well Logging with California-252", SPWLA 10th Annual Logging Symposium, Nouston, Texas, (1969).

[99] CZUBEK, J.A., "Recent Russian and European Developments in Nuclear Geophysics Applied to Mineral Exploration and Mining", Log Analyst, Nov-Dec. (1971), 20-34.

[100] GJIBELS, S.R., "Neutron Activation Analysis of Ores and Minerals, Miner. Sci.Eng., 5(4), (1973), 304-348.

[101] EISLER, P.L., HUPPERT, P., WYLIE, A.W., "Logging of Copper in Simulated Boreholes by Gamma Spectrometry", Geoexploration, 9, (1971), 181-194.

[102] HOYER, W.A., LOCKE, G.A., "Logging for Copper by In situ Neutron Activation Analysis", AIME Annual Meeting, San Francisco, Reprint No. 72-L-28 (1972).

[103] LANDSTRÖM, O., CHRISTELL, R., KOSKI, K., "Field Experiments on the Application of Neutron Activation Techniques to In situ Borehole Analysis", Geoexploration, 10, (1972) 23.

[104] MOXHAM, R.M., SENFTLE, F.E., BOYNTON, G.R., "Borehole Activation Analysis by Delayed and Capture Gamma Rays Using a ^{252}Cf Neutron Source", Economic Geology, 67, (1972), 579-591.

[105] NARGOLWALLA, S.S., KUNG, A., LEGRADY, O.J., STREVER, J., CSILLAG, A., SEIGEL, H.O., "Nuclear Metalog Grade Logging in Mineral Deposits", in International Symposium on Nuclear Techniques in Exploration, Extraction and Processing of Mineral Resources, IAEA, Vienna, (1977), in press.

[106] LAUBER, A., LANDSTRÖM, O., "A Ge(Li) Borehole Probe for In situ Gamma-ray Spectrometry", Geophysical Prospecting, 20, (1972), 800-813.

[107] TANNER, A.B., MOXHAM, R.M., SENFTLE, F.E., BAICKER, J.A., "A Probe for Neutron Activation Analysis in a Drill Hole Using ^{252}Cf and a Ge(Li) Detector Cooled by a Melting Cryogen", Nucl. Inst. Meth., 100, (1972), 1-7.

[108] DIBBS, H.P., "The Application of Neutron Activation Analysis to the Determination of Copper in Minerals", CIM Trans., 73, (1970), 102-108.

[109] KRISHNAN, S.S., J. Radioanalytical Chem., 15, (1973), 165-172.

[110] Proc. of a Symposium on "Use of Nuclear Techniques in the Measurement and Control of Environmental Pollution", Salzburg, IAEA, Vienna, (1970).

[111] Proc. of the Second International Conference on Nuclear Methods in Environmental Research", University of Missouri, ERDA, Rep. CONF-740701, (1974).

RECENT ANALYTICAL APPLICATIONS OF NEUTRON-CAPTURE GAMMA-RAY SPECTROSCOPY

J.A. LUBKOWITZ, M. HEURTEBISE,
H. BUENAFAMA
Instituto Venezolano de Investigaciones
   Científicas,
Centro de Petróleo y Química,
Caracas, Venezuela

Abstract

RECENT ANALYTICAL APPLICATIONS OF NEUTRON-CAPTURE GAMMA-RAY SPECTROSCOPY.
   The measurement of the prompt photons released from the compound nucleus following the absorption of a neutron is a complementary technique to activation analysis. The technique used with reactors requires the design of a well-collimated and thermalized neutron beam. A high-resolution Ge(Li), high-efficiency detector is required in conjunction with a sample gamma-ray collimator. The technique has been applied to the analysis of Cr, Mn, Fe, Ni, Ag, Aw, Ti, Zn, V, Co and Cu in alloys at concentrations ranging from 5–65%. The relative standard deviations ranged from 1.9–8.2%. Accuracy errors ranged from 0.8–8.9%. Prompt gamma rays have been found useful in the analysis of Co, Mo and Ni in desulphurization catalysts. Titanium oxide is used as internal standard. Standard deviations of 5% were typical for concentration ranges of 2–15%. Comparison of the results with those of activation analysis showed that the two methods differed by 0–13% for 11 samples. The water content can be obtained simultaneously with the determination of Co, Mo or Ni in the same sample. A relative standard deviation of 3.6% was obtained for a water content of 12.65%.

INTRODUCTION

   In neutron-capture gamma-ray spectroscopy, the prompt gamma-ray intensity is dependent on the capture cross-section and independent of the radioactive properties of the product nucleus. These two facts differentiate the technique from radioactivation analysis which relies on the ground-state properties of the product nucleus. In numerous cases the product nucleus may have a short half-life, a long half-life, or a stable composition or it may not be a gamma-emitter at all. It is also possible that the target nucleus has a small isotopic abundance. In these cases it is advantageous to use prompt gamma rays.

Since the lifetime of the excited composite nucleus is small ($10^{-9} - 10^{-12}$ s), the characteristic gamma-radiation must be measured during neutron bombardment. The withdrawal of neutrons through a beam port is an essential part of the system.

This paper deals with the application of neutron-capture gamma-ray spectroscopy to three types of problems:

    I. Analysis of alloys [1, 2]
    II. Analysis of Co, Mo and Ni in desulphurization catalysts [3]
    III. Determination of moisture in catalysts [4–6]

THEORY

During the irradiation of a matrix composed of elements A, B, C, ... i, the number of reactions per second, dN/dt, for a given element is expressed as:

$$\frac{dN}{dt} = \Phi \sigma N \qquad (1)$$

where Φ is the thermal neutron flux,
    σ is the microscopic cross-section and
    N is the number of target atoms of the element in question.

Thus, the number of prompt-gamma events measured per second is given by:

$$C = \Phi \sigma N I \epsilon \qquad (2)$$

where I is the absolute intensity, and
    ϵ is the efficiency.

Since the sensitivity factor S is defined as $S = I \sigma/M$, where M is the atomic mass of the element, and since the absolute efficiency ϵ can be redefined as $\epsilon = k \theta$, where k is a constant and θ is the relative efficiency, it is possible to rewrite Eq.(2) as

$$C = \Phi S W A \theta k \qquad (3)$$

where W is the weight of the element in question and
    A is the Avogadro's number.

If the matrix is composed of several elements A, B, C, ... i, then the sum of the masses of the constituents expressed in any units is given by:

$$1 = W_A + W_B + W_C + \ldots\ldots W_i \qquad (4)$$

The weight per cent of constituent A is given by:

$$W_A + \frac{W_A}{(W_A + W_B + W_C + ... W_i)} \; 100\% \tag{5}$$

By solving Eq.(3) for W and substituting it into Eq.(5), one gets:

$$W_A = \frac{C_A/S_A \theta_A}{(C_A/S_A \theta_A + C_B/S_B \theta_B + ... C_i/S_i \theta_i)} \; 100\% \tag{6}$$

Equation (6) was used for the determination of metals in alloys. The relative efficiency θ must be experimentally evaluated as a function of energy. The S values were obtained from two very complete references of prompt-gamma transitions [7, 8].

Equation (1) can also be expressed in terms of the accumulated counts in a given time interval due to nuclear reactions with an element in the catalyst. For example, for a catalyst containing hydrogen and an internal standard such as TiO_2, Eq.(1) can be expressed as:

$$H = k_1 \sigma_1 W_H \Phi \tag{7}$$

$$Ti = k_2 \sigma_2 W_{TiO_2} \Phi \tag{8}$$

where the constant k is a function of the branching ratio, detector efficiency and the atomic mass. H and Ti are the measured counts due to the hydrogen mass W_H and the titanium oxide mass W_{TiO_2} present in the sample. Dividing Eqs (2) and (3) yields:

$$\frac{H}{Ti} = \frac{KW_H}{W_{TiO_2}} \tag{9}$$

where $K = k_1 \sigma_1/k_2 \sigma_2$. This equation shows that the counts ratio for a homogeneous sample is independent of the flux and flux variations. If the ratio, Ti/W_{TiO_2}, is replaced by the symbol S, the titanium oxide specific counts, Eq.(4) can be rewritten as:

$$\frac{H}{S} = KW_H \tag{10}$$

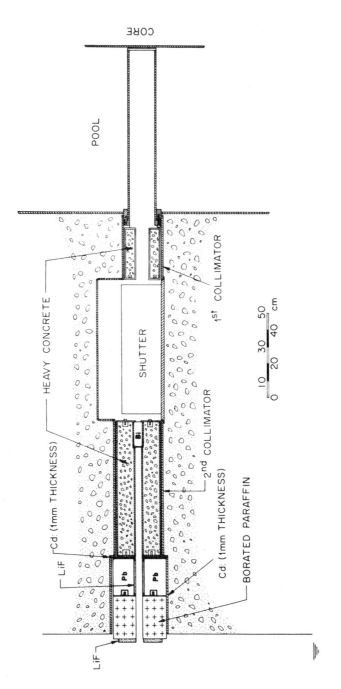

FIG.1. *Diagram of the collimation of the horizontal tube No.6 of the RV-1 reactor.*

The percentage of hydrogen in the sample is given by:

$$\%H = \frac{W_H 100}{W} \tag{11}$$

where W is the sample weight. Replacing W_H in Eq.(11) by its value in Eq.(10) yields:

$$(\%H)K = \frac{H \times 100}{SW} \tag{12}$$

This equation is used for calculating the hydrogen content in the sample after first evaluating the constant K from a standard sample containing a weighed amount of titanium oxide.

Equation (9) can be used to calculate the weight of Co, Mo and Ni present in the sample when standards are prepared with different metal concentrations but with a constant weight of TiO_2. Equation (7) can be used to determine %H or % metal when known amounts of a standard are added to the sample. These equations were used in the analysis of catalysts and in the determination of moisture in catalysts.

NEUTRON COLLIMATION

The horizontal beam used was formed in the horizontal tube T-6 of the RV-1 reactor. The neutron beam was collimated by means of two heavy concrete collimators, as shown in Fig.1. In the external collimator, a single bismuth crystal (15-cm long, 5.0 cm in diameter) placed in the central orifice reduced the direct fission-gamma radiation, epithermal and fast neutrons to an acceptable level. Behind this neutron collimator, a lead disc and a borated-paraffin disc (each 23-cm long and 23 cm in diameter) were used to reduce the remaining gamma radiation and fast neutrons. At the exit-end of the beam a compressed LiF slab (2-cm thick) served as a final neutron absorber. At 56 cm from the reactor wall a beam-catcher was placed which contained 2 m^3 of borated paraffin. Hydrogen prompt-gamma rays were eliminated by a concrete safety shield (40-cm wide).

Neutron flux characteristics

In the centre of the beam a flux of $4.75 \cdot 10^7$ $n \cdot cm^{-2} \cdot s^{-1}$ and a Cd ratio of 30.5 were measured using gold detectors. The flux deviation measured using

5-mm gold discs was < 10% in an area of 6.1 cm². These characteristics permit
1- to 5-g samples to be irradiated without subjection to flux inhomogeneity.

Measurement of prompt gamma rays

To eliminate the gamma rays arising from the neutron collimator and
neutron beam-catcher, a gamma collimator was built to ensure the measurement
of the sample gamma rays under conditions where little of the background
radiation reaches the detector. This collimator consists of two individual lead
bricks having a circular entrance slit smaller than the exit slit. The exit slit of
one block coincides in size with the entrance slit of the next block. Each
block is 5 cm in length, and at 2 cm of the last block a lead castle was built to
house the Ge(Li) detector. The lead castle has a circular opening of 4.4 cm
and the end facing the detector has a circular opening of 5.4 cm. At the end of
this opening the diameter of the hole remains constant at 6.0 cm, thus coinciding
with the detector dimensions. The position and diameter of the openings of
these blocks has been calculated so that a spherical cone arises having its apex at
12.7 cm from the first block while the other end has a surface coinciding with
the active detector surface. The sample is placed at a point of this cone having
a 1.0-cm diameter.

A 0.1-mm-thick aluminium tube, which can accommodate circular pellets,
powders and liquids, was used as sample holder. The sample holder was suspended above the beam path by a steel wire covered with shielding clay. Deflected
and diffracted neutrons reaching the Ge(Li) detector were avoided by covering
the lead collimators with compressed LiF slabs of 2-cm thickness. The detector
was located 23 cm from the centre of the beam. The neutron and gamma
collimators were located in the same plane but at 90° of each other. The gamma
radiations were measured with a 96-cm³ Ge(Li) detector (P/C 44:1, 1.76 keV
resolution at 1.33 MeV) coupled to a 4096-channel pulse height analyser. The
multichannel analyser was calibrated in the range of 5.0 – 9.096 MeV which is
the region of interest for the elements studied. Iron, with its three twin peaks,
nickel with its high-energy gamma rays and mercury with its high sensitivity,
permit the region of interest to be calibrated to the nearest keV by using the
tabulated values of Duffey et al. [7]. A diagram of the arrangement is shown
in Fig.2. The pulse height analyser was calibrated in the range of 0.5 – 2.5 MeV
for Co and Mo determinations and in the range of 5.0 – 9.096 MeV for Ni
determinations. Cobalt was measured at 556, Mo at 778, and Ti at 1381 keV.
Cobalt and Mo counts were obtained with good precision during a 10-min
irradiation-counting time. For Ni analysis, a 30-min irradiation-counting time
permits the accumulation of sufficient counts by integrating the counts of the
8999-keV gamma transition with the corresponding single- and double-escape
peaks. In this case the Ti is measured at 6249 keV (single-escape peak). Hydrogen

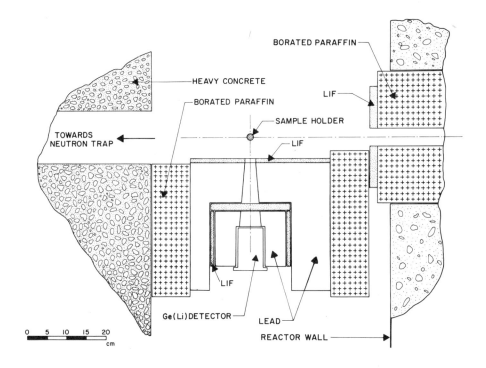

FIG.2. Top view of the detector shield and gamma collimator.

was measured by integrating the counts obtained at 2232 keV. The Ti peak at 1381 keV was used in the moisture determinations. Experimental details on sample treatment and mode of standardization are given in Refs [1–6].

I. APPLICATIONS IN THE ANALYSIS OF ALLOYS

A. Development of the model

It is possible to quantitate the elemental composition of alloys without the use of a standard. The method is based on the observation that a relationship exists between the peak areas representing different macro-constituents and their concentration [9]. The method is independent of flux, depressions and self-shielding neutron effects of a homogeneous matrix. We have successfully

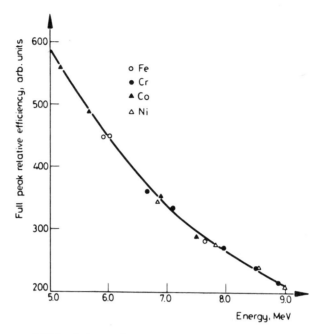

FIG.3. Relative efficiency of the detector for full peak.

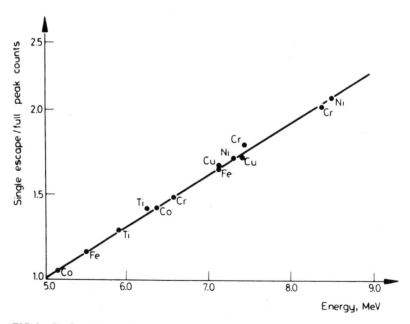

FIG.4. Single-escape peak counts to full peak counts ratio as a function of energy.

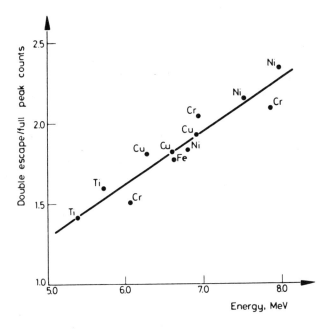

FIG.5. Double-escape peak counts to full peak counts ratio as a function of energy.

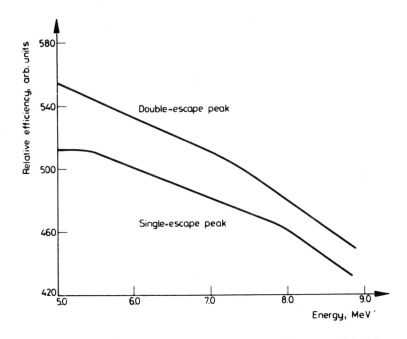

FIG.6. Relative efficiency curve of the Ge(Li) detector for single-escape and double-escape peaks.

applied this method to the analysis of gold alloys in conventional neutron activation analysis.

As shown by Eq.(6), it is necessary to measure accurately the relative efficiency θ for the elements being analysed. Figure 3 shows the full peak relative efficiency as a function of energy in the range of 5.0 – 9.0 MeV. Since single-escape and double-escape peaks are generally more intense than the corresponding full peak, their use is effective in obtaining an analysis with greater sensitivity. To establish the relative efficiency of single-escape peaks, a study of the ratio of single-escape peak counts to full peak counts at different energies was made, as shown in Fig.4. The function is linear in this region and is described by the equation, R = 0.3037 (±0.0057)E – 0.5023 (±0.0392). The correlation coefficient is 0.9981, indicating a good fit. A similar experiment was performed to measure the ratio of double-escape peak counts to full peak counts as a function of energy, and the relationship is shown in Fig.5. The equation obtained was R = 0.3219 (±0.0335)E – 3113 (±0.227). The correlation coefficient obtained was 0.9498. The large uncertainties obtained in the slope, intercept and correlation coefficient made the use of the double-escape peak not amenable to analytical purposes. Further study is in progress to understand the reasons for the poor correlation between the double-escape peak and the full peak. It has to be pointed out that the double-escape peak/full peak ratio was measured at the same time as the single-escape/full peak ratio. One possible explanation lies in the fact that the statistical distribution for the occurrence of a more complex phenomenon such as double-escape should have a larger standard deviation than that of a relatively simpler process such as single-escape.

By utilizing the full peak relative efficiency, the single-escape/full peak ratio and the double-escape/full peak ratio as a function of energy, it was possible to construct the relative efficiency as a function of energy for the double- and single-escape peaks. This function is shown in Fig.6. It is important to observe that the arbitrary units used in expressing the relative efficiencies of the double-escape and single-escape peaks versus energy are the same as those used in expressing the full peak relative efficiency versus energy curve. It is thus possible to compare measurements made by using the full peak, single-escape and double-escape peaks. It can be observed that as the full peak relative efficiency as a function of energy varies by a factor of 3 in the range of 5.0 – 9.0 MeV, the variation of the efficiency of the single-escape peak versus energy is only 20% under the same experimental conditions. Consequently, the use of the single-escape peak increases overall sensitivity of the determination particularly when high-energy gamma rays are measured. The poor results obtained with the double-escape peaks led us to use them only in qualitative measurements. An estimation of the absolute detector efficiency was obtained by using calibrated sources of ^{88}Y, ^{60}Co, ^{24}Na and ^{49}Ca. From the calibration curve, a detector efficiency of $2.8 \cdot 10^{-5}$ was obtained at 5 MeV by interpolation.

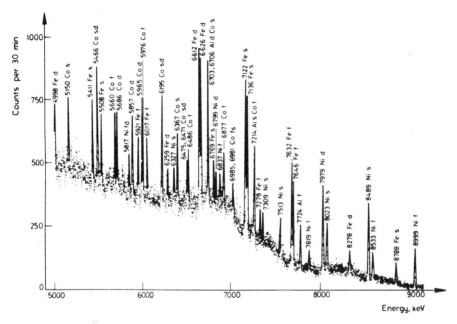

FIG. 7. Typical spectrum of a Maraging steel sample.

TABLE I. RESULTS OF THE ANALYSIS OF SYNTHETIC SAMPLES

Sample No.	Ni (%)	Cu (%)	Co (%)	Fe (%)	Ti (%)	Zn (%)	V (%)	Mn (%)
1	18.08	28.79	14.68	32.14	6.31			
1	17.88	29.28	15.56	31.31	6.45			
1	18.21	27.05	15.14	33.76	5.85			
1	18.00	29.81	14.92	31.74	6.13			
1	18.77	28.57	14.97	32.40	5.29			
1	17.98	28.67	14.39	32.63	6.32			
1	18.67	30.79	15.92	29.35	5.28			
						57.32	9.72	32.95
Average	18.23	28.99	15.08	31.90	5.95			
Standard deviation	0.35	1.16	0.52	1.36	0.49			
True composition (%)	20.02	28.66	14.72	30.70	5.90	58.76	9.35	31.88
Error (%)	−8.94	+1.15	+2.45	+3.91	+0.85	−2.4	+3.96	+3.36

TABLE II. RESULTS OF THE ANALYSIS OF A Ni-10-Cr-20.5 ALLOY SAMPLE

Trial No.	Ni (%)	Cr (%)	Fe (%)	Mn (%)
1	10.46	20.80	64.10	4.71
2	9.13	19.11	66.77	4.99
3	9.19	21.26	64.19	4.65
4	11.23	18.76	65.09	4.91
Average	10.18	20.00	65.04	4.81
Standard deviation	0.89	1.25	1.24	0.16
NBS reported composition (%)	10.10	20.51	62.90	4.62
Error (%)	−0.80	−2.40	+3.40	+4.11

B. Quantitative analysis

Figure 7 shows a spectrum obtained from the irradiation of a Maraging steel containing Fe (69.8%), Ni (19.0%) and Co (7.3%).

Table I shows the results obtained for the analysis of the artificial samples. The data indicate that good reproducibility is obtained. Table I also shows the results obtained for the Zn, V and Mn sample prepared. The accuracy obtained is good. Table II shows the results obtained for a Ni-10-Cr-20.5 steel sample and characterizes the ability of the method to analyse steel samples. It should be pointed out that there is 1.88% of impurities in the steel.

The method described can be successfully applied to the quantitative analysis of macro-constituents in alloys provided that they can all be measured by neutron-capture gamma-ray spectroscopy. This implies a good compromise between the relative composition of the sample and the S values of the components. When this technique can be applied, it has the advantage of being simple and it does not require any type of correction for self-absorption shielding, flux inhomogeneity differences between sample and standard, etc. Moreover, the method can generally be applied to determine mass ratios of components. Once the concentration of one of these constituents is known by applying any method, it is then possible to quantitate the other measurable components. Furthermore, the method is amenable to computer treatment of data and calculations of composition.

II. APPLICATION TO THE ANALYSIS OF Co, Mo AND Ni IN HYDRO-DESULPHURIZATION CATALYSTS

A. Importance of the analysis

The most important catalysts in the petroleum industry are those which contain Co-Mo, Mo-Ni and Co-Mo-Ni. The carrier for these metals is usually a mixture of synthetic γ-alumina and small amounts of silica. The ratio of these three elements in the catalyst determines to a great extent, among other variables, the catalytic activity. A great interest exists in the preparation of the catalysts by impregnation, co-precipitation, and other techniques with varying concentrations of these three elements in the γ-alumina-silica matrix. These catalysts are important for the hydro-desulphurization and hydrotreating processes. Furthermore, it is of economical interest to study the poisoning of the catalysts after processing of crudes at high temperatures and pressures. These requirements must be met for accurate analysis of Co, Mo and Ni.

B. Prompt gamma-ray spectra

Figure 8 shows the spectrum of a typical alumina catalyst sample containing CoO and MoO_3. It can be observed that a 10-min irradiation is sufficient to obtain the precision for the counts of Co, Ti and Mo. It should be pointed out that the Mo peak at 778 keV is well resolved from the Co peak at 786 keV. The relatively high counts accumulated at 2223 keV are due to the hydrogen present in the sample. This peak, as will be shown later, is used to determine the moisture content of the sample. The delayed gamma transition of aluminium at 1779 keV can also be observed. Figure 9 shows a typical spectrum of an aluminium catalyst containing Ni. Peaks arising from the sample due to Ni, Ti, Al and Co are observed in conjunction with Pb peaks arising from the Ge(Li) shielding.

A 10-min irradiation of sample No.9 yields 2810 ± 145, 5274 ± 134 and 6587 ± 118 counts for Co, Mo and Ti, respectively. After a 30-min irradiation, the sample yields 1071 ± 63 counts obtained at 8999, 8488 and 7977 keV due to nickel. The Ti counts obtained were 3629 ± 58 at 6249 keV.

Titanium oxide is a good internal standard because it can be easily and homogeneously mixed with the sample. Furthermore, its nuclear properties are also advantageous. Its high cross-section, coupled with the fact that it presents three prominent peaks at 6760 keV (54%), 6418 keV (36%) and 1381 keV (66%) yielding a good spectrum contrast, made this element a judicious choice as internal standard. Thus, it can be used in both energy ranges studied.

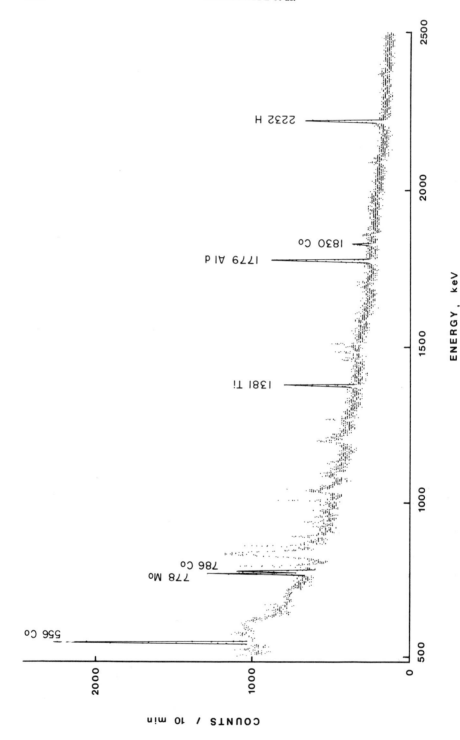

FIG.8. *Typical catalyst spectrum measured from 500 to 2500 keV.*

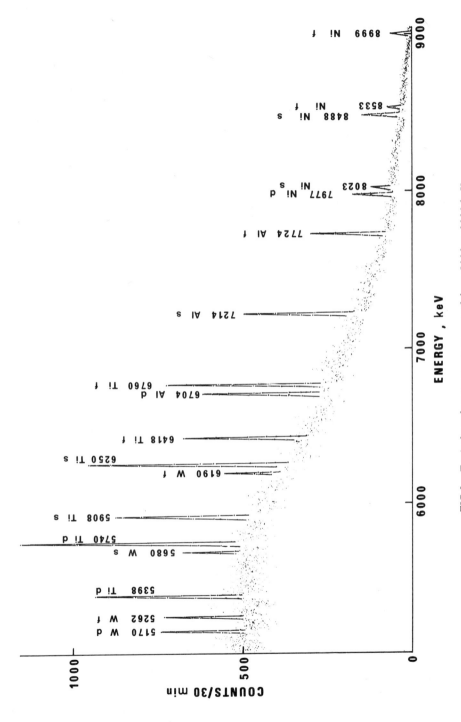

FIG. 9. *Typical catalyst spectrum measured from 5000 to 9000 keV.*

TABLE III. LINEARITY OF CALIBRATION CURVES FOR COBALT, MOLYBDENUM AND NICKEL

Cobalt		Molybdenum		Nickel	
Mass ratio, Co/TiO$_2$	Counts ratio, Co/Ti	Mass ratio, Mo/TiO$_2$	Counts ratio, Mo/Ti	Mass ratio, Ni/TiO$_2$	Counts ratio, Ni/Ti
0.1801	0.5259	0.4923[a]	0.2637	0.1237	0.2271
0.3055[c]	0.8752	0.6066[a]	0.3079	0.2658	0.5103
0.4909[a]	1.5396	0.6630	0.3219	0.3962	0.7585
0.5873	1.7040	1.0495[a]	0.5068	0.5299	0.9742
0.6523[a]	2.1259	1.2280[a]	0.6372	0.6824	1.2406
0.9018[c]	2.7630	2.1110[c]	0.9894		
0.9074	2.6820	2.3593	1.0780		
1.0500	3.3078	5.0000[c]	2.4088		
Slope	3.1312 (0.1077)[b]	0.4748	(0.06966)	1.802	(0.04266)
Intercept	−0.04334 (0.07501)[b]	0.08787	(0.01535)	0.02184	(0.01897)
Correlation coefficient	0.9965		0.9993		0.9992

[a] Added as oxide; remaining ones added as metal.
[b] Values in parenthesis are standard deviations.
[c] Matrix is powdered aluminium.

TABLE IV. STANDARD ADDITION ANALYSIS

	Cobalt counts[b]	Mass of CoO added (mg)		Molybdenum counts[c]	Mass of MoO_3 added (mg)
	2071	0		4414	0
	5749	23.29		5599	48.55
	9197	39.07		8556	151.13
	12978	62.27		11304	255.45
Cobalt mass intercept		11.02 mg	MoO_3 mass intercept		160.3 mg
% CoO[a] (dry basis)		1.22	% MoO_3^a (dry basis)		13.93

[a] Sample weight 1151 mg on dry basis (sample No.9).
[b] Molybdenum counts constant.
[c] Cobalt counts constant.

C. Standard calibration curves

The data obtained for the application of Eq.(9) are shown in Table III. The data were obtained from prepared artificial alumina samples spiked with Co, CoO, Mo, MoO_3 and Ni. The data show that the calibration is linear, indicating that the metal/TiO_2 mass ratio is directly proportional to the metal/Ti counts ratio. These calibration curves were reproducible from day to day so that it was acceptable to prepare one standard when analysing samples. Standards prepared by using the metal or an oxide form in an alumina or aluminium matrix yielded counts ratios that did not depart from linearity, indicating that no matrix effects are observed.

D. Standard addition study

The standard addition method was employed to check the previous results obtained by using the standard calibration curves as well as the possible interferences in the Co and Mo determinations. Moreover, the use of a TiO_2-matrix combination may cause shadowing effects. Since the activation cross-section is a function of neutron energy, resonance peaks (when present) may interfere with the neutron-element interaction. This situation, on the other hand, can be complicated by the fact that the water-moderated reactor employed in this work has a daily cyclic operation of 6 h during which the water temperature may increase causing a shift in the cross-section-neutron-energy function and thus varying the neutron-sample element interaction. The standard addition method should confirm the presence or absence of these effects. Table IV shows the results for CoO and MoO_3 obtained for a typical commercial catalyst. A comparison of the % CoO and % MoO_3 obtained by the calibration technique, as shown in Table III, with the values obtained in Table IV, shows an absolute difference of 0.10% for the MoO_3 content and a difference of 0.11% for the CoO content. This indicates that the two modes are in agreement. Moreover, the choice of TiO_2 as internal standard is appropriate for the analysis. Nickel prompt-gamma transitions used occur in the range of 7.977–8.999 MeV and are free of interferences; thus the standard addition technique was not employed.

E. Comparison of prompt gamma analysis with neutron activation analysis (NAA)

The results of analysing twelve commercial catalysts by prompt gamma-ray spectroscopy and NAA are presented in Table V. A standard reference catalyst is at present not available so it is difficult to assess accuracy. However, the two methods do show a satisfactory agreement. In the Co determination, four samples show a difference of greater than 10% by the NAA analysis as reference for the calculation. In the MoO_3 determination, three samples show a relative difference of 10% or more. It is interesting to note that the average of these relative differences with regard to sign is +1.83% for the CoO determinations and +1.32% for the MoO_3 determinations, indicating that neither method showed any bias. Sample No.13 is a synthetic catalyst which was prepared by impregnation of about 100 g of γ-alumina. The concentration of the impregnating MoO_3 solution was determined by NAA and the wash solutions were analysed by atomic absorption. Thus, a MoO_3 content could be calculated. The percentage differences of the MoO_3 determination on this sample were 6.38% and 5.17% as determined by prompt gamma-ray spectroscopy and NAA, respectively. Only three of the commercial catalysts analysed contained Ni, and the NiO content of the catalysts was compared with the corresponding analysis by the gravimetric technique using dimethylglyoxime. The differences between the nickel determinations by both methods and by the gravimetric method are between 4 and 5%. The prompt gamma-ray standard deviation experimentally obtained by statistical treatment of the data of several different analyses of the sample is quite similar to that obtained by the statistical treatment of counting data. This would seem to imply that the use of the internal standard is effective in correcting for geometry effects from sample to sample, flux gradients within the sample, and flux variations with time. Calculated standard deviations on counting statistics in NAA are about 2% as compared with 5.6% for the real

TABLE V. ANALYSIS OF CATALYST SAMPLES BY PROMPT GAMMA-RAY SPECTROSCOPY (P-γ) AND NAA

Sample	% CoO P-γ	% CoO NAA	% MoO$_3$ P-γ	% MoO$_3$ NAA	% NiO P-γ	% NiO NAA	Gravimetric
1	3.32 (0.20)[1]	3.06 (0.05)[2]	13.21 (0.65)[1]	12.81 (0.12)[2]	a		
2	1.74 (0.09)[1]	1.87 (0.06)[2]	8.59 (0.53)[1]	9.04 (0.13)[2]	4.21 (0.23)[2]	3.82 (0.17)[2]	4.04 (0.07)
3	<0.05	0.02	15.03 (0.75)[1]	13.23 (0.12)[2]	3.71 (0.21)[2]	3.39 (0.12)[2]	3.58 (0.04)
4	3.61 (0.17)[1]	3.39 (0.06)[2]	8.79 (0.55)[1]	8.66 (0.13)[2]	a		
5	2.71 (0.12)[2]	2.47 (0.05)[2]	13.23 (0.66)[2]	11.85 (0.13)[2]	a		
6	3.19 (0.13)[2]	2.71 (0.05)[2]	14.49 (0.58)[2]	14.61 (0.12)[2]	a		
7	2.89 (0.14)[2]	2.59 (0.06)[2]	16.14 (0.61)[2]	15.29 (0.11)[2]	a		
8	3.70 (0.15)[2]	3.58 (0.05)[2]	11.22 (0.53)[2]	10.20 (0.13)[2]	a		
9	1.33 (0.07)[1]	1.51 (0.07)[1]	14.03 (0.60)[1]	14.88 (0.63)[1]	2.68 (0.11)[1]	2.78 (0.17)[1]	2.59 (0.03)
10	4.78 (0.17)[2]	4.78 (0.05)[2]	14.27 (0.58)[2]	14.21 (0.11)[2]	a		
11	3.85 (0.15)[2]	3.54 (0.05)[2]	13.01 (0.62)[2]	13.86 (0.12)[2]	a		
12	3.36 (0.15)[2]	2.79 (0.05)[2]	13.91 (0.62)[2]	15.44 (0.11)[2]	a		
13[b]	a		11.78 (0.60)[2]	11.92 (0.12)[2]	a		

[a] Indicates metal is not reported as a constituent by manufacturer.
[b] Synthetic catalyst of 12.57% MoO$_3$.
[1] Experimental standard deviation.
[2] Standard deviation calculated on the basis of counting statistics.

TABLE VI. DETERMINATION OF HYDROGEN BY NEUTRON-CAPTURE GAMMA-RAY SPECTROSCOPY

Sample name	H counts	Ti counts	TiO$_2$ weight (mg)	S ratio (counts Ti/mg TiO$_2$)	Sample weight, W (g)	$\dfrac{HW100}{S}$	% H theoretical	% H found	% H$_2$O in catalyst
Potassium biphthalate	2179	3461	85.8	40.3	0.6927	7798	2.468	a	
Potassium biphthalate	2015	2551	64.7	39.4	0.6273	8146	2.468	a	
Cupferron	2397	5171	95.1	54.4	0.2336	18873	5.847	5.84	
Zirconium oxychloride 8H$_2$O	2604	4188	155.3	27.0	0.5987	16127	5.005	4.99	
Sodium barbital	2775	5688	141.6	40.2	0.3990	17314	5.378	5.36	
Sodium molybdate 2H$_2$O	1226	4391	109.7	40.0	0.5676	5396	1.666	1.67	
Benzoic acid	3977	5419	148.5	36.5	0.6482	16814	4.953	5.20	
Na$_2$ EDTA 2H$_2$O	2849	2555	51.5	49.6	0.3665	15669	4.874	4.85	
Catalyst 1	1255	3370	82.6	40.8	0.7310	4208	b	1.30	11.6
Catalyst 1	1375	2940	71.1	4.13	0.7608	4374	b	1.35	12.1

a Reference material. b Not determinable.

standard deviation. This may imply that this method is affected by flux variations in sample and standard during irradiation by counting geometry, and by physical differences between standard and samples.

F. Conclusions

The prompt gamma-ray analysis of catalysts is a precise, rapid and non-destructive analysis of hydro-desulphurization catalysts. The time required for a Co and Mo determination is 10 min, and an additional 30 min is required for a Ni analysis. The corresponding NAA analysis requires about 3–4 days. Furthermore, because of the alumina-silica matrix, the mineralization of this type of catalyst is also rather time-consuming. The applications of prompt gamma-rays for routine analysis utilizing a reactor neutron beam have been few, perhaps because of the complexity of obtaining a suitable collimated neutron beam whose characteristics depend on the neutron source used. However, once the beam became available, this work showed that prompt gamma rays can be useful in analysing samples containing elements whose irradiation and counting characteristics may make them laborious to analyse by conventional NAA. This work has also shown that the use of energy spectra in the range of 500–2500 keV is convenient for routine applications.

III. DETERMINATION OF WATER IN HYDRO-DESULPHURIZATION CATALYSTS

A. Introduction

The determination of hydrogen by neutron-capture gamma-ray spectroscopy was applied to coal by Rasmussen and Hukai [10]. We have used prompt gamma rays to determine moisture in highly sorptive surfaces such as aluminas, zeolites and impregnated alumina catalysts. Catalysts based on γ-alumina are known to absorb 6–20% of their weight in water, depending on surface characteristics and environmental factors.

B. Application to the determination of hydrogen in organic compounds

The use of Eq.(12) requires the determination of K which was determined by using potassium biphthalate as a standard. Once this constant was known, the percentage of hydrogen was determined in a selected number of compounds. The data are shown in Table VI. In all cases the counts of the hydrogen peak at 2232 keV and the counts of the TiO_2 peak at 1381 keV were used in conjunction with Eq.(12). Table VI shows that the agreement is excellent between

TABLE VII. STANDARD ADDITION METHOD FOR THE MOISTURE DETERMINATION OF CATALYST No.1 BY NEUTRON-CAPTURE GAMMA-RAY SPECTROSCOPY

Trial No.	H^a (counts)	Ti (counts)	TiO_2 weight (mg)	S ratio (counts Ti/mg TiO_2)	$\frac{H100}{S}$	H_2O added (mg)
1	1814	2130	45.4	46.9	3866	0
2	1986	1971	47.5	41.5	4786	18.7
3	2492	2202	49.8	44.2	5635	39.2
4	2933	2705	57.8	46.8	6267	58.5
5	3054	2835	62.6	45.3	6743	77.9
6	3392	2429	53.2	45.7	7429	97.1
			Slope	35.6 ± 1.99		
			Intercept	4057 ± 177		
			Correlation coefficient	0.9938		
			% H_2O	11.4%		

[a] Weight of catalyst was 1.000 g.

the % H found and the corresponding theoretical value calculated from the molecular weight and composition. The data also show that TiO_2 is a suitable internal standard for the hydrogen determination.

C. Standard addition studies

The standard addition method was used to check the result obtained by direct comparison with standards. It can be observed from Eq.(12) that a plot of H × 100/SW versus weight of water added to identical aliquots of the sample should yield a straight line whose intercept permits the calculation of the original content of water in the sample. The data obtained show good linearity, as can be seen from Table VII. The content of water in catalyst No.1 determined by this method is 11.4% which is in good agreement with the values obtained by standard comparison, as shown in Table VI.

D. Comparison of moisture determinations in catalysts by prompt gamma-ray spectroscopy, thermogravimetric analysis (TGA) and oven drying

Because it is not feasible to acquire a reference hydro-desulphurization catalyst with a standard water content and since in the prompt gamma-ray analysis it is assumed that the hydrogen measured is solely due to water, the prompt gamma-ray analyses were checked by thermogravimetric analyses (TGA). The catalysts subjected to TGA contained 10–15% MoO_3 and thus could not be heated to temperatures above 500°C since losses of Mo would occur. A comparison of the results obtained by prompt gamma-ray analyses and the weight loss obtained at 500°C is shown in Table VIII. The differences between the prompt-gamma determinations and the weight losses obtained at 500°C seem fairly constant and average 9.20% lower for the TGA study.

The lower values are due to the fact that water remains at 500°C and cannot be eliminated without decomposing the MoO_3 in the catalyst. This was confirmed by heating a non-impregnated catalyst support to temperatures of 700°C where, reportedly, γ-alumina loses all its water content. The result obtained is in good agreement with the prompt-gamma value. The 200°C difference yields a percentage difference weight loss of 9.05% which is similar to the relative percentage water difference as determined in the catalyst at 500°C and by prompt gamma-ray analysis. Furthermore, a naturally occurring alumino-silicate or zeolite also yields a weight loss which, expressed as per cent water, is also in good agreement with the prompt-gamma determination. Oven drying yields considerably lower results than those of the other two methods showing that this technique is unsatisfactory in totally removing the water content in the catalyst.

TABLE VIII. COMPARISON OF MOISTURE DETERMINATIONS BY NEUTRON CAPTURE GAMMA-RAY SPECTROSCOPY (NC), TGA AND OVEN DRYING

Sample	% H_2O determined		
	NC	TGA[a]	Oven dried
Catalyst 1	11.86	11.30	8.08
" 2	12.65	11.33	8.62
" 3	22.66	20.54	16.81
" 4	12.03	10.82	9.06
" 5	19.27	17.31	15.06
Zeolite	22.79	22.10	–
Catalyst support	16.88	15.08 $(16.58)^b$	–

[a] Weight loss at 500°C.
[b] Weight loss of catalyst support at 700°C.

IV. OTHER RECENT DEVELOPMENTS

Applications of neutron capture are not too numerous although they are becoming more numerous. Recently, Gladney et al. [11] determined B and Cd non-destructively. Owing to the large cross-section of these two elements (3840 and 2450 barns, respectively), these researchers were able to determine the two elements at concentrations of less than 1 ppm with a standard deviation of 3–7%. Henkelman and Born [12] have recently discussed the problematics of determining Boron using the $^{10}B(n,\alpha)^7Li$ reaction.

Pouraghabagher and Profio [13] recently reported the determination of V in crudes in a stream by utilizing a ^{252}Cf source. Vanadium prompt gamma-ray transitions were used to analyse the crude at concentrations of about 100 ppm.

Zwittlinger [14] reported the analyses of alloys of similar composition to the ones reported in this work. However, because of the characteristics of the beam, samples of 50 g and irradiation times of about 10 h were required. The method used standard comparison which requires that the total macroscopic cross-section of both the standard and the sample must be very similar. This implies a prior knowledge of the elemental composition of the sample.

With the advent of ^{252}Cf sources, it is possible that the applications of prompt gamma-rays will increase, since it is easier to obtain a thermalized beam relatively free of interfering gamma radiation.

REFERENCES

[1] HEURTEBISE, M., LUBKOWITZ, J.A., J. Radioanal. Chem. **31** (1976) 503.
[2] HEURTEBISE, M., LUBKOWITZ, J.A., Trans. Am. Nucl. Soc. **21** 2 (1975) 60.
[3] HEURTEBISE, M., BUENAFAMA, H., LUBKOWITZ, J.A,, Anal. Chem. **48** 13 (1976) 1971.
[4] HEURTEBISE, M., LUBKOWITZ, J.A., Anal. Chem. **48** 14 (1976) 2143.
[5] HEURTEBISE, M., LUBKOWITZ, J.A., IVIC, Center of Petroleum and Chemistry, Report No.16, Sep.1975.
[6] HEURTEBISE, M., LUBKOWITZ, J.A., Proc. Modern Trends in Neutron Activation Analysis, Munich, Sep.1975.
[7] DUFFEY, D., EL-KADY, A., SENFTLE, F.E., Nucl. Instrum. Methods **80** (1970) 149.
[8] SENFTLE, F.E., MOORE, H.D., LEEP, D.B., EL-KADY, A., DUFFEY, D., Nucl. Instrum. Methods **93** (1971) 425.
[9] HEURTEBISE, M., LUBKOWITZ, J.A., Anal. Chem. **45** (1973) 47.
[10] RASMUSSEN, N.C., HUKAI, Y., Trans. Am. Nucl. Soc. **12** (1967) 29.
[11] GLADNEY, E.S., JURNEY, E., CURTIS, D.B., Anal. Chem. **48** 14 (1976) 2139.
[12] HENKELMAN, R., BORN, H.J., J. Radioanal. Chem. **16** (1973) 473.
[13] REZA POURAGHABAGHER, PROFIO, E., Anal. Chem. **46** (1974) 1223.
[14] ZWITTLINGER, H., J. Radioanal. Chem. **14** (1973) 147.

APPLICATIONS OF POSITRON ANNIHILATION

P. HAUTOJÄRVI, A. VEHANEN
Helsinki University of Technology,
Espoo, Finland

Abstract

APPLICATIONS OF POSITRON ANNIHILATION.
The paper reviews various applications of the positron annihilation technique. When positrons from radioisotopes are injected into a condensed medium they rapidly slow down to thermal energies, live a relatively long time in equilibrium with the surrounding matter and finally annihilate with electrons almost always into two gamma quanta. By studying this annihilation radiation with methods developed in nuclear physics, information can be obtained about the state of the annihilating electron-positron pair and thus about the electronic structure of the medium. The lifetime distribution of positrons is measured with a coincidence technique using fast scintillators and time-to-pulse height conversion. The mean lifetime of free positrons varies from 0.1 to 0.5 ns depending on the electron density of the medium. The momentum of an electron-positron pair is studied by measuring the angular correlation of the annihilation quanta. The deviation of the two gamma rays from collinearity is determined almost totally by the momentum of the electron, especially if the positron is free. Thus, the angular correlation curve reflects the momentum distribution of electrons in the medium. This techique is powerful, e.g. in Fermi surface studies of metals and alloys. Positrons are found to be trapped in crystal lattice defects like metal vacancies, voids and dislocations. This phenomenon is widely utilized in the determination of vacancy formation energies and in the study of void formation. The annihilation radiation from the trapped positrons contains also unique information about the internal electronic structure of defects. In molecular substances the positron can capture an electron while slowing down and form a positronium atom. The formation probability and lifetime of positronium depends strongly on the properties of the medium. Because of its hydrogen-like structure the positronium atom takes part in chemical reactions with different kinds of atoms and molecules. Positronium chemistry deals with the chemical aspects of positron annihilation. The medical applications are concentrated on the development of positron cameras. With these instruments it is possible to image the distribution of positron emitters like ^{11}C, ^{13}N and ^{15}O and thus to follow various physiological processes in the human body.

1. INTRODUCTION

Positron physics is concerned with annihilation phenomena of low-energy positrons in matter. The existence of the positron was predicted by Dirac [1] in 1930 and it was experimentally observed by Anderson [2] in 1932. The birth and rapid growth of positron physics occurred in the early 1950s as the applicability of positron annihilation to the study of the electronic structure of matter was realized. Since then the field has grown very rapidly, as shown in Fig.1 [3]

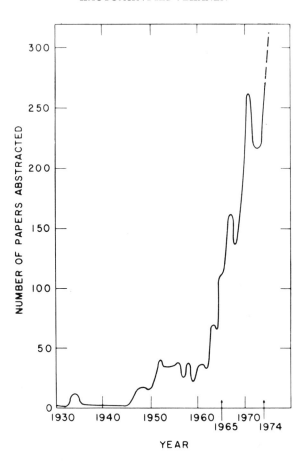

FIG.1. The number of annually published papers dealing with the study of low-energy positrons and positronium [3].

which indicates the number of annually published papers. The growth rate is at present about 20% a year. The reasons for the rapid growth lie in the unique information obtained by positrons, e.g. on crystal lattice defects. At the same time, commercial and inexpensive devices became available.

There have been several international conferences specially devoted to positron annihilation studies. The first was held in Detroit, USA, in 1965 [4], the second in Kingston, Canada [5], the third in Otaniemi, Finland [6] and the most recent in Helsingør, Denmark in 1976 [7]. The fifth conference will be held in Japan in 1979 and already the 1982 conference has been scheduled for the USA.

The positron annihilation technique provides several advantages in the study of matter. It provides a non-destructive testing of materials because the information is carried out of the material with penetrating annihilation quanta. The area scanned by positrons is typically ca. 30 mm². The layer thickness is ca. 30 − 150 mg/cm² depending on the positron emitter but being quite insensitive to the atomic number of the sample. The preparation of samples is relatively easy and in some applications also 'in situ' studies, for example on dynamic phenomena at elevated temperatures, are possible. Several reviews on positron physics in the study of condensed matter have been published [8−11].

2. POSITRON METHOD

2.1. Positron experiment

When energetic positrons from radioisotopes are injected into a condensed medium they first slow down to thermal energies in a very short time, of the order of 1 ps [12]. The positron then lives in the medium in thermal equilibrium for a somewhat longer period and the mean lifetime is in the range 0.1 to 0.5 ns, characteristic of each material. Finally, the positron annihilates with an electron of the surrounding medium preferentially into two 511-keV gamma quanta. The above picture is distorted in some media where positronium formation occurs during the slowing down of the positron. This phenomenon, however, is treated separately in Section 5.

Figure 2 shows the schematic course of the positron annihilation experiment in which the most commonly used radioisotope ^{22}Na is involved. Almost simultaneously with the positron, the radioisotope emits an energetic (1.28-MeV) photon used as the birth signal. The lifetime of the positron can thus be measured as the time delay between the birth and the annihilation gammas.

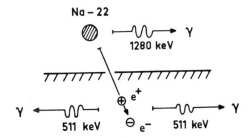

FIG.2. The positron experiment. Positrons from radioactive isotopes penetrate into the material, slow down and annihilate with electrons. The gamma radiation gives information on the structure of the material.

By measuring the angle between the annihilating quanta we can deduce the momentum of the electron-positron pair characterizing the electronic structure of the medium. The motion of the annihilating pair causes a Doppler-shift to the annihilation radiation and this is seen in an accurate energy measurement of one of the photons.

2.2. Annihilation of free positron

The annihilation process is a relativistic event where the masses of the positron and the electron are converted into electromagnetic energy, the annihilation photons. From the invariance properties of the electromagnetic interaction we can derive several selection rules for annihilation. One-gamma annihilation is possible only in an external field and has no practical meaning. The most important process is the two-gamma annihilation. Higher-order phenomena may be neglected unless some selection rule forbids the annihilation into two-gamma quanta. This situation is treated in Section 5.

From the non-relativistic limit of the two-gamma annihilation cross-section (Dirac [13]), we obtain the annihilation probability per unit time or the annihilation rate

$$\gamma_{2\gamma} = \pi r_0^2 c n_{e\varrho} \qquad (1)$$

which is independent of the velocity of the positron. Here, r_0 is the classical electron radius, c the velocity of light and $n_{e\varrho}$ is the electron density at the site of the positron. By measuring the annihilation rate $\gamma_{2\gamma}$, one directly obtains the electron density $n_{e\varrho}$, i.e. the positron serves as a test particle for the electron density of the medium. However, because of their opposite charge, strong Coulomb attraction exists between electrons and the positron. Consequently, the electron density $n_{e\varrho}$ will be enhanced from the equilibrium electron density in matter due to the Coulomb screening of the positron. Calculation of these positron-electron correlations is a difficult many-body problem.

The kinetic energy of the annihilating pair is typically a few electron volts. In their centre-of-mass frame the photon energy in the two-gamma annihilation is $m_0 c^2 = 511$ keV, and the photons are moving exactly in opposite directions. Because of the non-zero momentum of the annihilating pair the photons deviate from co-linearity in the laboratory frame. Simple relativistic energy-momentum conservation rules yield a result

$$\theta \cong p_T/m_0 c \qquad (2)$$

where $180° - \theta$ is the angle between the two photons in the laboratory and p_T is the momentum component of the electron-positron pair transverse to the

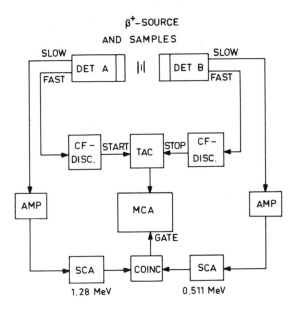

FIG.3. *The schematic diagram of the fast-slow coincidence system used in the positron lifetime measurements.*

photon emission direction. Usually, θ is very small ($\theta \lesssim 1°$) and Eq.(2) is valid. Because the momentum of the thermalized positron is almost zero [12], the measured angular correlation curve describes the momentum distribution of the annihilated electrons in matter.

As mentioned previously, the energy of the annihilation photon suffers a Doppler-shift, because the electron-positron pair was originally moving in the laboratory. With similar analysis as above the energy deviation from 511 keV equals

$$\delta E = \frac{1}{2} c p_L \qquad (3)$$

where p_L is now the longitudinal momentum component of the annihilating pair.

2.3. Lifetime measurement system

Figure 3 shows a schematic diagram of the positron lifetime spectrometer. The most common positron emitter is ^{22}Na. The lifetime is measured as the time delay between the 1.28-MeV and 511-keV photons shown in Fig.2. The lifetime spectrometer is a fast-slow coincidence system conventionally used in nuclear physics. The positron source is prepared by evaporating a few microcuries of aqueous ^{22}NaCl solution on a thin metallic or plastic foil (typically 1 mg/cm^2) and it is surrounded by two pieces of the sample material.

The detectors consist of fast plastic scintillators coupled to fast photomultiplier tubes. The fast signals taken from the anodes of the photomultipliers are then fed to constant-fraction timing discriminators to produce time signals. These are then guided to a time-to-amplitude converter (TAC), whose output amplitude is proportional to the time difference between the two input signals. The TAC pulses are then collected into the multichannel analyser.

The slow channels in Fig.3 exist only to check that the energies of the two photons detected were correct and that the two pulses originate from the same process. The corresponding energy regions are selected with the single-channel analysers.

In practice, there are random and systematic errors in the system and the measured lifetime spectrum becomes a convolution of the ideal spectrum and an instrumental resolution function. The latter can be measured by replacing the positron source by a ^{60}Co gamma source without touching other settings of the system. This source emits two almost simultaneous photons in a cascade. The measured resolution function turns out to be nearly Gaussian shaped. A time resolution of 300 ps (FWHM) is typically obtained with available commercial equipment. The resolution depends on the widths of the energy windows in the slow channels. The best resolution obtained is about 170 ps with 33% energy windows [14].

The resolution described above allows measurements of positron lifetime spectra in the form $\exp(-\lambda t)$, where the mean positron lifetime $\tau = \lambda^{-1}$ is ~ 100 ps or more. The measured curve can then be fitted with computer programs to analyse the lifetime value τ. Usually, complex spectra with many exponential terms are observed.

In certain cases described, e.g. in Section 4, there is some statistical advantage in measuring the mean lifetime of positrons as the time displacement between the centroids of the lifetime spectrum and the ^{60}Co resolution curve. This demands simultaneous measurement of these two curves realized with a router-mixer system [15].

2.4. Angular correlation apparatus

The angle between the annihilation photons is measured with a system described schematically in Fig.4. The photons are detected with usual NaI(Tl) detectors and the angle is determined by the position of the detectors and the lead collimators.

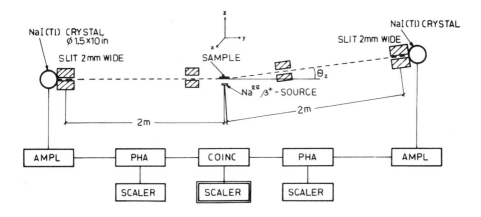

FIG.4. *The angular correlation apparatus with long-slit geometry.*

The positron source is usually ^{22}Na, ^{64}Cu or ^{58}Co with an activity of 10 mCi or more. The positrons penetrate into the sample and annihilate there. The lead shields hinder the photons annihilating in the source from reaching the detectors. The resolution of the equipment is determined almost totally by the geometry of the detectors and collimators. A typical slit width in front of the detectors is 0.5 mrad.

The discriminators are tuned at 511-keV photon energies and the device simply counts the coincidence pulses as a function of the angle θ_z. To minimize the errors one usually measures several runs over the total angular range, typically ± 15 mrad.

The geometry in Fig.4 is called the long-slit geometry, according to the shape of the collimators. The device cannot resolve either the angular deviation in the y-direction or the Doppler-shift in the x-direction and consequently the coincidence counting rate becomes

$$N(\theta_z) = C \int\int_{-\infty}^{\infty} dp_x dp_y \rho(p_x, p_y, \theta_z m_0 c) \qquad (4)$$

where $\rho(p_x, p_y, p_z)$ is the momentum distribution of the annihilation positron-electron pairs in the studied medium. Typically, an angular correlation measurement takes several days with the coincidence counting rate about 10 s^{-1} at $\theta_z = 0$.

Also point-slit geometries are used to measure two-dimensional momentum densities, but the accumulation rate of data is very slow due to geometrical conditions. The latest step in the development of angular correlation devices is the two-dimensional detector system consisting of a multicounter system or a pair of position-sensitive detectors [16, 17]. With such devices the two-dimensional momentum distribution can be measured in a reasonable time.

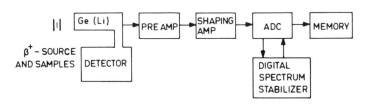

FIG.5. *The system for measuring the Doppler-broadened annihilation line.*

2.5. Annihilation line spectrometer

As described previously, the measurement of the 511-keV photon energy distribution coming from the studied medium is identical with the angular correlation measurement in the long-slit geometry. Figure 5 shows a typical installation. The source and samples are prepared in the same way as for lifetime measurements and the annihilation radiation is detected conventionally with a high-resolution Ge(Li) detector. The efficiency is roughly a hundred times better than that of the angular correlation system since no coincidence requirements exist. Typically, a one-hour measurement with a 5 μCi ^{22}Na source is enough for sufficient statistical accuracy. The disadvantage, however, is the resolution of the system. The best Ge(Li) detectors have an energy resolution of about 1.2 keV at the 514-keV line of ^{85}Sr, corresponding to an angular resolution of about 4 mrad, i.e. an order of magnitude worse than the resolution of the angular correlation device. Also, electronic stability problems arise, a relatively small drift during the measurement can severely destroy the information obtained. The digital spectrum stabilizer shown in Fig.5 is essential, if reliable results are required.

There is also a disadvantage in interpreting the resulting curve in physical terms because of the poor resolution. Some deconvolution procedures have been developed but their reliability is limited. The consequence of the above fact is that the Doppler-results can only be used to define a line-shape parameter characterizing the measured curve. One example of such a parameter $S = C/A$

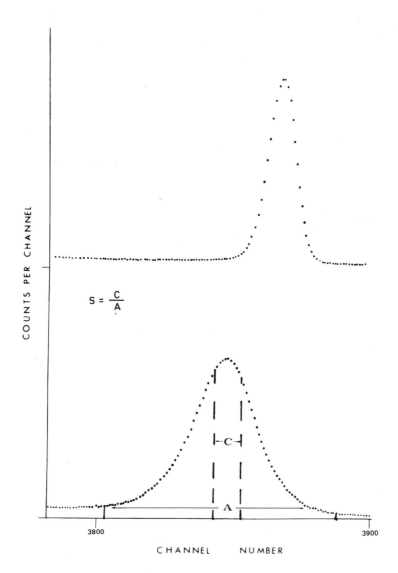

FIG.6. *The Doppler-broadened 511-keV annihilation line (lower) and the ^{85}Sr 514-keV line describing the instrumental resolution (upper)* [18].

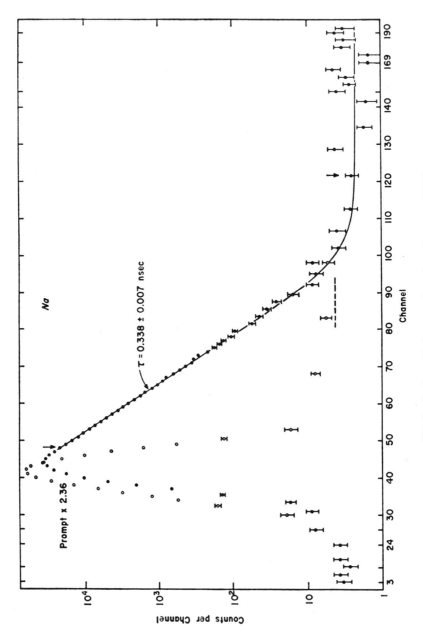

FIG.7. *Positron lifetime spectrum in sodium metal [19].*

is given in Fig.6 [18], which shows the Doppler-curve together with the instrumental resolution function obtained with the monoenergetic ^{85}Sr gamma source. Thus, the Doppler-technique can be used to follow various phenomena which require measurements of many stages of samples in a comparatively short time.

3. ELECTRONIC STRUCTURE OF IDEAL METALS

3.1. The annihilation rate

We shall now treat the problem of a thermalized positron in a perfect metal lattice. The electrons in simple metals can be roughly divided into nearly free valence electrons and tightly bound core electrons. The presence of the dense free electron gas totally prevents positronium formation. The repulsive interaction between the positron and the ions results in small overlapping of the positron and core electron wave functions. Thus, the annihilation rate λ is almost totally determined by the valence electron density and is also time independent. A lifetime measurement then produces a single-exponential spectrum $\exp(-\lambda t)$ analogously to the radioactive decay process. Figure 7 [19] shows a typical lifetime spectrum in a pure well-annealed metal.

The dependence of the annihilation rate λ on the valence electron density $n_{va\varrho}$ is conventionally expressed in terms of a density parameter r_s, which is defined by $4/3 \pi r_s^3 = n_{va\varrho}^{-1}$. Figure 8 shows the measured annihilation rates in various metals as a function of r_s. The theoretical curve 1 in the figure is calculated from Eq.(1) with n_e replaced by $n_{va\varrho}$. We see, however, that the resulting curve is far too low and its r_s-dependence is too steep. The reason for this discrepancy is partly the small contribution of core annihilations and mainly the many-body positron-electron correlations mentioned previously. Curve 2 in Fig.8 corresponds to the results of many-body calculations applied to a positron in an electron gas [20]. For simple metals the agreement is quite good, but in the noble and transition metals the measured annihilation rates are much higher because of the big overlap of the positron and the core electrons.

3.2. Angular correlation studies

The angular distribution of the annihilation photons $N(\theta_z)$ measured with the long-slit apparatus is an inverted parabola in the case of free electrons with a spherical Fermi surface because the area of the cross-section of the Fermi sphere at a plane $p_z = \theta_z m_0 c$ decreases quadratically. Two examples of angular correlation curves in aluminium [21] and copper [22] are given in Fig.9. The curves are seen to consist of two parts, the inverted parabola and a broader, roughly Gaussian-shaped contribution which is due to annihilations with the core electrons.

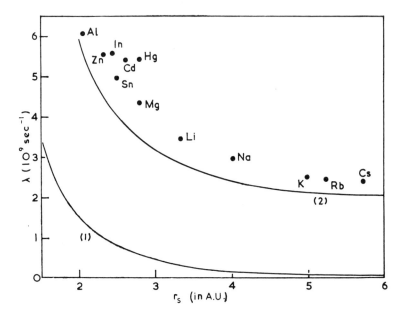

FIG.8. Experimental and theoretical annihilation rates in simple metals as a function of the electron density parameter r_s. Curve (1) is the one-electron approximation and curve (2) is a result of many-body calculations [20].

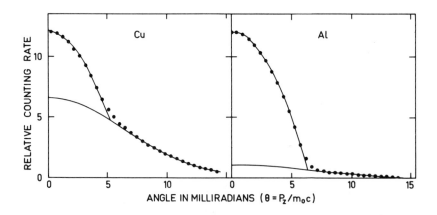

FIG.9. The angular correlation curves for Al and Cu [21, 22].

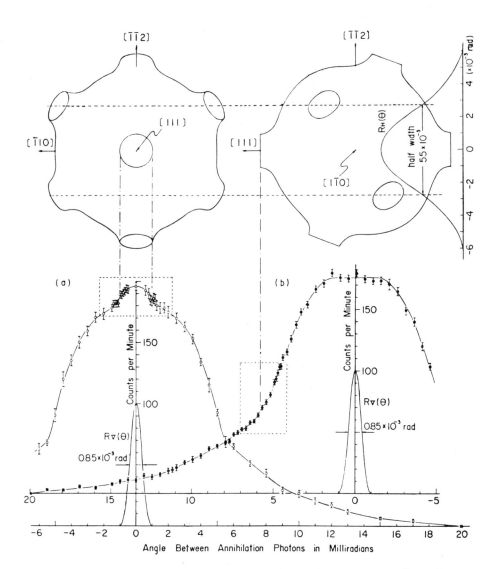

FIG.10. *The angular correlation curves in copper single crystals in various crystallographic directions. A special short-slit geometry is used where the p_y-integration in Eq.(4) is restricted inside the Fermi sphere* [23].

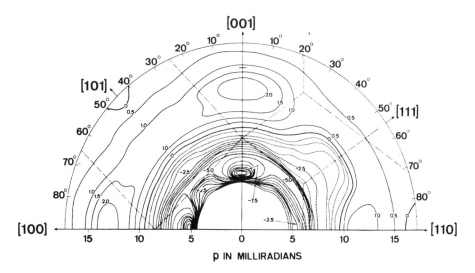

FIG.11. *The three-dimensional spin-polarization contour diagram in momentum space measured in ferromagnetic iron with polarized positrons* [24].

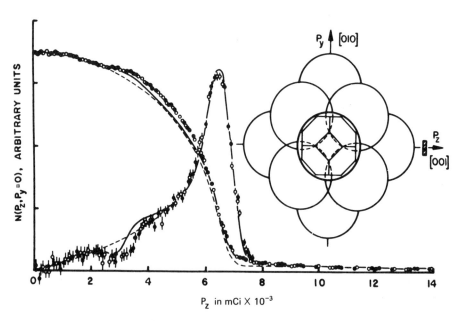

FIG.12. *Point-slit angular correlation in an aluminium single crystal measured with a multi-counter system. The solid curve represents an OPW calculation, the dashed curve the Wigner-Seitz model* [16].

The intersection of these two parts directly gives the Fermi angle θ_F corresponding to the Fermi momentum. Figure 9 shows also a significant difference in the core annihilation fraction and a small difference in the widths of the parabolas between the two metals. These effects are due to different core and valence electron densities in the metals.

In contrast to the lifetime distribution, the angular correlation curves seem to be surprisingly insensitive to many-body effects. Thus, a simple one-electron description quite adequately characterizes the measured curves. Therefore, positrons are particularly suitable in studies of anisotropies in Fermi surfaces, when single crystal samples are used. Such an example is given in Fig.10 [23] where the necks in the copper Fermi surface are clearly visible.

By combining several long-slit measurements one can reconstruct the three-dimensional momentum density $\rho(\vec{p})$. Figure 11 [24] shows a result for $\rho(\vec{p})$ in contour diagram representation. The additional pecularity in the figure is that it denotes the electron spin polarization diagram $\rho_{up}(\vec{p}) - \rho_{down}(\vec{p})$ in ferromagnetic iron obtained by using polarized positrons.

The development of two-dimensional angular correlation devices has enabled us to achieve significant details in the momentum density studies. Figure 12 gives a point-slit angular correlation curve of aluminium measured with a multi-counter system [16].

The positron studies of momentum densities have enlarged amply. Nowadays the Fermi radius can be measured with 0.1% accuracy [25]. The same accuracy is obtained with more conventional methods. However, with alloys or impure metals the other methods fail because of the short scattering mean free path of electrons. In these materials the positron method offers unique information and the subject is under extensive study [26].

4. STUDIES OF METAL LATTICE DEFECTS

4.1. Trapping of positrons

The behaviour of positrons in simple metals is quite well understood as discussed in the previous chapter. However, it was found that various physical states of the metal could drastically change annihilation properties. As the metal was mechanically deformed, the mean lifetime increased [27, 28] and the angular correlation curve became significantly narrower [29, 30]. These effects are clearly visible in Figs 13, 14 [22] and 15 [31]. In a more detailed study on deformed aluminium [28] the lifetime spectrum was found to contain two exponential lifetime components. The effect of crystal defects on positron behaviour was seen to saturate at relatively low degrees of deformation, the lifetime spectrum became one-exponential again, but the lifetime value was higher.

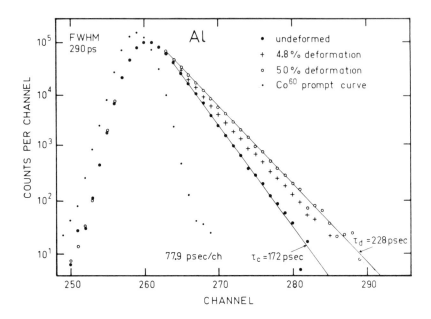

FIG.13. *Positron lifetime spectra in aluminium with different degrees of deformation.*

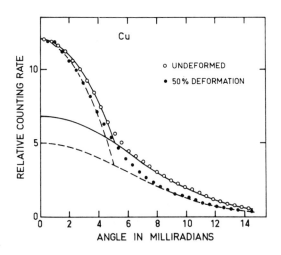

FIG.14. *The effect of deformation on the angular correlation curve in copper. The fitted Gaussian and parabolic parts are shown in both cases* [22].

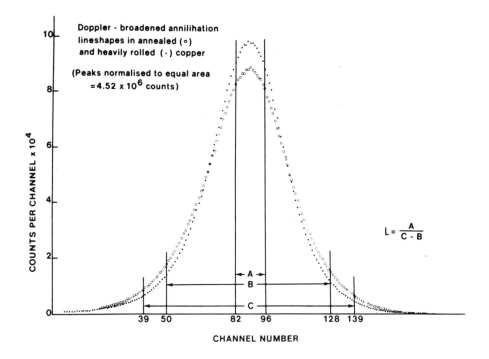

FIG.15. The Doppler-curves in deformed and annealed copper [31].

The explanation for these phenomena is the trapping of positrons by crystal defects. At low-defect concentrations only some positrons get trapped and in this region the annihilation characteristics depend sensitively on the number of defects. When all positrons are trapped, saturation is reached. The defects that have been found to trap positrons are vacancies, dislocations, vacancy clusters or voids and surfaces. In the disordered regions the positive charge of ions is reduced and the resulting redistribution of free electrons leads to a net negative electrostatic potential attractive to the positron. If the potential is strong enough the positron suffers a transition from the Bloch-like state to the trapped, localized state. The electron density at the site of the defect is smaller and consequently the positron lifetime increases. The same reason also leads to the narrowing of the angular correlation curve, since the local Fermi momentum decreases [21]. These features can be also quantitatively understood on the basis of the electronic structure of lattice defects [32, 33].

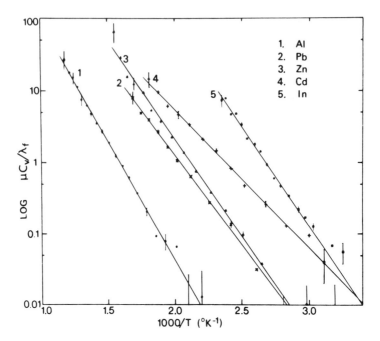

FIG.16. *Arrhenius plots for the determintion of vacancy formation energies with the positron method for several metals* [37].

4.2. Determination of defect concentrations

The positron trapping can be utilized to study the density of defects in the sample. According to a simple trapping model [34–36] there are two types of positrons, either free or trapped. Thus, any annihilation characteristic (peak counting rate in angular correlation curve, mean lifetime, the shape of annihilation line etc.), which depends linearly on the fraction of positrons in the trapped state, can be used to study the densities of the relative defects. The positron technique has proved to be a very sensitive and powerful method and it has been extensively applied especially for the determination of vacancy formation energies in metals. This is simply done by measuring some positron parameter as a function of temperature. Figure 16 [37] shows the resulting Arrhenius plot for different metals. Each slope gives directly the vacancy formation energy.

The advantages of the positron method lie in the fact that its sensitivity starts already from the vacancy concentrations of about 10^{-6} and thus the contribution of divacancies is quite small. Even reasonable estimates for the divacancy binding energy have been obtained [38]. For more details on the applications to the determination of the vacancy formation energies see, for example, West [39].

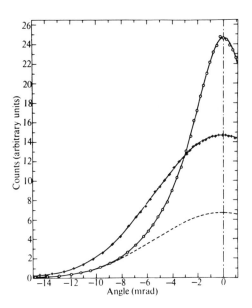

FIG.17. *The effect of voids on the angular correlation curve of molybdenum. The voids were produced during high-dose fast-neutron irradiation* [41]. *(Open circles with voids, crosses without voids.)*

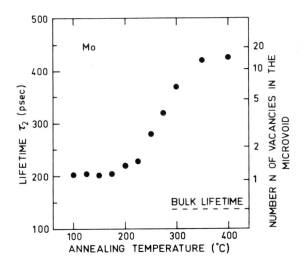

FIG.18. *The void formation during isochronal annealing of electron-irradiated molybdenum. The experimental points are from Ref.*[45] *and the void size was calculated in Ref.*[46].

4.3. Clustering of vacancies into voids

Positron annihilation in irradiated and annealed metals has shown remarkably different behaviour from annihilation of positrons trapped in vacancies or dislocations. On the other hand, small cavities or voids are known to agglomerate from vacancies under high-dose fast-neutron irradiation [40]. Drastic changes in the angular correlation and lifetime curves have been observed in molybdenum under similar conditions [41, 42]. The lifetime of positrons increased by a factor of four and the angular correlation curve narrowed to about one half, as shown in Fig.17 [41]. The voids with a diameter of about 30 Å were also seen in an electron microscope. It was suggested [43, 44] that in such big voids the positron is trapped in a surface state.

The next step in the void studies was to reduce the irradiation dose and to use electrons instead of fast neutrons. Eldrup et al. [45] revealed that the motion of vacancies during stage III annealing caused submicroscopic void formation at relatively small irradiation doses. After annealing at higher temperatures the void structure coarsened and became visible in the electron microscope, too. The lifetime results were correlated with the void size [46]; the resulting annealing curve predicts the number of vacancies clustered in the void, as shown in Fig.18.

Thus, positrons can reveal voids from the very beginning, long before they become visible in the electron microscope. In this sense the positron annihilation method is a unique tool. In practice, void formation is a serious problem in the construction of nuclear reactors and positrons are now being used for these studies, see for example Ref.[47].

4.4. Studies of deformed metals

As positrons are sensitive to various defects produced during plastic deformation of metals, the dynamic properties of different defects and their interactions can be studied. The topics in this field are the recovery of metal defects during annealing and related phenomena. Numerous annealing studies have recently been made and new features are found with positrons.

Figure 19 shows an isochronal annealing study of two plastically deformed irons with different degrees of impurities [48]. The positron parameters, lifetime and an annihilation parameter S defined in Fig.6, change remarkably at about 600°C, which is the recrystallization temperature of iron. Thus, the main recovery at this temperature shows that the trapping of positrons is caused by dislocations. Another interesting feature is the difference between the two curves. The purer iron shows partial recovery at about 250°C, which is absent in the iron containing more impurities, mainly carbon. The explanation here is the reordering of dislocation structure (polygonization process) in the pure iron.

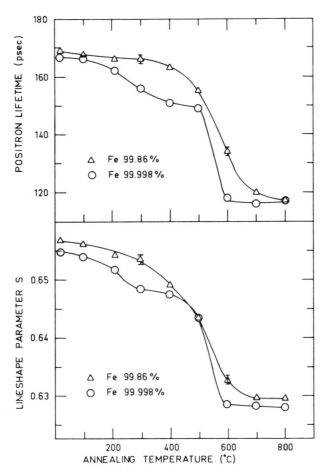

FIG.19. The recovery of positron lifetime and annihilation lineshape parameter S (defined in Fig.6) in two deformed irons during isochronal annealing [48].

The carbon interstitials are known to migrate into dislocation surroundings and this causes the blocking of the dislocation structure and consequently the absence of the low-temperature recovery.

Dlubek et al. [49] have studied the behaviour of two nickels with different purities during the plastic deformation. The interesting result was the observation that the presence of impurities catalysed clustering of vacancies.

Other studies more oriented to the metallurgical field are shown in Figs 20 and 21 [50] dealing with fatigue damage in commercial steels. In Fig.20 the steel was expected to show fatigue hardening and in Fig.21 fatigue softening. The mean positron lifetime indeed showed systematic behaviour with respect to these effects. Thus, positrons can be used to predict failures in commercial materials.

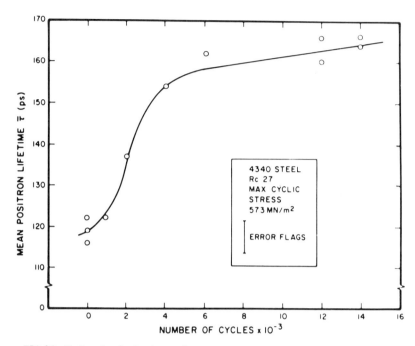

FIG.20. Fatigue hardening in a soft commercial steel studied by positrons [50].

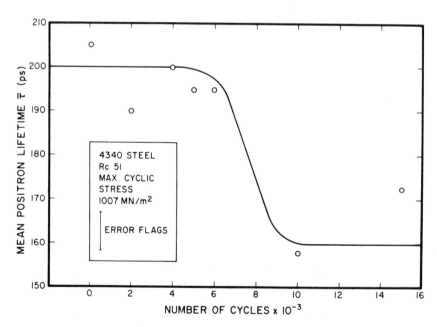

FIG.21. Fatigue softening in a hard commercial steel studied by positrons [50].

It should be emphasized that the choice of parameters used to describe the changes during the kinetic studies is essential. The parameters should be statistically accurate, rapidly obtainable and linear with respect to the studied effect. The parameters obtained after tedious mathematical manipulations from the measured curves, such as multicomponent lifetimes and the relative intensities, etc., should not be used.

4.5. Internal structure of lattice defects

Because the positron annihilates in a localized state inside a defect such as a vacancy, the annihilation characteristics reflect the electronic structure of the defect. A vacancy is not an empty hole in the lattice, but the electron density in the centre is about 30% of the bulk electron density. This calculation [33] is confirmed by comparing the resulting annihilation rates with the experimental values.

The electron density inside the voids is, however, calculated to reach zero very fast as the void size increases [46]. This gives considerable support to the ideas described previously that the positron interacts strongly with the void surface [51].

The positron states in dislocations are theoretically more complex. It has not been verified whether the positron can move freely along the dislocation line or whether it is localized at the jogs or vacancies inside the dislocations. More experimental and theoretical work in this field is needed.

5. POSITRONS IN MOLECULAR SUBSTANCES

5.1. Positronium formation

Positrons in molecular substances can capture an electron from the surrounding medium and a positronium atom, the bound state of the positron-electron pair, is formed. As the size of the positronium is about two times that of the hydrogen atom, the positronium formation occurs in a molecular medium where comparatively large empty regions exist.

The formation process of positronium is at present under discussion. Two simple models have been presented. The "Ore gap" model [52] states that positronium formation is most probable when the positron energy lies within the so-called Ore gap, where no other energy transfer process is possible. According to the radiochemical "spur" model [53], positronium formation is a reaction between the free positron and the electrons in the spur produced during the slowing down of the positron itself.

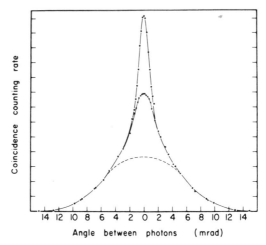

FIG.22. *Parapositronium peak in the angular correlation curves of SiO_2 powders with different grain size* [54].

The ground states of positronium are the singlet ^1S state or parapositronium and the triplet ^3S state or orthopositronium. The lifetime of parapositronium in two-photon annihilation is 125 ps, about the same as the free positron lifetime in metals. However, the two-photon annihilation is forbidden by the selection rules in the case of orthopositronium. Thus, it decays via three photon emissions with a lifetime of 140 ns, more than three orders of magnitude longer than the parapositronium lifetime. A competing mechanism called the pick-off-annihilation is always present when the orthopositronium atom moves in a medium. In the pick-off process the positron in the positronium atom annihilates with another electron of opposite spin and two-gamma annihilation results. Consequently, orthopositronium lifetimes in materials are of the order of a few nanoseconds.

As is quite evident, the angular correlation curve and the lifetime spectrum are complex in the presence of positronium. The self-annihilation of parapositronium causes a peak at small angles in the angular correlation curve, because the centre-of-mass of the parapositronium atom has a small momentum in the laboratory frame. One example of the parapositronium peak is shown in Fig.22, where positronium formation occurs in Si_2O powder [54].

Because of the anomalously long lifetime of orthopositronium, other competing annihilation mechanisms may further decrease the orthopositronium pick-off lifetime. Such quenching processes are the ortho-para conversion and chemical reactions of the positronium. The ortho-para conversion may be due to electron exchange between the positronium and surrounding molecules while chemical reactions are more complex including the formation of e^+A^- type molecules. Positronium chemistry studies these phenomena.

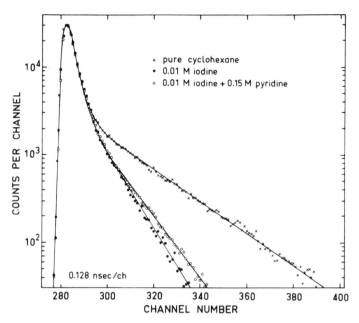

FIG.23. Positron lifetime spectra in cyclohexane, where positronium is formed. The long tails represent the orthopositronium decay. The chemical reaction of positronium with added iodine molecules is seen to decrease the orthopositronium pick-off lifetime [57].

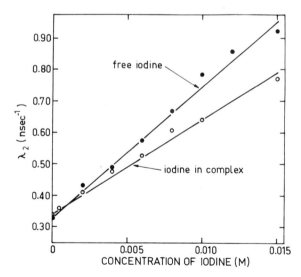

FIG.24. Orthopositronium decay rates as a function of the iodine concentration. The slopes in each case denote the chemical reaction rate constants [57].

5.2. Positronium chemistry

The orthopositronium decay rate and its relative intensity depend strongly on the properties of the medium. Because its hydrogen-like nature and relatively long lifetime, the positronium can take part in chemical reactions. This topic has been discussed in more detail in Refs [10,55,56].

Figure 23 shows an example of the studies of orthopositronium reactions with iodine molecules [57]. The chemical reaction between positronium and iodine causes a considerable decrease in the measured lifetime. The measured decay rate is described by

$$\lambda_{meas} = \lambda_{pick\text{-}off} + \lambda_{react} \tag{5}$$

where λ_{react} is the reaction rate of orthopositronium. At low reagent concentrations C, we have

$$\lambda_{react} = k \cdot C \tag{6}$$

where k is the reaction rate constant, an important parameter in chemistry.

We see from Fig.23 that as iodine is dissolved in cyclohexane solvent the orthopositronium lifetime decreases. A further addition of pyridine to the solution causes the binding of iodine to a complex molecule and the reaction properties change slightly. Figure 24 shows the measured decay rates as a function of iodine concentration in both cases. The linear dependence predicted by Eq.(5) is seen and the slopes give the chemical reaction rate constants k in absolute time scale.

5.3. Study of glasses with positrons

The first glass studies concerned quartz, where positronium formation occurred in the amorphous state but not in the crystalline form [58]. Positrons have been applied to various problems like glass transition temperatures [59], positron behaviour as a function of glass composition in binary glasses [60] and studies of dynamic processes like liquid-liquid phase-separation and crystallization [61–65].

In the lithium silicate ($Li_2O\text{-}SiO_2$) glass the phase-separation is a fast process where nearly pure SiO_2 droplets are segregated from the bulk glass. The lifetime spectra can be divided into three exponential components and the long-lifetime intensity associated with the orthopositronium pick-off annihilation is seen to be proportional to the relative volume of the SiO_2 droplets, as shown in Fig.25 [64]. Thus the origin of the long-lifetime component is the SiO_2 droplet phase.

As the glass is heat treated at moderate temperatures, the droplet phase remains the same and the lithia-rich bulk phase crystallizes. Consequently,

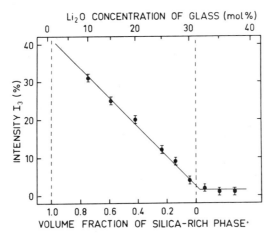

FIG.25. *Pick-off annihilation intensity in the phase-separating Li_2O-SiO_2 glass as a function of the relative volume of the separated SiO_2-droplets* [64].

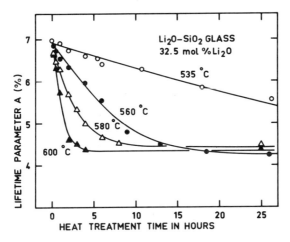

FIG.26. *Kinetic curves of crystallization in a Li_2O-SiO_2 glass at different heat-treatment temperatures. Changes in the lifetime parameter "A" are proportional to volume crystallinity* [65].

changes in the lifetime spectrum are seen only as a decrease of the intermediate lifetime intensity. Thus, the intensity can be used to detect changes in the crystallinity. However, a statistically more accurate parameter "A" [63], which was shown to depend linearly on the crystallinity by comparative X-ray diffraction measurements, was used. Figure 26 [65] shows the kinetic curves measured with the "A"-parameter at different temperatures, and thus the activation energy of crystallization can be determined.

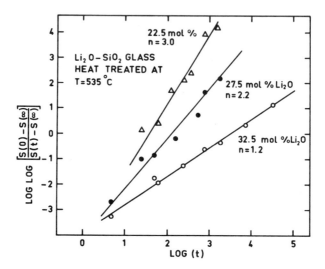

FIG.27. *Kinetic curves of crystallization in different Li_2O-SiO_2 glasses at constant temperature. The changes in the Doppler-parameter "S" (see Fig.6) are proportional to volume crystallinity. The scales are chosen in such a way that the slopes of the fitted straight lines give the morphology index of crystallization [65].*

The phase-separation occurring before crystallization has a strong influence on the parameters describing the crystallization process. This is seen in Fig.27, which shows crystallization kinetics at constant temperature measured in three different lithium silicate glasses with different fractions of the SiO_2 droplets. The scale is chosen in such a way that a mathematical model, the Johnson–Mehl-Avrami equation, describing the crystallization [65] is a straight line and the slope denotes the morphology index process of crystal growth. This is seen to vary with the glass composition.

The phase-separation has a strong effect on the activation energy of volume crystallization as well as on the morphology index. On the other hand, the activation energy associated with the linear crystal growth stays constant. Thus, positrons can reveal information on the connection between phase-separation and the crystallization process, which is important in the production of glass-ceramic materials.

6. MEDICAL APPLICATIONS OF POSITRON ANNIHILATION

Positrons can be used in cancer localization, in metabolic studies and in meson beam collimation inside the human body. The basic problem is to deduce the distribution of positron emitters from the annihilation radiation. Figure 28

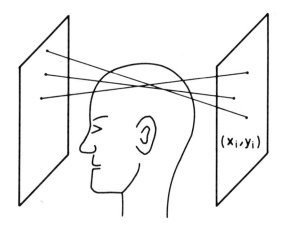

FIG.28. *Schematic representation of the positron camera.*

shows the principle of the positron camera for this determination. Two position-sensitive gamma detectors are used in coincidence to determine the two annihilation quanta emitting in opposite directions. Thus, the pair of detection coordinates determines the straight line, where the positron emitter existed. The advantage of the positron camera is the lack of collimators needed in conventional gamma cameras. The three-dimensional distribution of the positron emitter can be calculated. Some mathematical and instrumental questions are discussed in Refs [17, 66, 67].

The most useful application of positron cameras is perhaps the metabolic studies in the human body. A superior method for brain metabolism studies is to label a suitable chemical (CO, glucose etc.) with positron emitters. The reason is that the only suitable radioactive isotopes among carbon, oxygen and nitrogen are the positron-active isotopes ^{11}C, ^{13}N and ^{15}O. Thus, the labelling technique can be utilized with any organic compounds. The practical limitation is the lifetime of these isotopes (20 min or less). A cyclotron for the isotope production must be situated in close connection with the positron camera.

7. FUTURE DEVELOPMENT

As seen in Fig.1, the amount of research on positron physics shows continuous growth. The instruments needed are cheap, commercially available and normally exist in well-equipped nuclear physics laboratories. A further increase in the work in this field is expected.

The new two-dimensional angular correlation devices provide further possibilities in electronic structure studies of alloys and non-metallic compounds. The lattice defect studies have enabled us to understand the positron behaviour in disordered regions. Thus, it is possible to apply the method to more technological problems like deformed metals, radiation damage, fatigue and other metallurgical applications. In molecular substances the positronium formation and its interaction with the surrounding medium are still, to some extent, open questions. A totally new area of applications, the use of positrons in the study of surfaces, is at present breaking out [68].

In conclusion, we may state that although the field of positron annihilation is rather wide, it still keeps growing. There are plenty of opportunities and interesting applications to anyone who wants to enter this field.

REFERENCES

[1] DIRAC, P.A.M., Proc. R. Soc. **126** (1930) 361.
[2] ANDERSON, C.D., Phys. Rev. **43** (1933) 1056.
[3] LAMBRECHT, R.M., Antimatter-matter Interactions, 1. Positrons and Positronium, A Bibliography, Brookhaven National Laboratory Report 50510, New York (1975).
[4] Positron Annihilation (STEWART, A.T., ROELLIG, L.O., Eds), Academic Press, New York (1967).
[5] Proc. 2nd Int. Conf. on Positron Annihilation, Kingston, Ontario, 1971 (unpublished).
[6] Proc. 3rd Int. Conf. on Poistron Annihilation, Otaniemi, Finland, 1973 (HAUTOJÄRVI, P., SEEGER, A., Eds) Springer-Verlag, Heidelberg (1975).
[7] Proc. 4th Int. Conf. on Positron Annihilation, Helsingør, Denmark, 1976 (unpublished).
[8] WEST, R.N., Adv. Phys. **22** (1973) 263.
[9] DEKTHYAR, I.Ya., Phys. Rep. **9C** (1974) 243.
[10] GOLDANSKII, V.I., At. Energy Review **6** (1968) 3.
[11] Positrons in Solids (HAUTOJÄRVI, P., Ed.), Springer-Verlag, Heidelberg (1977), to be published in Topics in Current Physics.
[12] BERGERSEN, B., PAJANNE, E., Appl. Phys. **4** (1974) 25.
[13] DIRAC, P.A.M., Proc. Camb. Philos. Soc. Math. Phys. Sci. **26** (1930) 361.
[14] MYLLYLÄ, R., KARRAS, M., MIETTINEN, T., Appl. Phys. **13** (to be published).
[15] CRISP, V.H.C., MACKENZIE, I.K., WEST, R.N., J. PHys. E **6** (1973) 1191.
[16] MADER, J., BERKO, S., KRAKAUER, H., BANSIL, A., Phys. Rev. Lett. **37** (1976) 1232.
[17] JEAVONS, A.P., CHARPAK, G., STUBBS, R.J., Nucl. Instrum. Methods **124** (1975) 491.
[18] CAMPBELL, J.L., JACKMAN, T.E., MACKENZIE, I.K., SCHULTE, C.W., WHITE, C.G., Nucl. Instrum. Methods **116** (1974) 369.
[19] WEISBERG, H., BERKO, S., Phys. Rev. **154** (1967) 249.
[20] BHATTACHARYYA, P., SINGWI, K.S., Phys. Rev. Lett. **29** (1972) 22.
[21] HAUTOJÄRVI, P., Solid State Commun. **11** (1972) 1049.
[22] HAUTOJÄRVI, P., JAUHO, P., Acta Polytech. Scand. **98** (1973) 1.
[23] FUJIWARA, K., SUEOKA, O., J. Phys. Soc. Jpn **21** (1966) 1947.
[24] MIJNARENDS, P.E., Physica **63** (1973) 248.
[25] KUBICA, P., STEWART, A., Phys. Rev. Lett. **34** (1975) 852.

[26] MIJNARENDS, P.E., in Positrons in Solids (HAUTOJÄRVI, P., Ed.), Springer-Verlag, Heidelberg (1977), to be published in Topics in Current Physics.
[27] GROSSKREUTZ, J.C., MILLET, W.E., Phys. Lett., A **28** (1969) 621.
[28] HAUTOJÄRVI, P., TAMMINEN, A., JAUHO, P., Phys. Rev. Lett. **24** (1970) 459.
[29] DEKTHYAR, I.Ya., LEVINA, D.A., MIKHALENKOV, V.S.M., Soviet Phys. Dokl. **9** (1964) 492.
[30] BERKO, S., ERSKINE, J.C., Phys. Rev. Lett. **19** (1967) 307.
[31] MACKENZIE, I.K., LICHTENBERGER, P.C., Appl. Phys. **3** (1974) 393.
[32] ARPONEN, J., HAUTOJÄRVI, P., NIEMINEN, R., PAJANNE, E., J. Phys. F **3** (1973) 2092.
[33] MANNINEN, M., NIEMINEN, R., HAUTOJÄRVI, P., ARPONEN, J., Phys. Rev. B **12** (1975) 4012.
[34] BRANDT, W., in Positron Annihilation (STEWART, A.T., ROELLIG, L.O., Eds), Academic Press, New York (1967) pp. 155–182.
[35] BERGERSEN, B., STOTT, M.J., Solid State Commun. **7** (1969) 1203.
[36] CONNORS, D.C., WEST, R.N., Phys. Lett., A **30** (1969) 24.
[37] McKEE, B.T.A., TRIFTSHÄUSER, W., STEWART, A.T., Phys. Rev. Lett. **28** (1972) 358.
[38] DLUBEK, G., BRÜMMER, O., MEYENDORF, N., Appl. Phys. **13** (1977) 67.
[39] WEST, R.N., in Positrons in Solids (HAUTOJÄRVI, P., Ed.), Springer-Verlag, Heidelberg (1977), to be published in Topics in Current Physics.
[40] CAWTHORNE, C., FULTON, E.J., Nature (London) **216** (1967) 575.
[41] MOGENSEN, O., PETERSEN, K., COTTERILL, R.M.J., HUDSON, B., Nature (London) **239** (1972) 98.
[42] COTTERILL, R.M.J., MACKENZIE, I.K., SMEDSKJAER, L., TRUMPY, G., TRÄFF, J.H.D.L., Nature (London) **239** (1972) 99.
[43] HODGES, C.II., STOTT, M.J., Solid State Commun. **12** (1973) 1153.
[44] NIEMINEN, R., MANNINEN, M., Solid State Commun. **15** (1974) 403.
[45] ELDRUP, M., MOGENSEN, O.E., EVANS, J.H., J. Phys. F **6** (1976) 499.
[46] HAUTOJÄRVI, P., HEINIÖ, J., MANNINEN, M., NIEMINEN, R., Phil. Mag. **35** (to be published).
[47] GAUSTER, W.B., J. Nucl. Mater. **62** (1976) 118.
[48] HAUTOJÄRVI, P., VEHANEN, A., MIKHALENKOV, V.S., Appl. Phys. **11** (1976) 191.
[49] DLUBEK, G., BRÜMMER, O., HENSEL, E., Phys. Status Solidi A **34** (1976) 737.
[50] LYNN, K.G., BYRNE, J.G., Metall. Trans., A **7** (1976) 604.
[51] NIEMINEN, R., MANNINEN, M., in Positrons in Solids (HAUTOJÄRVI, P., Ed.), Springer-Verlag, Heidelberg (1977), to be published in Topics in Current Physics.
[52] ORE, A., Univ. i Bergen, Arbok Naturvitenskap, Rekke No.9 (1949).
[53] MOGENSEN, O.E., J. Chem. Phys. **60** (1974) 998.
[54] STELT, F.R., VARLASHKIN, P.G., Phys. Rev. B **5** (1972) 4265.
[55] ACHE, H.J., Angew. Chem. Int. Ed. Engl. **11** (1972) 179.
[56] GOLDANSKII, V.I., SHANTAROVICH, V.P., Appl. Phys. **3** (1974) 335.
[57] LEVAY, B., HAUTOJÄRVI, P., J. Phys. Chem. **76** (1972) 1951.
[58] BELL, R.E., GRAHAM, R.L., Phys. Rev. **90** (1953) 644.
[59] SINGH, K.P., SINGRU, R.M., RAO, C.N.R., J. Phys. C **5** (1972) 1067.
[60] SINGH, K.P., WEST, R.N., PAUL, A., J. Phys. C **9** (1976) 305.
[61] HAUTOJÄRVI, P., PAJANNE, E., J. Phys. C **7** (1974) 3817.
[62] JAMES, P.F., PAUL, A., SINGRU, R.M., DAUVE, C., DORIKENS-VANPRAET, L., DORIKENS, M., J. Phys. C **8** (1975) 393.
[63] HAUTOJÄRVI, P., LEHMUSOKSA, I., KOMPPA. V., PAJANNE, E., J. Non-Cryst. Solids **18** (1975) 395.

[64] HAUTOJÄRVI, P., VEHANEN, A., KOMPPA, V., PAJANNE, E., Solid State Commun. **18** (1976) 1137.
[65] HAUTOJÄRVI, P., KOMPPA, V., PAJANNE, E., VEHANEN, A., in The Physics of Non-Crystalline Solids (FRISCHAT, G.H., Ed.) Trans. Tech. Publications, Germany (1977) pp.129–134.
[66] BROWNELL, G.L., BURNHAM, C.A., in Instrumentation in Nuclear Medicine, Vol.2 (HINE, G.J., SORENSON, J.A., Eds), Academic Press, New York (1974) pp.135–159.
[67] IEEE Trans. Nucl. Sci. **21** 3 (1974), Special Issue on Physical and Computational Aspects of Three-Dimensional Image Reconstruction.
[68] BRANDT, W., Adv. Chem. **158** (1976) 219.

APPLICATIONS OF THERMAL NEUTRON SCATTERING

G. KOSTORZ
Institut Laue-Langevin,
Grenoble, France

Abstract

APPLICATIONS OF THERMAL NEUTRON SCATTERING.
　　Although in the past neutrons have been used quite frequently in the study of condensed matter, a more recent development has lead to applications of thermal neutron scattering in the investigation of more practical rather than purely academic problems. Physicists, chemists, materials scientists, biologists, and others have recognized and demonstrated that neutron scattering techniques can yield supplementary information which, in many cases, could not be obtained with other methods. The paper illustrates the use of neutron scattering in these areas of applied research. No attempt is made to present all the aspects of neutron scattering which can be found in textbooks. From the vast amount of experimental data, only a few examples are presented for the study of structure and atomic arrangement, "extended" structure, and dynamic phenomena in substances of current interest in applied research.

INTRODUCTION

　　The use of thermal neutrons in scattering experiments started more than thirty years ago [1], and many outstanding and unique results have been obtained on structural and dynamical properties of a great number of substances [2–6]. As the neutron carries no net charge, scattering of thermal neutrons by an assembly of atoms is mainly controlled by the interaction of neutrons with the nuclei which can be described by the Fermi pseudopotential, and the interaction of the neutron magnetic moment with magnetic moments of electrons. The theory of thermal neutron scattering is based on the use of the first Born approximation, and the scattering cross-section can be written (see, e.g., Marshall and Lovesey [5] for details)

$$\frac{d^2\sigma}{d\Omega dE} = \frac{k}{\hbar k_0} \cdot \mathscr{A} \cdot \mathscr{S}(\vec{Q}, \omega) \tag{1}$$

where Ω is a solid angle, E is energy, $\vec{k}_0$ is the wave-vector of the incident neutrons, and $k_0 = |\vec{k}_0|$, $\vec{k}$ is the wave-vector of the scattered neutrons, and $k = |\vec{k}|$; $\hbar$ has its usual meaning. The quantity $\mathscr{A}$ describes the neutron-target interaction. It

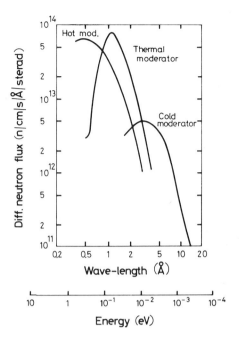

FIG.1. Differential neutron flux as a function of wave-length for different moderators at the High Flux Reactor of the Institut Laue-Langevin, Grenoble, France [7].

has the dimension of area, and for purely nuclear scattering $\mathscr{A}$ is related directly to the appropriately averaged single-nucleus cross-section which is independent of the scattering vector $\vec{Q}$ defined by

$$\vec{Q} = \vec{k}_0 - \vec{k} \qquad (2)$$

If magnetic scattering is considered, $\mathscr{A}$ depends on the direction and the magnitude of $\vec{Q}$ and on the polarization of the incident neutron beam. The quantity $\mathscr{S}(Q,\omega)$ is called scattering function (or scattering law) and has the dimension of time. The angular frequency ω in the argument is determined by

$$\hbar\omega = E_0 - E \qquad (3)$$

where E_0 and E are the energies of incident and scattered neutrons. $\mathscr{S}(\vec{Q},\omega)$ describes the physical processes of the sample probed in the scattering experiment and is related, by Fourier transformation, to the time-dependent correlation functions of the scattering system. Equation (1) implies that correlations between the processes contained in $\mathscr{A}$ and in $\mathscr{S}(\vec{Q},\omega)$ can be neglected which is in fact permitted in many cases of interest [5].

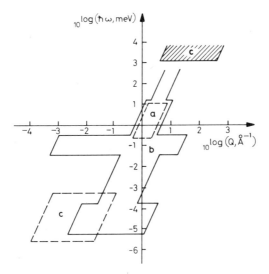

FIG.2. *The energy-momentum plane for neutron scattering (a) in 1970, (b) in 1976, (c) expected ranges accessible with neutrons from pulsed sources (top) and with ultra-cold neutrons (bottom). Replotted from White* [8].

Most of the neutrons used in scattering experiments today are produced in nuclear reactors, as described in textbooks (e.g. Bacon [6]). Beam tubes serve to extract neutrons in thermal equilibrium with the moderator (at a temperature slightly above room temperature). For special applications, smaller moderators for different temperatures are incorporated to modify the available spectrum of neutrons to be used for scattering studies. As an example, Fig.1 shows the three different spectra obtained from the High Flux Reactor of the Institut Laue-Langevin [7]. A hot source (graphite at about 2000 K) and a cold source (D_2 at 25 K) provide "hot" and "cold" neutrons of a mean wave-length of roughly 0.9 and 6 Å, in addition to the thermal neutrons (~2 Å) from the D_2O moderator. The range of useful wave-lengths thus extends from 0.6 to 20 Å (and higher for special applications), with corresponding incident energies from 230 to 0.2 meV. As these wave-lengths cover the range of interatomic distances and the neutron energies are similar to typical energies of excitation in condensed matter, thermal neutrons offer ideal possibilities to study the structure and dynamical properties of solids, liquids, etc. The recent progress in expanding the range of accessible values of scattering vector and energy transfer is documented in Fig.2, adapted from White [8]. This progress is the combined result of more powerful sources and more sophisticated instrumentation for scattering experiments. Figure 2 also indicates the expected ranges of Q and ΔE for ultra-cold neutron sources [9] and pulsed spallation sources [10] which are currently being developed.

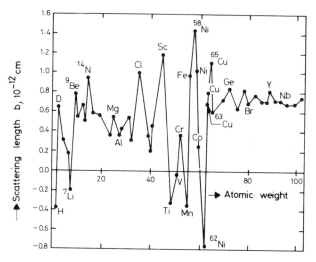

FIG.3. Coherent scattering length as a function of atomic weight for some elements and isotopes. Replotted from Schmatz et al. [11].

Neutron sources have a lower luminosity than X-ray or gamma sources, and larger samples are usually necessary. As absorption is normally small, large sample volumes can in fact be studied, and this may also be advantageous if average bulk properties are to be investigated. For the same reason, neutrons are particularly suited for scattering experiments under extreme conditions (temperature, pressure, etc.) where samples have to be contained and surrounded by other materials.

Although neutron scattering experiments may be expensive and time-consuming, they offer information that cannot (or only indirectly or with greater difficulty) be obtained from other experiments. Inelastic neutron scattering is the most direct and detailed method to study dynamical phenomena as documented in a series of conference proceedings published by the IAEA.

The magnetic interaction yields unique details on magnetism on a microscopic scale [5], and the non-systematic variation of the nuclear scattering cross-sections with isotopic mass can be used to produce scattering contrast between otherwise indistinguishable atoms. Figure 3 [11] shows the coherent scattering length b, which is related to the coherent scattering cross-section, σ_{coh}, by

$$\sigma_{coh} = 4\pi b^2 \qquad (4)$$

for bound nuclei and atoms (i.e. averaged according to the natural abundance of isotopes of one chemical species) of the lighter elements. The large difference of b for hydrogen and deuterium has led to a variety of applications in chemistry and biology, as is discussed later. Incoherent scattering studies also benefit from the high incoherent scattering cross-section of H, whereas D has a much lower incoherent cross-section and can be used to replace H where coherent phenomena are investigated. Similar use can be made of the widely different scattering length of, e.g. nickel isotopes, and neighbouring elements (e.g. Al-Mg), which show no contrast with electromagnetic radiation, can be distinguished.

As stated already, neutron scattering experiments have been performed for over thirty years, and it is not the purpose of this paper to review any of these activities. A bibliography is available covering the literature from 1932 to 1974 [12], and recent references may be consulted for more current information (e.g. Ref. [1] and all subsequent papers). With the availability of more neutron scattering instruments, neutron scattering has become a more frequently used method in applied research or research leading to applications. In the following, an attempt is made to illustrate how neutrons are used in the fields of physical chemistry, materials science, and biology. The applied aspect of neutron scattering is emphasized, and examples are given for experiments relating to structure (atomic arrangements), "extended" structure (large-scale inhomogeneities), and dynamics (excitations and atomic or molecular motion). As in any developing field, a coherent picture will only emerge after the situation has been reviewed from several different points of view and, quite naturally, only one can be offered at a time.

STRUCTURE AND ATOMIC ARRANGEMENTS

Coherent elastic scattering yields information about the time-averaged local arrangement of scattering centres with a spatial resolution corresponding to $\delta \cong 2\pi/Q_{max}$, where Q_{max} is the maximum scattering vector used in the experiment. For any spectrometer, the upper limit of Q_{max} is $2k_0$, and a large range of Q values may be desirable for precise structural studies, especially for amorphous materials and liquids.

For ideal crystalline substances, coherent elastic scattering occurs at Bragg peaks only, i.e. for $\vec{Q} = \vec{\tau}$, where $\vec{\tau}$ is a vector of the reciprocal lattice. This condition can be expressed by the Bragg equation

$$n\lambda = 2d_{hk\ell} \sin \theta \qquad (5)$$

where $d_{hk\ell}$ is the interplanar distance for planes $\{hk\ell\}$, n is the order of the reflection, and θ is half the scattering angle. For simple systems, a few Bragg peaks are sufficient to determine the crystallographic structure, but several thousands of them may be necessary for a complete analysis of molecular crystals. Neutron crystallography is specifically indicated if the position of light atoms (of small scattering cross-section for electromagnetic radiation) is to be studied, or if the magnetic structure is to be determined.

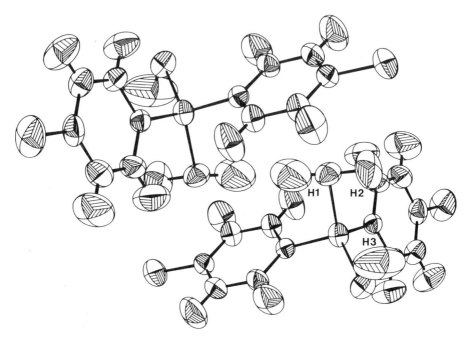

FIG.4. Stereoscopic drawing of the dimeric unit of di(p-chlorophenyl)dithiophosphinic acid. The two $C_{12}H_8Cl_2P(S)SH$ molecules are linked by one S-H ... S hydrogen bond (from Krebs et al. [13]*).*

As an example for a structural investigation to reveal the position of protons in an organic molecule, the work of Krebs et al. [13] is quoted on a small (3 mm³) crystal of di(p-chlorophenyl) dithiophosphinic acid. This acid crystallizes in the space group $P\bar{1}$ with two molecules per unit cell. Figure 4 shows the dimeric unit, two $C_{12}H_8Cl_2$ P(S)SH molecules linked by one S-H ... S hydrogen bond. The proton positions were obtained from a difference Fourier synthesis using X-ray results for the heavier atoms and 1798 recorded neutron reflections. Whereas all hydrogen atoms in the aromatic ring are well localized as expected, the protons of the sulfhydryl groups occupy three partially filled positions which are cystallographically different.

Some of the structural work in materials science has recently been summarized [14], both for nuclear and magnetic scattering, and a major review is forthcoming [15]. Here, less common efforts in applied research are emphasized.

A large amount of work on materials concerns multi-phase systems obtained by phase separation or compaction. As neutrons sample large volumes, it is possible to investigate the structure of minority phases. For example, the binary alloy Al-Mg decomposes near room temperature if the magnesium concentration exceeds about 8 at.% [16]. The coherent Guinier-Preston zones of about 50 Å radius form an ordered structure which could not be determined unambiguously from X-ray and electron diffraction results. Neutron diffraction measurements on a set of samples after room temperature aging and reversion indicate L12-type (Cu_3Au) order [17]. This leads to Mg concentrations of 25 at.% in the zones and 7.5 at.% in the matrix after complete ageing. In another diffraction study, the interlayer spacing c_o of graphite nodules in a high-carbon steel were measured after different stages of graphitization [18]. From the change of c_o upon stress relief in the surrounding matrix, the pressure inside the nodules can be estimated.

Structure studies of minority phases represent only one example of the general possibilities of structure determinations under extreme conditions mentioned earlier. As a variety of materials are available which are very transparent for neutrons, cryostats, furnaces [19], pressure cells [20], sample containers for liquids etc. can be made and used without major difficulties in many cases. Other examples where the large sample volume is the key point are texture studies (extensively reviewed by Szpunar [21]), structure determinations of surface layers [8,22], and kinetic measurements in intercalation [23] and adsorption processes.

Monolayers, multilayers or clusters (droplets) of crystalline or amorphous structure may be formed upon physiorption. Neutron diffraction (and also quasielastic scattering, see later) can be used in all these cases if sufficient surface area is exposed to the neutron beam. The structure of deuterated ammonia adsorbed on precooled graphon (graphitized carbon of about 300 Å mean particle diameter and a surface area of 86 m^2/g) was recently studied [8,24] as a function of adsorbed amount and temperature (between 80 and 200 K). The diffraction patterns showed that ammonia forms large crystallite sheets with the structure of bulk ND_3 above a certain critical coverage (~0.25, expressed in hypothetical monolayer units) but a fraction always remains amorphous. Figure 5 [8] shows the diffraction pattern from ammonia adsorbed on graphon, measured at 100 K and at several higher temperatures. A change of the height and width of the diffraction peaks indicates annealing phenomena, while the integrated coherent scattering decreases with increasing temperature until, at a temperature T_{ms} (150 to 180 K, depending on the coverage), all Bragg peaks have disappeared. The ammonia crystallites melt at $T_{ms} < T_m$ where T_m = 197 K is the melting point of bulk ammonia.

The structure of nitrogen adsorbed on grafoil (exfoliated graphite, surface area 20 m^2/g) has been investigated by Kjems et al. [22]. In this case, continuous monolayers are formed and, despite the low intensities, very significant results, e.g. on the first monolayer structure and on the transition from one to two monolayers, were found.

Although crystallographic structure determination provides essential information for many applied problems, the deviations from the ideal crystal structure control many practical properties as well, and neutron scattering has

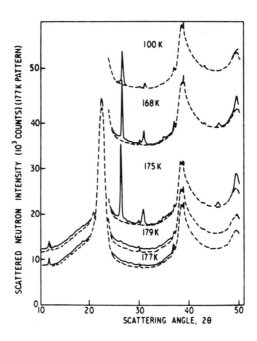

FIG.5. Diffraction patterns at different temperatures from graphon after adsorption of deuterated ammonia. The sharp lines are the diffraction peaks of ND_3 (from White [8]).

been used successfully to study short-range order or clustering phenomena in alloys as well as strain fields around substitutional and interstitial solute atoms. Under simplified conditions, the diffuse coherent elastic scattering for a substitutional binary alloy A-B can be written as [25–27]

$$\left(\frac{d\sigma}{d\Omega}\right)_{diff} = N_s \overline{|C(\vec{Q})|^2} |b_A - b_B + i\bar{b}\,\vec{s}(\vec{Q})\cdot\vec{Q}|^2 \tag{6}$$

where N_s is the number of atoms, $C(\vec{Q})$ is the Fourier component of the compositional fluctuations as introduced by Krivoglaz [25], $\bar{b}$ is the average scattering length of the alloy, and $\vec{s}(\vec{Q})$ is the Fourier transform of the displacement field around one solute atom (Debye-Waller factors have been neglected). With neutrons, it is possible to study compositional fluctuations alone by preparing alloys with $\bar{b} = 0$. This has been done by several authors [28–30] for the system Cu–Ni, and the state of local clustering has been characterized for different heat treatments.

A knowledge of the displacements around solute atoms may be interesting for several reasons, e.g. for an understanding of the solute atom-dislocation interaction. If the solute site is unknown, its symmetry and lattice distortion strength

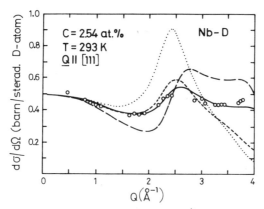

FIG.6. *Diffuse scattering of Nb-2.54 at.% D at 293K for $\vec{Q}$ parallel [111]. The open circles are interpolated values obtained from the measured data in the $(01\bar{1})$ plane of the crystal. Four theoretical curves are shown:*

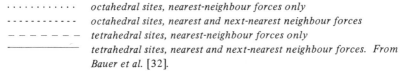

can be obtained from diffuse scattering on single crystals (obviously, dilute solutions and $b_A = b_B$ are desirable). Bauer et al. [31] have shown that deuterium (used instead of hydrogen which would contribute too much incoherent scattering) in niobium occupies tetrahedral interstitial sites, and the displacement scattering can be described by a model with forces on nearest and next-nearest neighbours. Figure 6 shows one example of more recent results [32] including larger scattering vectors to confirm the earlier conclusion. Similar work on Nb-O is in progress [33]. The {001} plane was investigated for pure Nb and two different oxygen concentrations. Pure Nb showed constant (incoherent) scattering for Q below the Bragg condition, whereas for the Nb-O samples diffuse peaks at (1/2 1/2 0), (1/2 0 0) and (1 0 0) were found.

It should be stressed that diffuse scattering is a well-established method with X-rays, and that neutrons are not always advantageous, especially if there is a large amount of incoherent scattering. If the coherent contrast is high enough, however, an evaluation of neutron diffuse scattering should be more straightforward as there are no form factor problems, and inelastic scattering can be separated.

In concluding this section on atomic arrangements, the possibilities of neutron scattering in research on amorphous and vitreous systems (metallic and non-metallic), molten salts and liquid alloys are mentioned only briefly. Neutron diffraction is particularly useful in the analysis of amorphous ferromagnets [34,35],

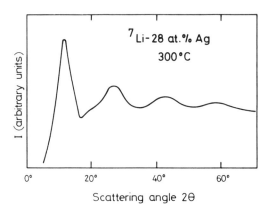

FIG.7. *The scattered intensity due to compositional fluctuations $S_{cc}(Q)$ for a liquid "null-matrix" alloy ($\bar{b} = 0$), ^{7}Li-28 at.% Ag at 300°C. The peak at about 12° and subsequent oscillations of $S_{cc}(Q)$ are characteristic for short-range order, i.e. a preference for unlike nearest neighbours (from Ruppersberg [38]).*

especially if polarized neutrons are available [36]. Diffraction measurements on liquid alloys are of special value for the study of compositional fluctuations. The formalism is not essentially different from Eq.(6) although other notations are commonly used. For a binary liquid, Bhatia and Thornton [37] have introduced the scattering functions S_{NN}, S_{CC} and S_{NC} where N stands for total number density and C for compositional fluctuations. With these scattering functions, the coherent elastic scattering is

$$\left(\frac{d\sigma}{d\Omega}\right)_{coh} = (\bar{b})^2 S_{NN}(Q) + 2S_{NC}(Q)(b_A - b_B)\bar{b} + S_{CC}(Q)(b_A - b_B)^2 \quad (7)$$

Again, a "null-matrix alloy" with $\bar{b} = 0$ gives the compositional fluctuations directly. Ruppersberg [38] has used the isotope ^{7}Li with a negative scattering length (see Fig.3) to measure $S_{CC}(Q)$ in ^{7}Li-28 at.% Ag. Figure 7 shows the short-range order in the liquid state at 300°C. More recent measurements at 400°C and 500°C do not reveal any major changes of the scattering pattern [39]. Systems with segregation tendency have also been studied, e.g. Bi-Sb [40], and more work is in progress (Li-Na [41,42], Al-Sn [43] etc.). In these alloys, neutrons also offer the possibility of following directly the process of mixing at different temperatures.

As clusters lead to an increase of scattering at small Q, small-angle scattering studies may be necessary in some of these segregation systems. This particular technique is the basis for the results discussed in the next section.

"EXTENDED" STRUCTURE

If we restrict ourselves to small Q values in a scattering experiment, the spatial resolution may no longer be adequate for a complete determination of all atomic positions. In fact, this detailed information is not absolutely necessary or not of primary interest for a large variety of problems. The term "extended" structure is to characterize this situation where we seek a more global information on the distribution or arrangement of scattering centres. Typical problems of this nature are: the shape, global structure and mutual arrangement of large molecules in solution and in the bulk (biological macromolecules, fibres, synthetic polymers, liquid crystals) of precipitates in liquid, crystalline and amorphous substances, and the distribution of magnetization in magnets and superconductors. If the desired spatial resolution is, e.g. $\delta \gtrsim 20$ Å, the Q range to be investigated can be limited to $Q \lesssim 0.3$ Å^{-1}. For Cu-K$_\alpha$ X-rays, the corresponding full scattering angle would be $2\theta_{max} = 4.2°$, and $16.5°$ for neutrons of 6 Å wave-length. For such small angles, special theoretical and experimental requirements have led to the development of particular techniques which constitute the field of small-angle scattering (SAS). In comparison with X-ray SAS, neutron SAS is still a relatively new method but among all neutron scattering techniques it is undoubtedly the one which is most relevant and promising for applied research. Instrumentation and application of neutron SAS have been reviewed by Schmatz et al. [11,44], and descriptions of instruments are available for high [45], medium [11] and low flux [46] installations.

As the discrete arrangement of scattering centres can be disregarded, the scattering length per site can be replaced by an (appropriately averaged) scattering length density $\rho(\vec{r})$, and the observable small-angle cross-section is

$$\frac{d\sigma}{d\Omega} = \frac{1}{N_s} \left| \int_V \left[\rho(r) - \bar{\rho} \right] \exp(i \vec{Q} \cdot \vec{r}) d^3\vec{r} \right|^2 \tag{8}$$

where V is the sample volume and $\bar{\rho}$ is the scattering length density averaged over a volume larger than the resolution volume of the instrument (determined by the minimum observable Q value). For a given scattering length distribution, the cross-section can be calculated analytically or numerically. However, it is in general impossible to evaluate $\rho(\vec{r})$ from measured values of $d\sigma/d\Omega$.

In the case of N_p widely separated identical particles of volume V_p each and of homogeneous scattering length density ρ_p, imbedded in a homogeneous matrix with scattering length density ρ_m, very simple expressions are obtained for $d\sigma/d\Omega$:

$$\frac{d\sigma}{d\Omega}(\vec{Q}) = \frac{V_p^2 N_p}{N_s} (\rho_p - \rho_m)^2 |F_p(\vec{Q})|^2 \tag{9}$$

$F_p(\vec{Q})$ is the single-particle form factor. For spherical particles or a random orientational distribution of non-spherical particles, the isotropic Guinier approximation yields

$$\frac{d\sigma}{d\Omega}(Q) \propto \exp(-R_G^2 Q^2/3) \qquad (10)$$

where R_G is the average radius of gyration. This approximation is valid for small Q, ($R_G Q \lesssim 1$) depending on the shape of the particles. In cases of orientational correlations between particles, the Guinier approximation is still valid but may yield different inertial distances for different directions of $\vec{Q}$. More theoretical details can be found in the X-ray literature [47–49], in particular form factor calculations for various particle shapes and approximations of Eqs (8) and (9).

There is a wide range of applications of neutron SAS, e.g. in the study of polymers, macromolecules, fibres and membranes, catalysts, colloids, minerals, metals and alloys, glasses, etc. Only a few characteristic examples are presented here.

In inorganic materials "extended" inhomogeneities control many important properties, e.g. the mechanical strength of alloys, critical currents in hard superconductors, coercive forces in magnetic materials, optical properties of glasses, etc. With neutron SAS, the decomposition in alloys, compounds and glasses, clustering of point defects and spins, dislocation arrangements, pores and microcracks, voids and displacement cascades, critical fluctuations, interfaces and surfaces etc., can be investigated. In crystalline substances, double Bragg scattering can easily be avoided by working with wave-lengths above the Bragg cut-off.

The binary system Al-Zn has been the subject of many neutron SAS experiments [14, 50–55] in recent years, as it presents an interesting case for the study of different decomposition mechanisms (starting from "homogeneous" Al-rich solid solutions). Whereas the decomposition (formation of coherent zones) proceeds by nucleation and growth at elevated temperatures, e.g. at $T \leqslant 150°C$ for 6.8 at.% Zn, all experimental techniques seem to indicate that spinodal decomposition [56] takes place at lower aging temperatures ($T < 80°C$). For this composition of 6.8 at.% Zn, magnetic susceptibility, electrical resistivity and X-ray SAS measurements have been combined [57–59], and a temperature of $T_S = (129 \pm 2)°C$ was proposed for the transition from one mechanism to the other. One of the criteria suggested for the existence of spinodal decomposition [58] was the appearance of a SAS "ring", i.e. a maximum of SAS intensity at $Q \neq 0$ as shown in Fig.8 for an alloy of the same concentration aged at room temperature [50]. In this case, a peak of the neutron SAS at $Q_{max} = 5.6 \times 10^{-2} \text{ Å}^{-1}$ is clearly revealed and can be attributed to interparticle interference of a system of spherical particles of a radius $R \cong 30$ Å. Figure 9 shows that after aging at 133°C

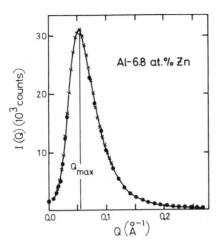

FIG. 8. *Neutron SAS intensity as a function of scattering vector Q for Al-6.8 at.% Zn aged for several weeks at room temperature (from Raynal et al. [50]).*

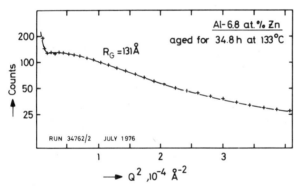

FIG. 9. *Neutron SAS intensity (logarithmic scale) as a function of the square of the scattering vector, Q^2 ("Guinier plot"), for Al-6.8 at.% Zn aged at 133°C for 35 h. Note that the single-particle scattering function for any homogeneous particle cannot produce a plateau below $Q^2 \lesssim 5 \times 10^{-5}$ Å^{-2} (from Laslaz et al. [53]).*

an indication of a peak is also visible [53], but the value of Q_{max} is much smaller ($\sim 7 \times 10^{-3}$ Å^{-1}) and could in fact not be observed with common X-ray scattering instruments. It was concluded [53] that the existence of a SAS peak cannot be claimed to give evidence for the operation of any particular decomposition mechanism but simply represents the more or less pronounced interference of an assembly of particles. Furthermore, transmission electron microscopy performed on samples of the same composition indicates [60] that at the temperature of 129°C, no abrupt change of the precipitate morphology occurs.

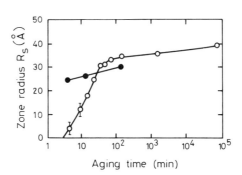

FIG.10. Zone radius R as a function of aging time at room temperature for an Al-12 at.% Zn alloy quenched from 310°C (open circles) and 380°C (full circles) as determined from neutron SAS (and diffuse scattering) measurements (from Allen et al. [52]).

From the slope of a Guinier plot, as shown in Fig.9, the radius of gyration can be determined if there is a well-defined particle size, and if interference effects do not modify too much the slope at lower Q. For the case shown, one finds $R_G \cong 130$ Å, whereas the results shown in Fig.8 yield $R_G \cong 23$ Å. The zones formed at room temperature are spherical but the larger zones found after long aging times at 133°C are platelets of an intermediate phase α'_R, as is known from electron microscopy (e.g. [60]).

The analysis of SAS curves as a function of aging time can be useful in the study of the kinetic aspects of decomposition, its dependence on annealing temperature, quench rate, aging temperature, impurity concentration, etc. Figure 10 [52] shows an example for the change of zone radius R as a function of aging time at room temperature for Al-12 at.% Zn alloys quenched from 310 and 380°C. The aging was interrupted at the indicated times, and measurements were performed in a liquid-helium cryostat. Although it is not possible to decide whether there is an incubation time for the growth of zones or not, the change of growth rate for the sample quenched from 310°C after 80–100 min of aging is quite drastic and can be attributed to the elimination of quenched-in excess vacancies [61].

An integration of Eq.(9) over all $\vec{Q}$ space allows one to determine the scattering length density of matrix and precipitate (see, e.g. Gerold [62]). As other scattering phenomena intervene for larger $\vec{Q}$, an analytical extrapolation has to be used to complete the integration for $Q \to \infty$, and this is sometimes difficult. Also, absolute values for the cross-sections and careful calibrations are necessary. Despite these limitations, the integrated intensity can often yield the composition of matrix and precipitates in binary systems, and it is possible to establish stable and metastable miscibility gaps. X-ray and neutron SAS results have recently been combined [63] to determine the metastable miscibility gap for ternary

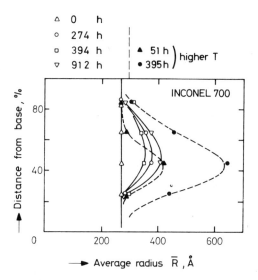

FIG.11. *Average radius of γ' precipitates in Inconel 700 turbine blades at different distances from the blade base (23, 43, 63 and 83 mm) after different service times under normal and elevated temperature conditions (from Pizzi et al. [67]).*

Al-rich Al-Mg-Zn alloys aged at room temperature [64]. An alternative way for such studies in ternary systems would be the variation of the isotopic composition of alloying elements as the total error could be reduced by using only one type of radiation.

Decomposition and reversion have been studied in Al-Mg [65] where the contrast is favourable. In Al-rich Al-Si, both low contrast and low solubility of Si lead to small scattering cross-sections, and surface irregularities have to date masked any possible zone scattering from the bulk [66].

The large sample volume for neutron SAS allows one to use this method nondestructively, and it is even conceivable to apply neutron SAS in routine testing operations. Pizzi and Walther [67] have studied these possibilities in detail using a multidetector SAS instrument at a small reactor [46]. In Fig.11 [68], neutron SAS results on the radius of γ' precipitates in turbine blades made from the alloy Inconel 700 (Ni, Co, Al, etc.) are shown for different positions along the blade axis. Two sets of blades, one used at normal temperature, the other at a more elevated temperature, have been examined after various service times. It can be seen that there is a considerable increase of the average precipitate radius with increasing service time in the middle of the blades, whereas precipitates in the base and the tip of the blades remain essentially unchanged. Operation at a higher temperature leads to more rapid growth of precipitates and degradation of mechanical properties. It is suggested [68] to use neutron SAS for a nondestructive evaluation of residual life-times of such components.

Other nickel alloys and steels were also studied [68] after thermal treatment, creep and fatigue. In magnetic materials, multiple refraction by domains of different magnetization has been observed [68]. This effect could be used to estimate domain dimensions and to follow changes thereof. (In this context it is briefly mentioned that neutron topography [69] and simultaneous measurements of symmetry-related magnetic reflections using the Laue technique [70] provide alternatives for the observation of magnetic domains.)

Cavities, pores, voids etc. give rise to rather strong SAS (as their scattering length density is zero). An increase of SAS intensity near creep rupture surfaces was attributed to the formation of microcracks [68]. Neutron SAS may be useful in fracture studies, e.g. by monitoring the formation of a microstructure leading to rupture during fatigue and creep. Dislocations also contribute to SAS, and Schmatz [71] has reviewed this field in detail. The enhancement of dislocation scattering by a coupling of the local magnetization vector to the elastic stress field of dislocations has been the basis for systematic investigations on the SAS scattering in deformed Ni [72] and Fe [73] single crystals near magnetic saturation. The scattering from Fe (deformed to the end of stage I, single glide) can be attributed to dislocation groups in the primary slip plane [73].

Voids in Al crystals after fast neutron irradiation have been studied by neutron SAS by Mook [74] and by Hendricks et al. [75]. It was found [75] that the single crystals (six of them, irradiated to doses between 0.3 and 2.0×10^{21} fast neutrons/cm², were studied) showed "isotropic" SAS at small Q (i.e. the scattered intensity depends on the magnitude of $\vec{Q}$, but not on its direction), whereas considerable anisotropy was observed for larger Q values as can be seen in Fig.12. The scattering patterns are interpreted by truncated octahedral voids (there are thus {111} and {001} faces). For anisotropic scattering patterns, a two-dimensional position-sensitive detector as used, for example, at the Institut Laue-Langevin [45], is most appropriate as it allows a complete contour map to be obtained for a given crystal orientation. Figure 13 shows lines of equal intensity for a single crystal of stoichiometric β'-NiAl containing facetted voids. In these crystals, voids form from thermal vacancies quenched-in from higher temperature. The details of void growth and shape are currently being studied as a function of aging time and aging temperature [76].

Other fast neutron irradiation effects have been investigated with neutron SAS, e.g. in GaAs [77], and in different types of graphite [78]. Most graphites contain a significant volume fraction of pores, and the porosity may change as a function of fast neutron dose. As the pore size distribution is very broad and other large defects may also contribute, the SAS curves do not follow any simple analytical relationship. It is nevertheless possible to follow the evolution of certain ranges of pore sizes as a function of fast neutron dose, and the authors find [78] that there is a decrease of pores with $R_G \gtrsim 100$ Å and an increase of

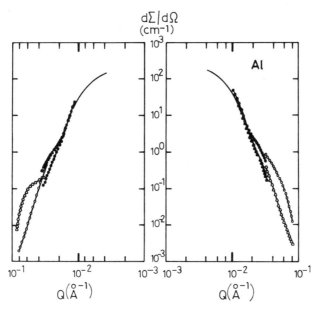

FIG.12. Neutron small-angle scattering curves for an aluminium single crystal irradiated with $\sim 1.7 \times 10^{21}$ fast neutrons/cm^2 at 55°C. Detectors D2 and D3 are linear position-sensitive detectors placed 7 cm below (D2) and 5 cm above (D3) the beam centre line at different distances L.

——	$D2, D3 : L = 12\ m$
▲	$D2 : L = 5\ m$
●	$D3 : L = 5\ m$
△	$D2 : L = 2\ m$
○	$D3 : L = 2\ m$

$d\Sigma/d\Omega$ is the macroscopic differential cross-section. As the voids are facetted but no appropriate symmetry axis is parallel to the detectors, the two sets of data are not identical (from Hendricks et al. [75]).

those with $R_G \gtrsim 25$ Å upon irradiation. The defect structures of different types of graphite seem to become more similar as a result of irradiation. Similar empirical studies are in progress on irradiated metals [79] and steel samples (surveillance samples) of reactor pressure vessels [80]. In the latter case, a considerable change of the SAS intensity after exposure to 2×10^{19} neutrons/cm^2 intensity was found for a weld containing minute amounts of copper whereas the bulk material did not show any measurable changes under similar conditions.

To conclude this brief review of applied SAS in inorganic materials, one very applied example from the field of soil science is presented. The rate of diffusion of water in clay minerals is of great practical importance. A programme has been started [81] aiming at an understanding of the properties of water in

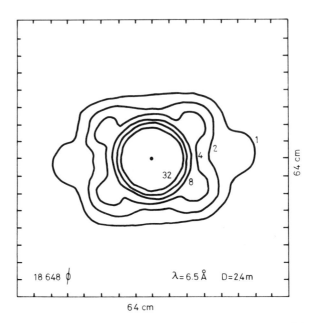

FIG.13. Contour lines on the two-dimensional position-sensitive detector of the neutron small-angle scattering instrument D11 at the Institut Laue-Langevin, Grenoble. The sample is a β'-NiAl single crystal quenched from $1600°C$ and aged for 22 h at $400°C$. The incident beam of a wave-length $\lambda = 6.5$ Å is parallel to a $\langle 110 \rangle$ direction in the cubic single crystal. The crystallographic symmetry of the voids contained in the sample is revealed by the anisotropy of the scattering pattern at larger scattering vectors (from Epperson et al. [76]).

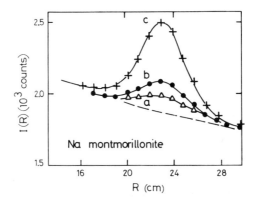

FIG.14. Scattered intensity in Debye-Scherrer cone for Na Montmorillonite prepared at 78% relative humidity, as a function of cone radius on the plane multidetector: (a) no compression, (b) 1 compression, (c) 2 compressions (from Thomas et al. [81]).

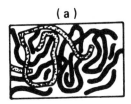 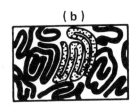 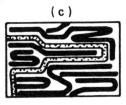

FIG.15. Models of polymer conformation in the bulk. A single tagged polymer molecule is shown in an untagged environment. (a) Gaussian coil, (b) ball, (c) meander concept (from Schelten et al. [85]).

clay. Clay samples are prepared from sols in water, with platelet diameters of 0.3 to 0.5 μm and so-called small material. The water may then be removed by different techniques, e.g. by compressing the sol above a semi-permeable and porous plate. Neutron SAS has been used to look at the coherent scattering from the small-particle fraction. Without compression, there is almost no Bragg scattering, as shown by the absence of a well-defined diffraction ring in Fig.14. One and two compressions lead to a "Debye-Scherrer ring" with marked "poles". The compression has apparently reduced the randomness of the platelet arrangement but a large amount of "polycrystalline" material remains.

Results in other fields like glasses and liquid alloys, liquid crystals, superconductors, magnetic systems, etc. have recently been summarized [14, 44, 54, 82]. Among the current applied studies, the continuing work on Cu-rich precipitates in $Ce(Co, Cu)_5$ compounds (high coercive force) [83] and on silicate glasses containing titania [84] should be mentioned.

In polymer science, neutrons offer particular possibilities for the bulk states as it is possible to dissolve some deuterated (protonated) chains in a protonated (deuterated) matrix and to determine the chain configuration (conformation) from neutron SAS experiments, subtracting the scattering of the pure matrix. The Flory model predicts that a polymer chain in the non-crystalline bulk should have a Gaussian coil shape with the chain segments oriented at random (see Fig.15), as is the case for polymers dissolved in a θ solvent (i.e. in an ideal solution where the second virial coefficient is zero). These predictions have been confirmed completely for several polymers, e.g. polystyrene [85, 86] in the amorphous bulk and polymethyl methacrylate (PMMA) [87]. Figure 16 illustrates the method which is based on the Zimm expansion of the scattering function as a function of concentration leading to

$$\frac{d\sigma}{d\Omega} = cK'|F(Q)|^2 \left(1/M + 2A_2 c + \ldots \right)^{-1} \qquad (11)$$

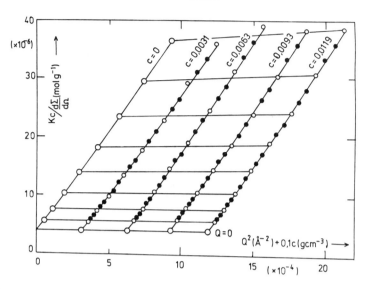

FIG.16. Zimm plot for protonated PMMA dispersed in deuterated PMMA (from Schelten et al. [87]). See text for details.

where c is the concentration (in weight %) of tagged molecules, K' is a constant containing the contrast due to $|b_D - b_H|$, M is the molecular weight, and A_2 is the second virial coefficient. As $|F(Q)|^{-2}$ can be replaced by

$$|F(Q)|^{-2} = 1 + \frac{Q^2 R_G^2}{3} \qquad (12)$$

for $Q^2 R_G^2 \ll 1$, a plot of $cK'(d\sigma/d\Omega)^{-1}$ as a function of Q for different c yields the radius of gyration R_G by extrapolating the slope to $C = 0$, and the second virial coefficient A_2 can be obtained from the slope of the ordinate values extrapolated to $Q = 0$. In Fig.16, this slope is zero. Extrapolation to $Q = 0$ and $c = 0$ must yield — if measurements are done on an absolute scale — the molecular weight of the tagged molecules, which provides an independent check on the assumption of isolated tagged molecules. Complications are frequently encountered in crystalline polymers (see, e.g. King et al. [88] for a discussion).

Interesting applied problems such as conformation in stretched polymers and polymer networks, and phase separation effects in polymer mixtures are currently being studied. Picot et al. [89] have found that the anisotropy of SAS patterns of stretched polymers (from a two-dimensional pattern, R_G can be determined for directions parallel and perpendicular to the stretch direction), measured on a sample containing deuterated polystyrene was much less than expected from an affine deformation of the molecules. This indicates some relaxation on the molecular level, and it is hoped that the rate of relaxation can be studied directly by neutron SAS. More details on these rapidly progressing topics can be found in forthcoming review articles [90, 91].

In the field of biology, there is considerable interest in solving the structure of many large molecules and complex assemblies of molecules as in proteins, viruses, ribosomes, chromosomes, etc. Even if these substances can be obtained in crystalline form, the amount of data in a large-angle diffraction experiment may sometimes be prohibitively large. But in many cases, crystallization has not yet been achieved and the objects must be studied in solution. Neutron SAS offers two useful approaches, and their applications have recently been reviewed by Jacrot [92].

The first method is based on a variation of the contrast between solvent and molecules [93, 94] adopted from light and X-ray scattering and applicable, of course, to the study of polymers as well [95]. Generalizing Eq.(9), we can write for the SAS intensity I(Q), for dilute solutions,

$$I(Q) \propto \left\langle \left| \int_{V_p} (\rho(\vec{r}) - \rho_s) \exp(i\vec{Q}\cdot\vec{r}) d^3\vec{r} \right|^2 \right\rangle \tag{13}$$

where the integration is over the particle volume (or the part inaccessible to the solvent) and the brackets indicate an orientational average for non-spherical particles as we assume orientational disorder. The important difference from Eq.(9) is that scattering length variations in the large particles under discussion are admitted and can be revealed by SAS. One can write

$$\rho(\vec{r}) = \rho_p + \rho_F(\vec{r}) \tag{14}$$

where ρ_p is the average scattering length density of the particle as before and $\rho_F(\vec{r})$ is the local deviation from the average value. As the integral of $\rho_F(\vec{r})$ over the particle volume must be zero, the extrapolation of SAS curves to $Q = 0$ yields (per particle)

$$I(0) = (\rho_p - \rho_s)^2 V_p^2 \tag{15}$$

If ρ_s is varied (by H_2O/D_2O mixtures) around ρ_p, a plot of the type shown in Fig.17 [96] yields the volume V_p of the scattering particle. V_p is the dry volume as it corresponds to the volume inaccessible to the solvent. As the H_2O/D_2O ratio is varied in the solvent, the value of ρ_p can also change because of H-D exchange. In equilibrium, the number of exchanged protons in the particle should be proportional to the D_2O concentration in the solvent, and a straight line will

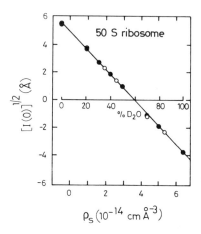

FIG.17. *Square root of SAS intensity extrapolated to Q = 0 as a function of D_2O concentration in H_2O/D_2O for the 50 S subunit of ribosome (RNA + 34 proteins). Based on Eq.(15), a dry volume of 1.5×10^6 Å^3 is obtained (from Stuhrmann et al. [96]).*

still result, but V_p can only be determined if the fraction of exchanged protons is known.

More interesting information can be obtained from the low Q portion of the SAS curve where the Guinier approximation, Eq.(10), is valid. In the presence of scattering length fluctuations $\rho_F(\vec{r})$, the apparent radius of gyration is

$$R_G^2 = R_{GV}^2 + \frac{1}{K_1} \int_{V_p} r^2 \rho_F(\vec{r}) d^3\vec{r} - \frac{1}{K_1^2} \left(\int_{V_p} r\, \rho_F(\vec{r}) d^3\vec{r} \right)^2 \qquad (16)$$

where

$$K_1 = (\rho_p - \rho_s) V_p \qquad (17)$$

and R_{GV} is the usual mechanical radius of gyration of a particle with homogeneous scattering length density. The second term in Eq.(16) is always present if scattering length fluctuations exist. It introduces a linear dependence of R_G on $(\rho_p - \rho_s)^{-1}$. The third term takes into account that the centre of mass of the particle and the centre of scattering length distribution may not be identical. Their separation will be contrast dependent. Stuhrmann and Fuess [97] have been able to observe this term in hen egg-white lysozyme. The results are shown in Fig.18. The separation is small in this case (2 Å for $\rho_p - \rho_s = 10^{10}$ cm^{-2}). It is somewhat easier to determine the separation of two parts of a particle which have

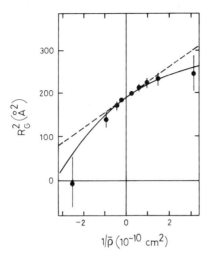

FIG.18. Variation of the square of the apparent radius of gyration as a function of $1/\bar{\rho} = \rho_p - \rho_s$, according to Eq.(16), for hen egg-white lysozyme (from Stuhrman and Fuess [97]).

FIG.19. Triangulation method applied to a complex particle in which a component can be deuterated (○) or protonated. The result of the operation shown in the diagram yields the distance R_{12} (drawing from Jacrot [92]).

drastically different scattering length densities, as demonstrated by Duval et al. [95] for a diblock copolymer of deuterated and protonated polystyrene in solution (D/H-cyclohexane). Careful separation of the different scattering terms over a wide Q range is necessary for testing models for the shape of large particles, e.g. proteins (recently reviewed by Stuhrmann [98]).

For very large assemblies, e.g. the ribosomes mentioned above, selective deuteration, the second important method in neutron SAS, is suitable. In these complex objects, the immediate goal is to understand the mutual arrangement of RNA and proteins. Since it is possible to reassemble the ribosome subunits from their components, it is easy to introduce deuterated proteins at identifyable positions. If one selects two components 1, 2 of the object which are either protonated or deuterated, the triangulation method [99, 100] sketched in Fig.19 yields

$$I_{pair} = I(1_d, 2_d) + I(1_p, 2_p) - I(1_d, 2_p) - I(1_p, 2_d) \tag{18}$$

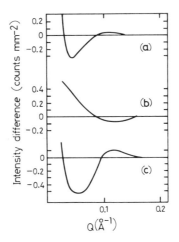

FIG.20. Triangulation experiment on the 30 S ribosomal subunit (RNA and 21 proteins). The intensity difference according to Eq.(18) is shown for three different pairs of proteins:
 (a)   pair S_2-S_5 (distance 105 Å)
 (b)   pair S_5-S_8 (distance 35 Å)
 (c)   pair S_3-S_7 (distance 115 Å)
(from Engelman et al. [101]).

where I is the intensity for the different combinations of protonation and deuteration and I_{pair} represents interference effects between 1 and 2. If both components can be approximated by points (within the experimental resolution),

$$I_{pair} \propto \frac{\sin(QR_{12})}{QR_{12}} \qquad (19)$$

where R_{12} is the distance between 1 and 2.

Figure 20 shows an example from Engelman et al. [101] on the other ribosome subunit, 30S. Much more work on proteins, chromatin, viruses, enzymes, etc. is discussed in the review by Jacrot [92] from which the above summary is drawn and where further references can be found.

Fibrous proteins, e.g. collagen [8, 102], muscles, and biological membranes and membrane components [103–105], represent somewhat more organized scattering systems than solutions of macromolecules, and low-angle diffraction peaks corresponding to the packing arrangement of fibrils and lamellar structures are analysed to determine molecular conformations.

Kinetic studies with neutrons on H/D exchange [106], on the incorporation of water and other species into biological substances and systems [92] are in an exploratory stage, and deserve further attention.

DYNAMICS

So far, we have only covered elastic scattering phenomena. As many practical properties of substances depend on structural features, this aspect of neutron scattering relates more directly to applied problems. Inelastic scattering of neutrons reveals dynamical properties of matter, either on individual scatterers or individual groups, mostly by incoherent scattering, or on collective excitations (phonons and magnons), mostly by coherent scattering.

Apart from the possibility of measuring elastic constants of substances for which other methods fail, by extrapolating phonon dispersion curves to long phonon wave-lengths, scattering by phonons offers very detailed information on dynamical properties of crystalline solids which may be important, e.g. for testing models for phase transformations, especially of the displacive type [107–109]. Much interest is currently devoted to the study of layer-type crystals (see, e.g., Stirling et al. [110]) which may have some practical importance. Phonons in metal-hydrogen (deuterium) systems [111, 112] and in crystals containing radiation-produced defects [113, 114] have been studied and provide some fundamental insight into symmetry and coupling of defects in a lattice.

However, much more practical information can be obtained from incoherent scattering studies related to the motion of individual atoms or molecules. The incoherent scattering cross-section is proportional to the Fourier transform of the self-correlation function $G_s(\vec{r}, t)$ where t is time,

$$\frac{d^2 \sigma_{inc}}{d\Omega dE} \propto S_{inc}(\vec{Q}, \omega) = \frac{1}{2\pi} \int G_s(\vec{r}, t) \exp(i\vec{Q} \cdot \vec{r} - \omega t) d^3\vec{r} \, dt \qquad (20)$$

For simple translational diffusion controlled by a rate equation, the Chudley-Elliott model [115, 116] yields for $S_{inc}(\vec{Q}, \omega)$ a sum of Lorentzians centred at energy transfer $\hbar\omega = 0$. The frequency width of this quasielastic scattering depends on the mean rest time and the site geometry in a solid. For small Q, S_{inc} can be expressed by a single Lorentzian.

$$S_{inc}(Q, \omega) = \frac{\Gamma/\pi}{\omega^2 + \Gamma^2} \exp(-Q^2 \langle u^2 \rangle) \qquad (21)$$

and

$$\Gamma = Q^2 D \qquad (22)$$

where D is the diffusion coefficient.

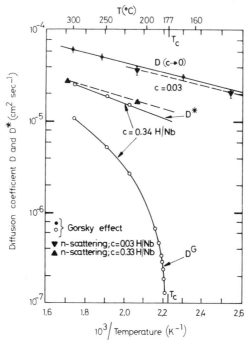

FIG.21. Diffusion coefficient for hydrogen in NbH as a function of reciprocal temperature, for dilute and near critical concentrations. Whereas the Gorski-effect results D^G show critical slowing down because of the thermodynamical factor, the neutron results show the diffusion coefficient in an equilibrium situation. If D^G is "reduced" to the same situation [123], there is good agreement between the neutron results and the reduced value D^* (from Springer and Richter [118]).

For diffusion in the solid state, high-resolution instruments are necessary as $D > 10^{-5}$ cm^2/s is required for standard time-of-flight instruments. With the backscattering technique [117], D values down to 10^{-7} cm^2/s are accessible. A high incoherent cross-section is nevertheless a prerequisite, and hydrogen has so far been studied most extensively [118, 119] although other nuclei have been used too (e.g. Ag in AgI [120], a superionic conductor with Ag as fast diffuser, Na self-diffusion near the melting point [121]).

As an example for hydrogen diffusion in the solid state, Fig.21 shows results for two different hydrogen concentrations in Nb [122] in comparison with results from Gorski-effect measurements [123]. More recent experiments with neutron quasielastic scattering [124] confirmed the change of activation energy near 250°K in NbH$_{0.012}$ which is probably related to a transition from classical jump diffusion to tunnelling motion and was first observed with the Gorski effect [125].

Another study concerns the trapping of hydrogen by nitrogen interstitials [126] which can be rationalized by a two-step random walk model involving trapped and free states for the proton with different life times [127].

First results have been obtained on the diffusion of hydrogen on surfaces of a nickel catalyst. At 150°C, the hydrogen surface diffusion coefficient for a 0.25 monolayer was found to be $\sim 0.8 \times 10^{-7}$ cm^2/s [128]. More experiments are planned. High energy transfer neutron spectroscopy has been used on similar material (Raney nickel) to study the vibrational properties of adsorbed hyrogen [129]. Hydrogen adsorbed on graphite-potassium intercalation compounds is currently being studied with the same techniques [130].

Whereas the above diffusion studies on hydrogen in the crystalline bulk or surface state are tedious and require high resolution, diffusive motion in the liquid state is easily measurable as long as the cross-sections are large enough. Diffusion coefficients of hydrogen in liquid Li have recently been determined by Sköld [131]. This information would be very difficult to obtain directly by other methods. In molecular systems, the dynamics can be decomposed into centre-of-mass motion, i.e. translational diffusion as above, and reorientation around the centre of mass. In addition, there are internal (vibrational) degrees of freedom. The latter can usually be separated quite easily but there is no general way of separating quasi-elastic scattering due to translational and rotational motion. The analysis of quasielastic scattering of liquid polymers, liquid crystals, etc. is therefore often based on simplified models, but well-planned experiments can also help to eliminate particular models. Review articles on polymers [132, 90], liquid crystals [133], and macromolecules [134] give complete accounts of these activities.

CONCLUSION

To review all the applications of neutron scattering in a single article is a formidable task, and the author apologizes for the lack of balance, profoundness, perspective and completeness that will certainly be discovered by the critical reader. It can only be hoped that some of the examples presented here have served their purpose of illustrating the possibilities of neutron scattering in applied research.

ACKNOWLEDGEMENTS

Thanks are due to many colleagues at I.L.L. for clarifying discussions and to Prof. B. Krebs, Bielefeld, for Fig.4.

REFERENCES

[1] SHULL, C.G., Proc. Conf. Neutron Scattering, Gatlinburg, Tenn., Vol.1 (MOON, R.M., Ed.), US Dept. of Commerce, Springfield, Va., CONF-760601-P1 (1976) 1.
[2] TURCHIN, V.F., Slow Neutrons, Israel Program for Scientific Translations, Jerusalem (1965).
[3] GUREVICH, I.I., TARASOV, L.V., Low Energy Neutron Physics, North-Holland, Amsterdam (1968).
[4] IZYUMOV, V.A., OZEROV, R.P., Magnetic Neutron Diffraction, Plenum Press, New York (1970).
[5] MARSHALL, W., LOVESEY, S.W., Theory of Thermal Neutron Scattering, Clarendon Press, Oxford (1971).
[6] BACON, G.E., Neutron Diffraction, Oxford University Press, Oxford (1975).
[7] MÖSSBAUER, R.L., Europhysics News 5 6 (1974) 1.
[8] WHITE, J.W., Physikalische Blätter 32 (1976) 633.
[9] AGERON, P., Private communication.
[10] CARPENTER, J., Draft Proposal for an Intense Pulsed Neutron Source, Argonne Nat. Lab. (1976).
[11] SCHMATZ, W., SPRINGER, T., SCHELTEN, J., IBEL, K., J. Appl. Crystallogr. 7 (1974) 96.
[12] LAROSE, A., VANDERWAL, J., Scattering of Thermal Neutrons — A Bibliography (1932—1974), Solid State Physics Literature Guide, Vol.7, IFI/Plenum, New York (1974).
[13] HENKEL, G., MASON, S.A., KREBS, B., Acta Crystallogr. (to be published).
[14] KOSTORZ, G., Atomkernenergie 28 (1976) 61.
[15] BROWN, P.J., Neutron Scattering in Materials Science (KOSTORZ, G., Ed., HERMAN, H., Series Ed.), Academic Press, New York (to be published).
[16] RAYNAL, J.M., ROTH, M., J. Appl. Crystallogr. 8 (1975) 535.
[17] DAUGER, A., BOUDILI, E.K., ROTH, M., Scr. Metall. 10 (1976) 1119.
[18] COWLAM, N.E., BACON, G.E., KIRKWOOD, D.H., Scr. Metall. 9 (1975) 1363.
[19] ALDEBERT, P., BADIE, J.M., TRAVERSE, J.P., BUEVOZ, J.L., ROULT, G., Rev. Int. Hautes Temp. Refract. 12 (1975) 307.
[20] BLOCH, D., PAUREAU, J., VOIRON, J., PARISOT, G., Rev. Sci. Instrum. 47 (1976) 296.
[21] SZPUNAR, J., At. Energy Rev. 14 (1976) 199.
[22] KJEMS, J.K., PASSELL, L., TAUB, H., DASH, J.G., NOVACO, A.D., Phys. Rev., B 13 (1976) 1446.
[23] RIEKEL, C., SCHÖLLHORN, R., Mater. Res. Bull. 11 (1976) 369.
[24] THOMAS, R.K., TREWERN, T., WHITE, J.W. (to be published).
[25] KRIVOGLAZ, M.A., The Theory of X-Ray and Thermal Neutron Scattering by Real Crystals (English translation), (MOSS, S.C., Ed.), Plenum Press, New York (1969).
[26] SCHMATZ, W., A Treatise on Materials Science and Technology, Vol.3 (HERMAN, H., Ed.), Academic Press, New York (1973) 105.
[27] BAUER, G., SEITZ, E., JUST, W., J. Appl. Crystallogr. 8 (1975) 162.
[28] MOZER, B., KEATING, D.T., MOSS, S.C., Phys. Rev. 175 (1968) 868.
[29] VRIJEN, J., VAN DIJK, C., Reactor Centrum Nederland Report RCN-75-054, Petten (1975).
[30] POERSCHKE, R., Private communication.
[31] BAUER, G., SEITZ, E., HORNER, H., SCHMATZ, W., Solid State Commun. 17 (1975) 161.
[32] BAUER, G., SCHMATZ, W., JUST, W., L'hydrogène dans les métaux, Int. Congr. Paris 1977 (to be published).

[33] DE NOVION, C., JUST, W., Annex to the Annual Report 1976, Institut Laue-Langevin, Grenoble (1977) 288.
[34] RHYNE, J.J., PICKART, S.J., ALPERIN, H.A., Phys. Rev. Lett. **29** (1972) 1562.
[35] BLETRY, J., SADOC, J.F., Phys. Rev. Lett. **33** (1974) 172.
[36] BLETRY, J., SADOC, J.F., J. Phys., F (London) **5** (1975) L110.
[37] BHATIA, A.B., THORNTON, D.E., Phys. Rev., B **2** (1970) 3004.
[38] RUPPERSBERG, H., Phys. Lett. **54A** (1975) 151.
[39] RUPPERSBERG, H., KNOLL, W., CHIEUX, P., Annex to the Annual Report 1976, Institut Laue-Langevin, Grenoble (1977) 219.
[40] LAMPERTER, P., STEEB, S., KNOLL, W., Z. Naturforsch. A **31** (1976) 90.
[41] RUPPERSBERG, H., KNOLL, W., CHIEUX, P., Annex to the Annual Report 1976, Institut Laue-Langevin, Grenoble (1977) 217.
[42] PAGE, D.I., RAINFORD, B.D., HOWELLS, W.S., ibid., p. 227.
[43] STEEB, S., KNOLL, W., ibid., p.219.
[44] SCHMATZ, W., Proc. Conf. Neutron Scattering, Gatlinburg, Tenn., Vol. 2 (MOON, R.M., Ed.), US Dept. of Commerce, Springfield, Va., CONF-760601-P2 (1976) 1037.
[45] IBEL, K., J. Appl. Crystallogr. **9** (1976) 296.
[46] GALOTTO, C.P., PIZZI, P., WALTHER, H., Nucl. Instrum. Methods **134** (1976) 369.
[47] GUINIER, A., FOURNET, G., Small-Angle Scattering of X-Rays, John Wiley, New York (1955).
[48] BEEMAN, W.H., KAESBERG, P., ANDEREGG, J.W., WEBB, M.B., Handbuch der Physik, Vol. 32 (FLÜGGE, S., Ed.), Springer, Berlin (1957) 321.
[49] GUINIER, A., X-Ray Diffraction, Freeman, San Francisco (1963).
[50] RAYNAL, J.M., SCHELTEN, J., SCHMATZ, W., J. Appl. Crystallogr. **4** (1971) 511.
[51] KOSTORZ, G., Atomic Structure and Mechanical Properties of Metals (CAGLIOTI, G., Ed.), North-Holland, Amsterdam (1976) 571.
[52] ALLEN, D., EPPERSON, J.E., GEROLD, V., KOSTORZ, G., MESSOLORAS, S., STEWART, R.J., Proc. Conf. Neutron Scattering, Gatlinburg, Tenn., Vol. 1 (MOON, R.M., Ed.), US Dept. of Commerce, Springfield, Va. CONF-760601-P1 (1976) 102.
[53] LASLAZ, G., KOSTORZ, G., ROTH, M., GUYOT, P., STEWART, R.J. (to be published).
[54] KOSTORZ, G., At. Energy Rev. (to be published).
[55] SCHWAHN, D., Dr. rer. nat. Dissertation, Ruhr-Universität, Bochum (1976).
[56] CAHN, J.W., Trans. AIME **242** (1968) 166.
[57] ALAIN, J., NAUDON, A., DELAFOND, J., JUNQUA, A., MIMAULT, J., Scr. Metall. **8** (1974) 831.
[58] NAUDON, A., ALAIN, J., DELAFOND, J., JUNQUA, A., MIMAULT, J., Scr. Metall. **8** (1974) 1105.
[59] JUNQUA, A., MIMAULT, J., DELAFOND, J., Acta. Metallogr. **24** (1976) 779.
[60] LASLAZ, G., GUYOT, P., Acta Metallogr. **25** (1977) 277.
[61] MESSOLORAS, S., Ph.D. Thesis, Reading (1974).
[62] GEROLD, V., Small Angle X-Ray Scattering (BRUMBERGER, H., Ed.), Gordon and Breach, New York (1967) 277.
[63] GEROLD, V., J. Appl. Cryst. **10** (1977) 25.
[64] GEROLD, V., EPPERSON, J.E., KOSTORZ, G., J. Appl. Crystallogr. **10** (1977) 28.
[65] RAYNAL, J.M., ROTH, M., J. Appl. Crystallogr. **8** (1975) 535.
[66] KOSTORZ, G., Z. Metallkde. **67** (1976) 704.
[67] PIZZI, P., WALTHER, H., J. Appl. Crystallogr. **7** (1974) 230.
[68] PIZZI, P., WALTHER, H., CORTESE, P., Eight World Conference on Nondestructive Testing, CECA, Brussels (1976) 3L7.

[69] SCHLENKER, M., BARUCHEL, J., PETROFF, J.F., YELON, W.B., Appl. Phys. Lett. 25 (1974) 382.
[70] MARMEGGI, J.C., HOHLWEIN, D., BERTAUT, E.F., (submitted to Phys. Status Solidi).
[71] SCHMATZ, W., Riv. Nuovo Cim. 5 (1975) 398.
[72] SCHEUER, H. (to be published).
[73] GÖLTZ, G. (to be published).
[74] MOOK, H.A., J. Appl. Phys. 45 (1974) 43.
[75] HENDRICKS, R.W., SCHELTEN, J., SCHMATZ, W., Philos. Mag. 30 (1974) 819.
[76] EPPERSON, J.E., KOSTORZ, G., RUANO, C. (to be published).
[77] GUPTA, S., Ph.D. Thesis, University of Reading (1976).
[78] MARTIN, D.G., CAISLEY, J., Some Studies of the Effect of Irradiation on the Neutron Small Angle Scattering from Graphite, UK Atomic Energy Authority, AERA-R8515, Harwell (1976).
[79] HOFMEYR, C., ISEBECK, K., MAYER, R.M., J. Appl. Crystallogr. 8 (195) 193.
[80] WEITKAMP, C., Technological and Industrial Application of Neutrons, GKSS 76/E/41, Geesthacht (1976).
[81] THOMAS, R.K., CEBULA, D., OTTEWIL, R.H., WHITE, J.W., ZACCAI, G., Annex to the Annual Report 1976, Institut Laue-Langevin, Grenoble (1977) 190.
[82] SCHELTEN, J., SCHMATZ, W., At. Energy Rev. (to be published).
[83] LEMAIRE, R., LABULLE, B., LAFORET, J., PETIPAS, C., ROTH, M., Annex to the Annual Report 1976, Institut Laue-Langevin, Grenoble (1977) 245.
[84] FENDER, B.E.F., KOSTORZ, G., TALBOT, J., WRIGHT, A. (to be published).
[85] SCHELTEN, J., WIGNALL, G.D., BALLARD, D.G.H., SCHMATZ, W., Colloid Polym. Sci. 252 (1974) 749.
[86] COTTON, J.P., DECKER, D., BENOIT, H., FARNOUX, B., HIGGINS, J., JANNINK, G., OBER, R., PICOT, C., DESCLOISEAUX, J., Macromolecules 7 (1974) 863.
[87] SCHELTEN, J., KRUSE, W.A., KIRSTE, R.G., Kolloid-Z. Z. Polym. 251 (1973) 919.
[88] KING, J.S., BAI, S.J., HIGGINS, J.S., SUMMERFIELD, G.G., TING, W.M., ULLMAN, R., J. Polym. Sci. (to be published).
[89] PICOT, C., DUPLESSIX, R., DECKER, D., BENOIT, H., COTTON, J.P., FARNOUX, B., JANNINK, G., OBER, R., Macromolecules 10 (1977) in press.
[90] HIGGINS, J.S., Neutron Scattering in Materials Science (KOSTORZ, G., Ed., HERMAN, H., Series Ed.), Academic Press, New York (to be published).
[91] HIGGINS, J.S., STEIN, R.S., Proc. Fourth International Conference on X-Ray and Neutron Small-Angle Scattering, Gatlinburg, 1977 (to be published).
[92] JACROT, B., Rep. Prog. Phys. 39 (1976) 911.
[93] STUHRMANN, H.B., J. Appl. Crystallogr. 7 (1974) 173.
[94] IBEL, K., STUHRMANN, H.B., J. Molec. Biol. 93 (1975) 255.
[95] DUVAL, M., DUPLESSIX, R., PICOT, C., DECKER, D., REMPP, P., BENOIT, H., COTTON, J.P., OBER, R., JANNINK, G., FARNOUX, B., J. Polym. Sci., Polym. Lett. Ed. 14 (1976) 585.
[96] STUHRMANN, H.B., HAAS, J., IBEL, K., DE WOLF, B., KOCH, M.H.J., PARFAIT, R., CHRICHTON, R.R., Proc. Natl. Acad. Sci. USA 73 (1976) 2379.
[97] STUHRMANN, H.B., FUESS, H., Acta Crystallogr. Sect. A 32 (1976) 67.
[98] STUHRMANN, H.B., Neutron Scattering for the Analysis of Biological Structures, Brookhaven Symposia in Biology, No.27, US Dept. of Commerce, Springfield, Va. (1976) IV-3.
[99] ENGELMAN, D.M., MOORE, P.B., Proc. Natl. Acad. Sci. USA 69 (1972) 1997.

[100] HOPPE, W., Israel J. Chem. **10** (1972) 321.
[101] ENGELMAN, D.M., MOORE, P.B., SCHOENBORN, B.P., Proc. Natl. Acad. Sci. USA **72** (1975) 3888.
[102] WHITE, J.W., MILLER, A., IBEL, K., J. Chem. Soc. (London), Faraday Trans., II **72** (1976) 435.
[103] WORCESTER, D.L., Neutron Scattering for the Analysis of Biological Structures, Brookhaven Symposia in Biology, No.27, US Dept. of Commerce, Springfield, Va. (1976) III-37.
[104] WORCESTER, D.L., Biological Membranes, Vol.3 (CHAPMAN, D., WALLACH, D.F.H., Eds), Academic Press, London (1976) 1.
[105] SCHOENBORN, B.P., Biochim. Biophys. Acta **457** (1976) 41.
[106] HAAS, J., Dissertation, Universität Mainz (1976).
[107] FLEURY, P.A., Ann. Rev. Mater. Sci. **6** (1976) 157.
[108] AXE, J.D., KEATING, D.T., MOSS, S.C., Phys. Rev. Lett. **35** (1975) 530.
[109] MORI, M., YAMADA, Y., SHIRANE, G., Solid State Commun. **7** (1975) 127.
[110] STIRLING, W.G., DORNER, B., CHEEKE, J.D.N., REVELLI, J., Solid State Commun. **18** (1976) 931.
[111] ROWE, J.M., RUSH, J.J., SMITH, H.G., MOSTOLLER, M., FLOTOW, H.E., Phys. Rev. Lett. **33** (1974) 1297.
[112] ALEFELD, G., STUMP, N., MAGERL, A., J. Phys., C (to be published).
[113] NICKLOW, R.M., COLTMAN, R.R., YOUNG, F.W., Jr., WOOD, R.F., Phys. Rev. Lett. **35** (1975) 1444.
[114] BÖNING, K., BAUER, G.S., FENZL, H.J., SCHERM, R., KAISER, W., Phys. Rev. Lett. **38** (1977) 852.
[115] CHUDLEY, C.T., ELLIOTT, R.J., Proc. Phys. Soc. **77** (1961) 353.
[116] SPRINGER, T., Springer Tracts in Modern Physics **64** (1972) 50.
[117] BIRR, M., HEIDEMANN, A., ALEFELD, B., Nucl. Instrum. Methods **95** (1971) 435.
[118] SPRINGER, T., RICHTER, D., At. Energy Rev. (to be published).
[119] MUELLER, M., SKOELD, K., Neutron Scattering in Materials Science (KOSTORZ, G., Ed., HERMAN, H., Series Ed.), Academic Press, New York (to be published).
[120] ECKOLD, G., FUNKE, K., KALUS, J., LECHNER, R.E., Phys. Lett., A **55** (1975) 125: J. Phys. Chem. Solids **37** (1976) 1097.
[121] GOELTZ, G., HEIDEMANN, A., MEHRER, H., SEEGER, A., Annex to the Annual Report 1976, Institut Laue-Langevin, Grenoble (1977) 295.
[122] GISSLER, W., ALEFELD, G., SPRINGER, T., J. Phys. Chem. Solids **31** (1970) 2361.
[123] VOELKL, J., Ber. Bunsenges. Phys. Chem. **76** (1972) 797.
[124] RICHTER, D., ALEFELD, B., HEIDEMANN, A., WAKABAYASHI, N., J. Phys., F **7** (to be published).
[125] WIPE, H., ALEFELD, G., Phys. Status Solidi A **23** (1974) 175.
[126] RICHTER, D., TOEPLER, J., SPRINGER, T., J. Phys., F **6** (1976) L93.
[127] RICHTER, D., KEHR, K.W., SPRINGER, T., Proc. Conf. on Neutron Scattering, Gatlinburg, Tenn., Vol. 1 (MOON, R.M., Ed.), US Dept. of Commerce, Springfield, Va., CONF-760 601-P1 (1976) 568.
[128] FOUILLOUX, P., RENOUPREZ, A.J., STOCKMEYER, P., Ber. Bunsenges. Phys. Chem. **81** (to be published).
[129] RENOUPREZ, A.J., FOUILLOUX, P., COUDURIER, G., TOCCHETTI, D., STOCKMEYER, R., J. Chem. Soc. (London), Faraday Trans., I **73** (1977) 1.
[130] WHITE, J.W., THOMAS, R.K., TREWERN, T.D., MARLOW, I., LESLIE, M., Annex to the Annual Report 1976, Institut Laue-Langevin, Grenoble (1977) 359.

[131] SKOELD, K. (to be published).
[132] ALLEN, G., HIGGINS, J.S., Rep. Prog. Phys. **36** (1973) 1073.
[133] VOLINO, F., DIANOUX, A.J., Proc. EUCHEM Conference: Organic Liquids, Dynamics and Chemical Properties, Schloss Elmau (1976).
[134] WHITE, J.W., Proc. R. Soc. (London), Ser. A **345** (1975) 119.

DEVELOPMENTS IN ION IMPLANTATION

D.W. PALMER
School of Mathematical and Physical Sciences,
University of Sussex,
Brighton, United Kingdom

Abstract

DEVELOPMENTS IN ION IMPLANTATION.
  Ion implantation is the deposition of impurity atoms into solids for the purpose of changing or studying the properties of the solids, this deposition being accomplished by bombardment with accelerated ion beams. The paper outlines the physical phenomena that have been the subjects of experimental and theoretical research in investigating the interaction of energetic ions with solids, and describes and discusses the important and recent applications in respect of the ion implantation of metals and alloys, of semiconductors and of insulating materials. The continuing expansion and progress in the applications of ion implantation owe much to the very close connections, as emphasized in the paper, between the many basic and applied aspects.

1. INTRODUCTION

"Ion implantation" is the term used to denote the irradiation of solid materials with fast atoms and molecules (of energies $\sim 10^3 - 10^7$ eV) for the purpose of changing or investigating the properties of the solids. The word "ion" in the name refers to the fact that the acceleration of the atoms or molecules is most conveniently accomplished by electric fields; the species to be implanted is thus usually formed into a beam in an ion accelerator comprising an ion source and the electric or electromagnetic acceleration and focussing systems. Equipment of the Cockroft-Walton, transformer or Van de Graaff types involving a high potential for the acceleration, or of the cyclotron type, have all been employed successfully to produce ion beams for implantation purposes. During the last ten years ion implantation has developed as an important field of academic and applied science, and its development can be considered as being one of continual interplay and innovation between basic research and an increasing variety of applications. It is this continuing connection, the variety of investigative techniques from solid-state, atomic and nuclear physics, chemistry, electronics and materials science, and the wide range of technological aspects that give ion implantation its great interest and value.
  The aim of the paper is to review recent developments in ion implantation and to emphasize the very close connections, or indeed the absence of distinction

in many cases, between the basic research and the applied aspects. The paper considers the experimental and theoretical investigations concerning the interactions of energetic ions with solids, and describes and discusses the important applications in respect of the ion implantation of metals and alloys, of semiconductors and of insulating materials.

2. GENERAL CONSIDERATIONS AND BACKGROUND

Many papers concerned with all aspects of ion implantation can be found in the proceedings of two series of international conferences. The emphasis in the conference series "Atomic Collisions in Solids" (of which the four most recent, [1–4], were held in 1969–1975) is upon all aspects of the dynamic behaviour (kinetic energy $E \gg kT$) of the ion as it traverses the solid (of temperature T), and includes consideration of the processes by which the ion loses energy, the atomic and electronic excitations in the solid, and the formation of radiation damage defects. The second conference series, [5–13], deals especially with the effects due to the presence and behaviour of the implanted atoms and the radiation damage defects after the ion has effectively come to rest ($E \cong kT$) in the solid; this is the realm of solid-state physics and chemistry, of materials science and of modern physical electronics. The dynamic, collision processes are by their nature (kinetic energy of ion $\gg$ atomic binding energy in the solid) somewhat insensitive to the detailed structure of any particular solid, and it has been found possible (see Sections 3 and 4) to formulate general theories of the processes which give reasonable or good agreement with experimental results in most cases. On the contrary, the non-dynamic ion-implantation effects depend strongly on the natures of the implanted atom and of the solid, and no completely general rules can be applicable; the experimental studies of the ion implantation effects of many different ions in different elements and compounds have in fact demonstrated how relatively meagre is our broad understanding of the properties of impurities in solids and of their interactions with lattice defects.

In addition to the two conference series [1–13] mentioned in the previous paragraph, various books [14–22] and review articles [23–26] have covered many aspects of ion implantation. A previous review article [26] by the present writer discussed ion implantation from a practical point of view, and described the equipment and techniques needed for developing experimental work in this field of academic and applied study. Two fairly recent bibliographies [27, 28] have given extensive lists of references to ion implantation work.

3. RATES OF ENERGY LOSS AND THE SPATIAL DISTRIBUTIONS OF IMPLANTED ATOMS

3.1. Distinction between energy loss rates due to electronic excitation and nuclear-nuclear collisions

The processes by which energetic ions lose their energy in traversing solids have been considered in various celebrated papers over many years. A recent review by Dearnaley [29] contains a concise but very clear description of all important aspects of this topic. The various treatments consider that when an energetic ion enters a solid it interacts with and can transfer energy to the electrons in the solid, and can also lose energy to the atoms by ion-atom forces arising from the screened-coulomb interaction potential between the nuclei of the ion and each of the target atoms. Lindhard [30] suggested that these two kinds of process, "electronic" and "nuclear", could be considered as independent contributions to the total rate of energy loss $(-dE/dR)_{total}$ with path distance R through the solids, and thus that

$$\left(\frac{-dE}{dR}\right)_{total} = \left(\frac{-dE}{dR}\right)_e + \left(\frac{-dE}{dR}\right)_n \tag{1}$$

The justification for this assumption is that the internuclear interactions are on average more violent but rarer collision events [31] than the ion-electron collisions and that the internuclear electronic screening is affected by excitation of only those electrons which are in orbitals at distances from each nucleus less than the internuclear separation, and not by the excitation of electrons at greater radii such as the valence electrons etc. (Excitation of non-inner shell electrons is almost always the major component of the electronic contribution to the total rate of energy loss.) It has been suggested by Thompson and Neilson [32] that under certain conditions, where there is a large probability for resonant inner-electron-shell excitation of target atoms, the screened-coulomb internuclear potential may be so significantly changed that the assumption of independent electronic and nuclear-collision energy loss rates may not be valid. However, recent experimental data [33–37], giving the projected ranges of various heavy ions in silicon, aluminium and aluminium oxide at ion energies where the dominant contribution to the energy loss rate is that of the screened-coulomb internuclear collisions, show smooth changes of projected range with ion energy and only small deviations from a common range-energy curve, in reduced coordinates ρ_p and ϵ (see Section 3.2), and give no evidence for the breakdown of the assumption of the separability of the electronic and nuclear energy loss contributions.

3.2. The LSS theory of ion energy-losses and ranges

3.2.1. Rates of energy loss in the LSS theory

The term "ion implantation" contains the word "ion" because in practice it is most convenient of course to accelerate atoms when they are electrically charged. Furthermore, the arrival of electric charge on the target solid is a straightforward way of measuring the dose of implanted atoms. In fact whether the species to be implanted arrives at the surface of the solid in a charged or neutral state is likely usually to have very little effect upon its behaviour in the solid or upon its effect on the solid. This is because the species will, in a time of $\sim 10^{-16}$ s (i.e. an electron orbital time), tend to take up an equilibrium charge state dependent mainly on its velocity but to some extent also on the electrons in the solid. Thus, for the ion energies usually used in implantation work the equilibrium charge state will be attained in a distance (e.g. of a few Å or a few tens of Å) small compared with the total range of the implanting atom. Ions moving at low velocities through solids will tend to capture electrons and become neutral atoms, and Bohr and Lindhard [38] in 1954 suggested that for atoms of atomic number Z_1 moving through a solid the critical velocity in respect of the charge-state could be considered as about $Z_1^{2/3} v_0$ (where v_0 is the Bohr velocity $c/137$), since this is the approximate value of the mean orbital electron velocity in the Thomas-Fermi statistical model of an atom [39a].

With this in mind and using the procedure of treating separately the electronic and nuclear energy loss contributions, Lindhard, Scharff and Schiøtt [39] in 1963 proposed a comprehensive theory ("LSS") for the energy loss rates of fast ions incident upon amorphous solids, valid for ion velocities up to about $Z_1^{2/3} v_0$. This upper validity limit corresponds to the velocities of 2.5-MeV B^+ ions, 15-MeV Na^+ ions and 200-MeV Kr^+ ions [29] and the regime of expected validity therefore includes the majority of uses and applications of ion implantation to the present date. It is certainly no exaggeration to say that the very considerable development of the subject and applications of ion implantation in the period 1967–77 would not have been possible without the availability of the LSS theory of 1963. Although (or even, perhaps, because) subsequent experimental results have shown some inaccuracies in the theory in respect of both the electronic and nuclear collision contributions, the existence of the theory has provided an extremely valuable stimulus to experimental and theoretical work.

The electronic contribution to the energy loss rate in the LSS theory is based on the work of Lindhard and Scharff [40] of 1961. Here, as also for considering the nuclear energy-loss contribution, the ion-target atom potential $V(r)$ is assumed to be screened-coulomb of the Thomas-Fermi (TF) type:

$$V(r) = \frac{Z_1 Z_2 e^2}{r} \phi_0 \left(\frac{r}{a_{TF}} \right) \qquad (2a)$$

where Z_1 and Z_2 are the atomic numbers of the projectile ion and target atoms respectively, a_{TF} is the Thomas-Fermi screening distance given by

$$a_{TF} = 0.885\, a_0/(Z_1^{2/3} + Z_2^{2/3})^{1/2} \tag{2b}$$

(where a_0 is the Bohr radius equal to 0.529 Å), and $\phi_0\,(r/a_{TF})$ is the Thomas-Fermi screening function (tabulated values are given in [39a]). The theory [40] treats the electrons as an electron gas of density and distribution calculated from the TF potential and gives the value of $(-dE/dR)_e$ as proportional to the ion velocity v for the condition $v < Z_1^{2/3} v_0$. The result is

$$\left(\frac{-dE}{dR}\right)_e \cong \frac{8 Z_1^{1/6}}{0.885} \cdot \left(\frac{Z_1 Z_2 e^2}{a_{TF}}\right) \pi\, a_{TF}^2\, N \cdot \left(\frac{v}{v_0 (Z_1^{2/3} + Z_2^{2/3})}\right) \tag{3}$$

where N is the number of atoms per unit volume in the solid. In terms of normalized energy and distance parameters ϵ and ρ, respectively, given [40, 41] by

$$\epsilon \equiv E \cdot \left(\frac{M_2}{M_1 + M_2}\right) \Big/ \left(\frac{Z_1 Z_2 e^2}{a_{TF}}\right) = \frac{32.5\, A_2\, E\, (\text{keV})}{(A_1 + A_2) \cdot Z_1 Z_2 \cdot (Z_1^{2/3} + Z_2^{2/3})^{1/2}}$$

and

$$\rho \equiv R \cdot \left(\frac{M_1 M_2}{(M_1 + M_2)^2}\right) \cdot 4\pi\, a_{TF}^2\, N = \frac{166.8\, A_1\, R(\mu g/cm^2)}{(A_1 + A_2)^2 \cdot (Z_1^{2/3} + Z_2^{2/3})} \tag{4}$$

where E is the energy of the projectile, and M_1 (and A_1) and M_2 (and A_2) are the masses (and atomic mass numbers) of projectile and target atom, respectively, the electronic energy loss rate can be expressed as

$$\left(\frac{-d\epsilon}{d\rho}\right)_e = k\, \epsilon^{1/2} \tag{5}$$

where $k \cong 3.396\, m_e^{1/2} \cdot \frac{Z_1^{2/3}\, Z_2^{1/2}}{(Z_1^{2/3} + Z_2^{2/3})^{3/4}} \cdot \frac{(M_1 + M_2)^{3/2}}{M_1^{3/2}\, M_2^{1/2}} \tag{6}$

where m_e is the electron mass. If the masses M_1 and M_2 in this expression are taken to be in atomic mass units then the factor $3.396\, m_e^{1/2}$ becomes equal to 0.0795. The value of k falls between 0.1 and 1.5 for most ion solid combinations.

Values of $(d\epsilon/d\rho)_e$ as a function of $\epsilon^{1/2}$ are shown in Fig. 1 for $k = 0.15$ and 1.5.

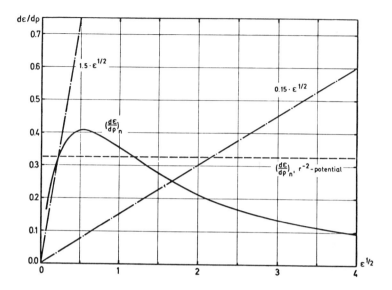

FIG.1. The rate of ion energy loss $(d\epsilon/d\rho)_n$ by elastic internuclear collisions (solid line) and two possible forms, $0.15\,\epsilon^{\frac{1}{2}}$ and $1.5\,\epsilon^{\frac{1}{2}}$, of the rate of ion energy loss by electronic excitation effects, according to the LSS theory [39]. The parameters ϵ and ρ are the reduced energy and distance values of expression (4). An energy-independent $(d\epsilon/d\rho)_n$ would be obtained for an r^{-2} internuclear potential (broken horizontal line). From Schiøtt [45].

The calculation of the nuclear collision contribution to the rate of energy loss in the LSS theory [39] is again based on the assumption of a Thomas-Fermi interaction potential, and uses the basic formula

$$\left(\frac{-dE}{dR}\right)_n = N \int_0^{T_m} T\left(\frac{d\sigma_n}{dT}\right) \cdot dT \qquad (7)$$

where $d\sigma_n/dT$ is the differential collision cross-section with respect to energy transfer T to a target atom and T_m is the maximum energy transferable (i.e. in a head-on collision). It turns out that a single universal curve, proposed to be valid for all ion-solid combinations, can be found giving the variation of $(d\epsilon/d\rho)_n$ with ϵ. This universal curve is shown in Fig. 1.

It is seen from Fig. 1 that the nuclear stopping contribution dominates at low energies and the electronic contribution at high energies. The cross-over energy is, for example, at 17 keV for B^+ ions in silicon and at 140 keV for P^+ ions in silicon [16]. At much higher energies ($v \gtrsim Z_1^{2/3} v_0$) the method of Lindhard and Scharff [40] for the electronic contribution is not applicable, and for $v \gg Z_1^{2/3} v_0$ the ion travels through the solid as a bare nucleus. For the latter condition the Bethe-Bloch expression [42, 43] for $(dE/dR)_e$ becomes valid; this causes $(dE/dR)_e$ to decrease with increasing ion energy.

3.2.2. Ranges and spatial distributions of implanted atoms

From the point of view of applications, the primary concern is the depth distribution of the implanted atoms when they have come to rest in the solid. This need has stimulated the development of calculational methods to provide theoretical depth distributions from the LSS energy loss data. The total path length R_{tot} of the ion in the solid can be found from the expression

$$R_{tot}(E) = \int_0^E \frac{dE}{(-dE/dR)_e + (-dE/dR)_n} \qquad (8)$$

Lindhard et al. [39] showed universal $\rho(\epsilon)$ range-energy curves obtained for different values of the electronic k parameter of expression (6); from these curves theoretical $R_{tot}(E)$ can be found by using expression (4). However, practical ion implantation requires knowledge of the projected range R_p parallel to the incidence direction and the perpendicular or lateral range $R_\perp$. Because the ion trajectory in the solid is zig-zag rather than straight, $R_p \neq R_{tot}$ and $R_\perp \neq 0$. Estimates of the ratio R_p/R_{tot} [16, 39, 41, 44] allow approximate values of R_p to be obtained. By using standard statistical formulae, Lindhard et al. [39] also estimated the spatial longitudinal spread of the implanted atoms by assuming in a first approximation that the distribution was of gaussian form defined by the mean range and the standard deviation (the "straggling") of the range distribution, these then being converted to estimates of the corresponding projected range parameters, $\langle R_p \rangle$ and $\langle \Delta R_p \rangle$.

However, in response to the need especially in semiconductor technology for greater accuracy in the knowledge of the spatial distribution of implanted atoms, more recent theoretical work has aimed to calculate directly the distribution with respect to the projected range coordinate. A convenient and effective method [45–49] is to calculate successive moments of the projected distance distributions (respectively, the mean range, straggling, skewness, kurtosis etc). The distributions themselves can be constructed from these moments by use of mathematical expressions of the Gram-Charlier or Edgeworth forms [50], or, if the skewness is not large, of the Mylroie-Gibbons (joined half-gaussian) form [51].

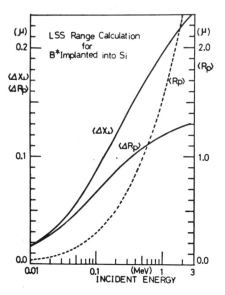

FIG.2. The results of an LSS calculation of the mean projected range $\langle R_p \rangle$, the standard deviation $\langle \Delta R_p \rangle$ of the projected range and the lateral spreading standard deviation $\langle \Delta X_1 \rangle$ for $^{11}B^+$ implanted into silicon. From Furukawa and Matsumura [54].

Tables of calculated values of the first, second and third projected distance moments for 20–1000 keV implantation of boron, phosphorus, arsenic and antimony ions into silicon have been given by Mylroie and Gibbons [51], and similar calculated data for many ions of energies 10–1000 keV in various solids have been published by Gibbons et al. [52]. Such tables are of great value in practical applications of ion implantation. It should however be emphasized that it has been found (see Section 3.2.3) that the use of Thomas-Fermi interaction potentials and of the associated electron density distributions gives energy loss and range data which differ to some extent from experimental results, and thus that the calculated TF data should be treated with some caution. This matter is considered briefly in Section 3.2.3. The projected distance tables of Mylroie and Gibbons [51] and of Gibbons et al. [52] mentioned above in fact make use of experimentally based electronic-energy-loss rates, where possible, instead of LSS values. It is also to be emphasized that the methods of calculation of these range statistics assume that there is no crystallographic atomic structure in the target solids (i.e. that the target is amorphous).

The same general method outlined in the previous paragraphs can be applied to find theoretical values for the extent of lateral spreading of implanted ions, and this has been done for example by Furukawa et al. [53, 54] for several ions in silicon. Figure 2 shows the results of their calculations for boron ions incident upon silicon. It can be seen that the lateral spreading is expected to be larger than the projected range straggling and to be a significant fraction of the mean projected range. In respect of the use of ion implantation in semiconductor technology for the fabrication of auto-registered FET devices (see Section 7.2), lateral spreading of the ions to regions inside the geometrical shadow of the mask can have deleterious effects on device characteristics, and lateral spreading as shown in Fig. 2 certainly needs to be taken into account. The tables of Gibbons et al. [52] also give theoretical values for the lateral standard deviations for all the ion-solid combinations considered. Comparison of the Furukawa et al. [54] and Gibbons et al. [52] data, where this is possible, shows that the two sets of calculations give very similar mean projected range values, fairly similar values for the standard deviation of the projected range, but rather different values for the lateral standard deviations.

3.2.3. Validity of the LSS energy-loss calculation

The LSS theory deals with energy losses by both electronic processes and screened nuclear collisions. For each of these kinds of energy loss the LSS theory predicts smooth changes of the rates of energy loss with respect to the atomic numbers or masses of the implanting ion and the target solid. For the energy region where the electronic energy loss rate $(-dE/dR_p)_e$ is dominant its dependence on Z_1 (the atomic number of the projectile) at constant Z_2 (i.e. in a certain element as target) is usually considered, in the context of the LSS theory, at a particular projectile velocity. Hvelplund and Fastrup [55] found that for various ions of the same velocity traversing amorphous carbon the electronic energy loss rate was an oscillatory function of Z_1, and that the smooth theoretical curve of the Lindhard-Scharff theory [39, 40] was a good mean of the observed oscillatory variation. Cheshire et al. [56] subsequently showed that the use of Hartree-Fock-Slater (instead of Thomas-Fermi) atomic wave functions in a Firsov-type [58] treatment of the energy loss to electrons gives theoretical results in very good agreement with the experimental data [55]; this indicates the effect of the shell-structure of the moving atom, the atoms of smaller size having lower electronic energy loss rates. Similar Z_1 oscillations in the electronic stopping regime have been found for other target solids.

Oscillations of electronic energy loss rate as a function of Z_2 at constant Z_1 are also found. Recently, Land and Brennan [57] have shown that the same kind of Firsov-type treatment, using HFS wave-functions as were employed by Cheshire et al. [56] for considering the Z_1 oscillation, gives very good agreement with the oscillatory energy-loss rates observed as a function of Z_2 when 800-keV nitrogen ions are incident on various target elements. In both calculations [56] and [57], ground-state HFS wave-functions for singly-ionised projectiles and neutral target atoms were used.

To complete this consideration of the electronic energy loss contribution it should be noted that the linear dependence on ion velocity predicted by both the basic Lindhard-Scharff [40] (expression (3)) and Firsov [58] treatments is not (e.g. [55, 59–61]) found to be in general agreement with experimental results. In the Firsov theory as used by Cheshire et al. [56] and by Land and Brennan [57] the linear dependence on ion velocity arises because of the rather arbitrary mathematical separation of the electron velocity due to the ion motion from the electron velocities within ion and atoms; it may be that better agreement of theory with experiment could be obtained if composite electron velocities could be used within a single integral in the calculation.

We now consider briefly the comparison of the nuclear energy loss data of the LSS theory with experimental results. Some recent experimental studies of the depth distributions of various heavy ions (Z_1 = 54–83; energies $\cong$ 20–100 keV) in silicon and aluminium [33–37] show that in the nuclear stopping region the actual projected ranges are systematically greater (by about 30%) than those predicted by the LSS theory. Thus, the LSS theory overestimates the nuclear energy-loss rate by about 30%. In contrast to electronic stopping, these nuclear stopping data seem to show no evidence for significant Z_1 oscillations. It is implied by Combasson et al. [37] that the Thomas-Fermi screened Coulomb potential (used in the LSS treatment) may not be accurate for these ion-atom collisions; they show that the measured projected ranges lie between those calculated using the Thomas-Fermi potential and the Lenz-Jensen potential [39a]. They also show that the relative range straggling (i.e. $\langle \Delta R_p \rangle / \langle R_p \rangle$) is rather insensitive to the choice of potential. From measurements of the nuclear energy loss rates for several light ions in silicon, Grahmann and Kalbitzer [62] have suggested that here also the LSS theory overestimates the loss rates; in their work the general difference between theoretical and experimental values was found to be much greater than 30%. However, Lee and Mayer [92] have presented graphical comparisons which show good agreement in the experimental and theoretical mean projected ranges and range straggling data for $^{11}B^+$, $^{31}P^+$ and $^{75}As^+$ of energies 20 keV to 1 MeV implanted into silicon (for boron in silicon the theoretical calculations use experimental electronic energy loss rates); it seems that for these not-very-heavy ions any error in the LSS nuclear energy loss rate is not significant when taken for the whole path of the ion over which the ion's energy changes from its initial value to zero.

3.3. Energy losses and ranges of very light ions

The LSS theory [39] of the energy loss rates of ions in solids deals with electronic-stopping in the low-velocity regime, and so is not applicable for considering the energy losses and ranges of very light ions (protons, deuterons and helium ions), of moderate or high energies. For protons and helium ions, however, empirical or semi-empirical energy-loss and range data are available [63, 64] in tabular form for various elemental target materials. Recent experimental data for particular elements include that for 1.5 to 60 keV protons in silicon [65], and for 250 keV to 2.5 MeV protons and ^{4}He ions in hydrogen, nitrogen, oxygen and erbium [66]. In the latter work [66] it is shown that the three-parameter theoretical formula of Brice [67] for electronic stopping (obtained by use of a modified Firsov [58] treatment) can be accurately fitted to the experimental results. The ranges of very light ions are much larger than those of moderate or heavy ions of the same energy because of the smallness of the nuclear stopping contribution.

4. DISPLACEMENT OF ATOMS DURING ION IMPLANTATION

It is well known that lattice defects in crystalline solids can strongly affect some properties of the solid, and theoretical solid-state physics can attempt to give quantitative explanations of these effects. Thus, it was clear from the beginning of the use of ion implantation that the ion-target-atom internuclear collisions (see Section 3) that transfer enough energy to displace target atoms from their usual sites could lead to large changes in the properties of the solid which might mask the required doping effect of the implanted atoms. It is therefore of importance to be able to make theoretical estimates of the depth, ion and target dependence of the lattice defects produced during ion implantation. (General discussions of radiation damage effects and of the properties of lattice defects can be found in Ref. [26].)

The most convenient approach in considering the irradiation damage effects of ion implantation is to say that the defect concentration as a function of projected distance R_p into the target should be proportional to $(-dE/dR_p)_n$, the rate of ion energy loss by nuclear collisions at the depth R_p [49, 68, 69]. If it is then assumed that it is possible to define, for each target material, a mean or effective displacement energy $\bar{E}_d$, which is the average energy needed to be imparted in ion-atom internuclear collisions for production of an interstitial defect and a lattice vacancy, the initial (i.e. pre-annealing) density n_d of displaced atoms per unit volume should be expressed as

$$n_d \cong 0.4 \, N_i \left(\frac{-dE}{dR_p}\right)_n / \bar{E}_d \qquad (9)$$

where N_i is the number flux of the incident ions. This expression (9) is based on theoretical ideas of Kinchin and Pease [70] modified slightly by Sigmund [71], and on consideration [49, 68, 69] of the collision processes themselves, often in this context called "atomic" or "elastic" collision processes. Expression (9) is expected to be only an approximation to the number of displaced atoms formed in any particular case because it ignores all the details of the recoil-energy spectrum of the displaced atoms, the likely anisotropy of the energy for displacement of an atom, the initiation of atomic collision sequences by recoiling atoms, the recombination of interstitials and vacancies and the formation of defect complexes. Nevertheless, when the spatial distribution of the energy into internuclear collisions, $(-dE/dR_p)_n$, calculated [47, 49, 68, 72] by the method of moments (see Section 3.2.2) is used in expression (9), the resulting depth distribution of lattice disorder is found to agree quite well in shape and depth scale with experimentally measured disorder distributions [49, 68]. Tabulations such as [72] of the rates of ion energy loss into nuclear collision processes for various ions, energies and target materials are thus of great value in ion implantation studies. It is difficult however to calculate absolute values for n_d in expression (9) because $\bar{E}_d$ is rarely known and because of the complexity of the defect processes as mentioned above. During electron irradiation the minimum energy transfer to an atom to displace it from a lattice site to an interstitial site is found to be ~ 10–25 eV for most crystalline solids; the values of $\bar{E}_d$ are expected in many materials to be significantly greater than the corresponding minimum electron-irradiation values.

Graphs and tabulations of the depth distribution of energy transfer into electronic excitation processes, calculated by the moment method, are also available [49, 72] for use in considering ion-implantation-related processes, such as the production of defects in insulators and possibly ionization-enhanced diffusion in insulators and semiconductors, which may be dependent on local ionisation densities. It has recently been emphasized [73] that the large ionisation density along the track of an ion moving through a covalently-bonded solid (such as the elemental and III-V compound semiconductors) may, by bond-breaking, significantly reduce the displacement energy value $\bar{E}_d$ to below the threshold value E_d found by electron-irradiation experiments. This would mean that $\bar{E}_d$ could not be taken in expression (9) as a constant, independent of ion-type, ion-energy and depth into the implanted solid.

5. PROPERTY CHANGES INDUCED BY ION IMPLANTATION: GENERAL ASPECTS

Before describing (in Sections 6, 7 and 8) the detailed applications of ion-implantation to metals, semiconductors and insulators, it is convenient in this section of the paper to discuss and list in a general way those properties of solids that have been found to be alterable by ion-implantation. Important features of such property changes are as follows:

(a) Because of the limited ranges of the implanted ions the effects will be confined to regions of the solid near the implanted surface. For ions of medium to high mass number the depth affected will be only about 0.01–1 μm for the ion energies usually used, unless channelling [1–4, 18] of the ions is employed to increase the ranges [74]. In practice, the use of channelling is not really convenient since it requires the solid being implanted to be crystallographically oriented with respect to the ion beam to an accuracy of better or much better than about 0.5°, and is somewhat difficult to accomplish in connection with scanning [75] of the ion beam across the solid surface for the purpose of achieving uniform implantation of the solid over reasonably large areas. Furthermore, for channelling to be effective the solid needs to be a single-crystal; although this is usually so for implanted semiconductors it is rarely so for technological metals and of course not so for glasses. Thus, the appropriate applications of ion implantation are those for which changing the **near-surface** properties of the solid can produce beneficial effects. When very light ions (especially protons, deuterons and helium ions) are used (Section 3.3), ranges of 1–5 μm or greater can be achieved at moderate ion energies of ~ 0.1–1 MeV (Table 1a in [26]).

(b) The greatest change of property in the as-implanted solid is likely to occur at a depth equal to the most probable projected range of the implanted atoms or to that of the maximum concentration of radiation damage defects, depending on whether the property change is due to the presence of the implanted atom itself or of the damage. It turns out that because both are dependent strongly on the nuclear energy loss rate the projected range and damage distribution are often not greatly different in shape, but that the peak damage concentration is somewhat closer to the surface than the peak implanted-atom concentration (see for example [75a] in connection with proton implantation of silicon). Heat treatment of the implanted solid will be needed to anneal the radiation damage if this damage causes deleterious property changes. The annealing of the disorder may itself then produce beneficial effects. Thus, crystalline silicon can be made amorphous by lattice defects formed during boron implantation; during subsequent heating at 600–650°C the implanted silicon layer recrystallises and at the same time almost 100% of the boron atoms become substitutional (and therefore electrically active) in the silicon lattice. On the contrary, if it is the lattice disorder rather than the implanted element that is producing the required property change then annealing of the defects is to be prevented; one can imagine the possibility of solids, such as potentially superconducting metals of certain kinds (see Section 6.4), being implanted at low temperature, and maintained at low temperature throughout their working life, in order to preserve the required lattice defects.

(c) The advantages of the use of ion implantation over other methods of introducing impurities into solids can be summarized by saying that ion implantation is a versatile, non-specific process allowing virtually any element to be introduced into any solid in accurately controllable quantities, and also to specified depth distributions by use of a sequence of ion doses at various energies

TABLE I. SUMMARY OF PROPERTY CHANGES AND APPLICATIONS ASSOCIATED WITH ION IMPLANTATION

Category of solid	Property changes and applications	References to reviews
Metals	General	[17, 24–26, 76–79]
	Electrochemical and chemical properties such as corrosion resistance and catalysis	[79–81]
	Hardness, frictional and wear properties	[76, 79, 82]
	Electrical resistance	See Section 6.3.
	Superconductivity	[83, 84]
	Simulation of neutron damage and void formation	[85–89]
	Implantation metallurgy (new phases, precipitation, gas-bubble formation, enhanced diffusion etc.)	[77, 144, 145]
Semi-conductors	General	[16, 17, 21, 23, 24, 26, 90–94]
	Electrical properties of silicon	[16, 17, 21, 90, 94]
	Silicon electronic devices and circuits	[91–102]
	Properties and devices of diamond	[109–110]
	Electrical properties and devices of germanium	See Section 7.3
	Electrical and optical properties of III-V compounds and device applications	[111–115]
	Electrical and optical properties of II-VI compounds and of SiC, and device applications	[111, 112, 114]
Glasses and other insulators	Refractive index changes	[116, 117, 186, 204]
	Electro-optical information storage	[117, 194]
	Properties of and information storage by magnetic bubbles	[118, 186, 187]

[75]. Provided that the temperature of the solid is never so high that diffusion of the implanted atom can occur then equilibrium solubilities can be exceeded without subsequent precipitate formation. For the production of semiconductor electronic devices and integrated circuit units the fact that the semiconductor need not usually be heated to very high temperatures can prevent many impurity contamination and semiconductor decomposition problems associated with doping by high temperature diffusion; even if for special reasons thermal diffusion

is needed, ion implantation can very effectively be employed to provide an accurately known quantity of implanted diffusant in the solid before the diffusion treatment.

This section concludes by giving in Table I a summary of property changes and applications associated with ion implantation in various kinds of solid. Subsequent sections of the paper review the recent developments in detail.

6. PROPERTY CHANGES AND APPLICATIONS FOR METALS

6.1. Electrochemical and chemical properties of metals including corrosion

The effects of ion-implantation on the chemical and electrochemical properties of metals have been recently reviewed by Dearnaley [79, 80] and by Grant [81]. It is clear that such properties can be affected by the implantation of foreign atoms, since these properties are determined by the elemental constitution of the metal and since the chemical or electrochemical reagent attacks the metal via its surface. The starting point for investigations of this kind is the knowledge of the behaviour of metal alloys formed by conventional metallurgical processes, and one can consider forming such alloys by implantation into a pure metal of ions of the other components of the alloy; this might for example be a cheaper way of producing an alloy surface for particular applications and components, especially as only small concentrations ($\lesssim 1\%$) of certain impurities may be required. Dearnaley [79, 80] has emphasized how the ease of incorporating various kinds of impurity element by implantation can allow experimental studies leading to improved understanding of corrosion behaviour; he has also emphasized that ion implantation can enable surface alloys of good corrosion properties to be formed without changing the chosen mechanical strength properties of the bulk metal.

The chemical property of metals and alloys that has, because of its technological importance, been studied most so far in respect of ion implantation is the oxidation rate. Dearnaley et al. [79, 80, 119, 120] have investigated the thermal dry oxidation of the technologically important materials, titanium, 18/8/1 stainless steel, zirconium and copper, after implantation of various ions at room temperature to depths of near 0.1 μm (ion energies up to 500 keV). It was found that both reductions and enhancements of oxidation rate occurred. Implantation of the less electronegative atoms calcium and europium reduced the oxidation rate of titanium while implants of the more electronegative atoms, ytterbium, bismuth, indium and aluminium, reduced the oxidation rate of the stainless steel. The implantation of argon, expected to be chemically inert, produced little effect on the oxidation of the titanium thus suggesting that radiation damage effects of the implantation were not significantly influencing

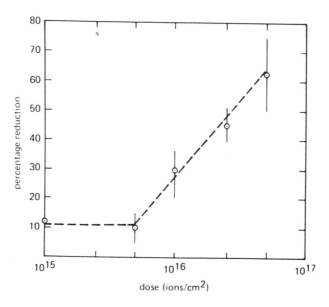

FIG.3. *The reduction in the rate of oxidation of copper (99.9%), in dry oxygen at 200°C and a pressure of 1 atm, as a result of implantation with B^+ ions; the effect seemed independent of ion energy for energies of 50–390 keV. From Naguib et al.* [121].

the oxidation. The use of inert gas implantation is often of great help in this way to distinguish doping and damage effects of ion implantation. For zirconium, Dearnaley et al. found that the oxidation rate could be decreased by implantation to particular doses ($\sim 3 \times 10^{15}$ ions/cm^2) of iron and nickel; they suggest that the smaller size of these atoms compared with that of zirconium itself reduces the mechanical stress within the ZrO_2 and thus decreases in the oxide the number of cracks, pores and grain boundaries which allow the oxygen to penetrate.

Interesting results on the effects of implantation of boron, carbon, nitrogen and neon ions (to a mean depth of about 700 Å) in copper have recently been presented by Naguib et al. [121]. They studied oxidation of the copper in dry oxygen at atmospheric pressure and 200°C and found that B^+ doses of 10^{16}–10^{17}/cm^2 were very effective in reducing the oxidation rate; their results for this case are shown in Fig. 3. They found that the oxidation rate was less reduced by the neon and nitrogen implants and increased by the carbon implantation. In contrast to the suggestions of Dearnaley et al. [79, 80, 119, 120] for titanium, Naguib et al. interpret their data as indicating that implantation-induced radiation damage in copper is detrimental to the beneficial effect of the chemical doping of the implanted atoms.

Concerning electrochemical effects, Ashworth et al. [81, 122] have studied the aqueous corrosion of iron implanted with several ions including chromium and iron itself. Their results show that the chromium-implanted layer behaves in respect of oxidation resistance in the same way as a bulk iron-chromium alloy of the same composition; the iron implantation did not significantly change the oxidation behaviour.

Dramatic effects of the implantation of 400-keV platinum ions to doses of $10^{15}-10^{16}/cm^2$ into tungsten and tungstic oxide have been found by Grenness et al. [123]; they report that such implanted surfaces, when used as cathodes during the electrolysis of water, can behave like platinum itself, and they point out the possible financial advantages of using platinum-implanted instead of platinum-alloy electrodes in electrolytic and fuel cells.

6.2. Hardness, frictional and wear properties of metals

It is clear that the hardness, friction and wear of metals and alloy surfaces are properties which are dependent upon the chemical and microstructural composition of the near-surface region of the material, and as such are likely to be altered by ion bombardment and implantation. Over many, many years surfaces with required properties, especially hardness, have been produced by conventional metallographic methods, the nitriding of steel being an important example of this. What ion implantation can do first is to allow studies of the mechanical properties of surfaces having compositions not previously obtained by traditional methods with the possibility then that large-scale non-implantation processes can be developed for manufacture of new surface alloys that are found to have valuable properties. Secondly ion-implantation itself can be usable on a production-line scale, such as for the surface treatment of small and crucial components of a larger mechanical assembly.

The theoretical aspects of how ion-implantation can affect surface mechanical properties have been recently carefully considered by Hartley [82] in an article which also reviews experimental results for implantation effects in metals. He points out that both hardness and friction are measures of the resistance of the surface to plastic deformation, and thus increase with yield stress. Ion implantation can introduce lattice defects and interstitial impurity atoms (such as boron, carbon and nitrogen in iron) which will hinder dislocation glide and therefore increase the yield stress (thus both damage and doping may contribute). At high implantation doses new alloys can be formed, and then the hardness and friction will be those of the new surface material. Kanaya et al. [123] and Gabovich et al. [125] have produced significant (e.g. at least 30%) increases in hardness in steel surfaces by implantation of nitrogen to doses of $\sim 10^{17}$ ions/cm^2. It is worth noting that these implantations were carried out using low-energy ions (up to only about 25 keV) and that this is an example of the fact that expensive high-energy ion accelerators are very often not needed for implantation purposes; indeed for a given dose, implantation at lower energies will produce a greater volume concentration of the implanted species.

In other studies of steel, Pavlov et al. [126] have related the increases in hardness and friction coefficient obtained by argon bombardment (40 keV, up to 10^{18} Ar$^+$/cm^2). Hartley et al. [127, 128] have reported experimental results showing both increases and decreases in friction coefficient brought about by ion implantation.

Of greater concern than friction, which with use of efficient liquid lubricators is not usually a problem, is the question of mechanical wear of lubricated surfaces. Most studies of this kind have been conducted at the AERE, Harwell by Hartley and Dearnaley. Hartley [82] suggests that ion implantation can increase the wear resistance by facilitating the smoothing of the surfaces brought about by the rubbing action between the moving parts; when the surfaces are smooth the load is taken more by the lubricant film, and the wear is reduced; the enhancement of smoothing may be due to the hardening of the surface (e.g. by implantation of light atoms) which causes rough contact points to break off more easily, or to the production of lateral stress in the surface (e.g. by implantation of large atoms). Figure 4 (from [82]) shows the very great improvement in the wear resistance of a steel surface as a result of implantation of 30-keV nitrogen ions.

Investigation of the effects of ion implantation on mechanical wear are to be considered as being mainly still in the research rather than the applications stage. But in view of the technological importance of this matter and the encouraging results mentioned above one can believe that this is a field of work of very great potential value.

6.3. Electrical resistivity

A number of investigations of the changes in the electrical resistivities of metals due to ion bombardment have been made in connection with studies of lattice defect production and of defect properties [26, 129]; lattice defects produced by ion-target nuclear collisions, as described in Section 4, scatter the conduction electrons in the metal or alloy and the electrical resistance is increased. There have, however, been several published reports of electrical resistivity changes due to the doping (rather than the damaging) effects of ion bombardment. Of particular interest is the work by Deery et al. [130] and Wilson et al. [131] who have made measurements of the resistivities and the thermal coefficients of resistivity of thin tantalum films (containing oxygen) as functions of implantation doses of argon, oxygen and nitrogen ions. The technological interest of these studies is that tantalum films can be used as resistors in integrated circuits (especially for use at high temperatures) and there is the possibility of optimizing the sheet resistivity of the films and of minimizing the thermal coefficient of resistivity by forming a film of the appropriate composition of Ta with Ta_2O_5, Ta_2N or TaN. The measurements showed [130, 131] that with nitrogen implantation thermal coefficients of resistivity

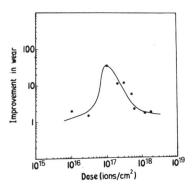

FIG.4. *The effect of 30-keV N^+ implantation on the wear rate of En 40B steel. From Hartley [82].*

of less than 10^{-4} per °C could be achieved, this being considerably better than that for conventionally produced tantalum oxide or tantalum nitride resistors.

6.4. Superconductivity

The subject of changing the electrical superconducting properties of metals and alloys by ion implantation has been reviewed recently in full articles by Meyer [83] and by Stritzker [84]. The proceedings of conferences held at Albuquerque [10] and at Warwick [12] contain additional research papers on this subject. The technological aim of course is to produce a stable or metastable alloy which stays superconducting to well above 20 K (liquid hydrogen temperature) and if possible to above 80 K. Among pure metals the superconducting/non-superconducting transition temperature T_c is highest (at 9.2 K) for niobium, and among alloys the highest T_c so far found is 23 K for Nb_3Ge.

The Bardeen-Cooper-Schrieffer (BCS) theory [132] relates the occurrence of superconductivity to interaction between electrons and phonons, and the value of T_c is greater when the density of electronic states at the Fermi energy is larger, when the average phonon frequency is smaller, and when there is more efficient electron-phonon coupling. Incorporation of impurities into a metallic solid (e.g. by implantation) may alter all three of these parameters in either direction and the transition temperature T_c may increase or decrease. Implantation as a doping process simultaneously introduces lattice disorder, and the disordered solid may well have a "softer" phonon spectrum (i.e. lower phonon frequencies) because of the presence of lattice vacancies, this in itself tending to raise T_c. However, the disordering may simultaneously reduce the density of electronic states, and so produce the opposite effect upon T_c; for example, Poate et al. [133] have shown that 2-MeV He^+-ion irradiation to doses of about 10^{17} ions/cm² near room temperature can reduce the T_c values of Nb-Ge films (of composition

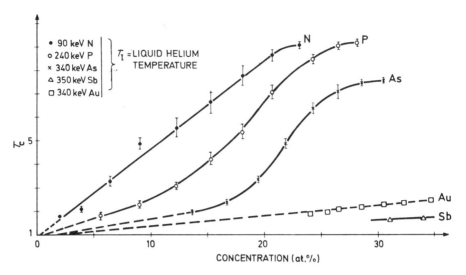

FIG.5. *The superconduction transition temperatures T_c of evaporated oxygen-free ($\leqslant 1\%$) molybdenum films implanted with N^+, P^+, As^+, Au^+ and Sb^+ ions to various concentrations. From Meyer [137].*

approximately that of Nb_3Ge) from their initial values of 8–22 K down to about 4 K, and they attribute the reduction in T_c to the presence of micro-strains in the irradiated samples.

A further fundamental problem involved in trying to increase T_c by softening the phonon spectrum is that the lattice structure consequently becomes less stable and is likely spontaneously to change to a more stable atomic arrangement of lower T_c. However, firstly, the thermodynamically unstable phase might be maintainable if kept at low temperatures, and secondly, as suggested by Stritzker [84], if a valuable new high-temperature superconductor is made, it might be possible to find a way to stabilize its lattice structure.

Let us now consider briefly examples of studies where significant increases in T_c have been brought about by ion implantation. Among these is work by Stritzker et al. [84, 134, 135] on palladium alloys. Although palladium itself is not superconductive it has been known since 1972 [136] that Pd-H alloys are superconductive if the H/Pd atomic ratio is greater than 0.8. Stritzker and Buckel [134] showed that T_c could be raised to 8.8 K by implanting H so as to give a H/Pd ratio of 1.0; furthermore, they showed [134, 135] that implantation of deuterium into palladium, so as to produce an alloy having a D/Pd ratio of 1.0, raised T_c to 10.7 K. In consideration of their results [135] with implants of other light impurities they suggest that the effectiveness of an impurity in palladium depends on its being interstitial and also able to donate electrons to the palladium 4d state, and on there being strong free-electron/phonon coupling via the coulomb potential of the impurity; they believe that the greater T_c value for deuterium implantation compared with that for implantation of hydrogen may be due to differences in the phonon spectra.

Further results where ion implantation raises the superconducting transition temperature concern molybdenum. Figure 5 (from Meyer [137]) shows the effect of implantation of N, P, As, Sb and Au at 4K upon the value of T_c for molybdenum (initial value for pure Mo, 0.9 K). It is seen that large increases in T_c resulted from the N, P and As implants. No increase in T_c was however produced by Ne, Xe or Al implantations, and it is clear that the considerable effects with N, P and As are impurity-specific and certainly not due just to disordering of the lattice. Meyer suggests however that the enhanced T_c values may be due to the presence of defect complexes involving the N, P or As atoms. In this wide study of implantation into molybdenum Meyer shows that if the molybdenum layer contains more than 1% of oxygen then Ne^+ and Xe^+ bombardments increase T_c, and he attributes this effect to the formation of defect-oxygen complexes. It is important to note that high-T_c properties of both the implanted pure and oxygen-containing molybdenum layers were retained even when the samples had been heated to 200°C.

The examples mentioned in this section show how ion-implantation, by its versatility in allowing the fabrication of very many kinds of alloy with a wide range of compositions, is leading to a better understanding of superconductivity in metal alloys and can produce very significant increases in critical superconducting temperatures. This seems to be a valuable field of basic and applied research, likely to lead to important technological advances.

6.5. Simulation of neutron radiation damage and void formation

As previously discussed in Section 4, ion implantation is always accompanied by the production of lattice defects in the implanted solids by collisions between the incident projectile and the atoms of the solid. This is part of the general field of irradiation damage studies [26, 139] and for metals and alloys many papers on recent investigations can be found in the proceedings of a conference held in 1975 in Gatlinburg [138]. However, because of its technological importance, the use of ion beam irradiation to simulate neutron irradiation deserves special mention in this paper. In particular, the formation at high temperatures of voids (large vacancy clusters), which can lead to macroscopic swelling of structural and fuel components by neutron irradiation in fast-breeder reactors, has been studied by means of irradiation with energetic ion beams; because of the high atomic displacement rates under ion irradiation, void formation effects that might take several years to develop in the fast-breeder reactor (despite the fast neutron flux being $\sim 10^{16}$ neutrons/cm^2·s) can be studied in a matter of hours using ion beam irradiations. This use of ion beams for void studies has been discussed in various review articles [85–89 and 139, 140].

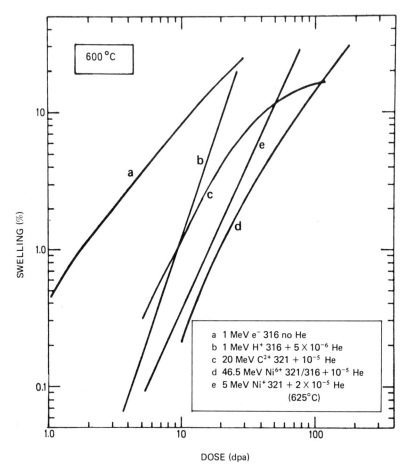

FIG.6. The experimentally-deduced volume swellings of types 316 and 321 stainless steels resulting from bombardment with electrons and various ions at 600°C; the bombardment doses are given in terms of calculated displacements per atom (dpa). The samples had been pre-implanted with helium to the atomic concentrations indicated. From Hudson [141].

In a recent paper [141] Hudson has reported measurements by transmission electron microscopy of void formation in 316 and 321 stainless steels under nickel ion bombardment and has compared the results with those obtained in other work for irradiation with charged-particle beams. Some of the results are shown here in Fig. 6, where the experimentally deduced swelling are plotted as functions of the theoretically estimated numbers of displacements per atom (dpa). Because very considerable and continuous annealing of defects occurs during the bombardments, each atom can be displaced many times and the dpa number is much greater than unity. As in the results of Fig. 6, in many experiments the metal sample being studied is pre-implanted with helium ions to concentrations of $10^{-6}-10^{-5}$ (atom/atom) of helium so as to represent that which would be formed by (n, α) reactions during actual neutron irradiation; in many materials the helium can play a rôle in the nucleation of voids. In order to use data such as those of Fig. 6 for prediction of neutron irradiation effects it is necessary to have a theoretical basis for comparing the damaging rates of $\sim 10^{-3}-10^{-2}$ dpa/s, that can be obtained by ion irradiation, with the much lower rates of $\sim 10^{-6}-10^{-5}$ dpa/s given by the fast neutron reactor fluxes. The first basic theoretical criterion [142] is that similar effects should be obtained if the ion beam and neutron irradiations are conducted with the samples at different temperatures such that the ratio of atomic displacement rate to vacancy diffusivity is the same for each. This means that the sample temperatures during ion irradiation experiments must be somewhat higher, for example by 150–200°C, than those of components in actual reactors in order to simulate void production by the reactor neutron irradiation. A further criterion for comparison of ion and neutron irradiations is whether the recoil energy spectra of knock-on target atoms are similar in the two cases. Hudson [141] points out that the void-induced swelling rates in the 316 and 321 stainless steels are almost the same for irradiations with nickel ions (of 5 MeV or 46.5 MeV) and fast neutrons when the expected ion/neutron temperature shift is taken into account; this result seems to be in agreement with conclusions that can be drawn from calculations of Marwick [143] concerning the recoil energy spectra for different kinds of irradiating species.

6.6. Implantation metallurgy (new phases, precipitation, gas bubbles blistering, enhanced diffusion)

This Section 6 concerned with ion implantation effects in metals and alloys is terminated by mentioning some general aspects which have been called "implantation metallurgy". The same refers to the fact that alloys of new compositions and micro-structures can be produced by implantation, and that these new alloys can have special metallurgical characteristics and properties. Some of the special properties have been discussed in previous parts of this paper, but in a recent review article [77] Picraux has suggested that questions concerning the substitutional or interstitial nature of the implanted atoms, and their diffusion, enhanced diffusion, solubility and precipitation, need particular attention from a metallurgical point of view. The subject of the behaviour (migration, trapping, precipitation into bubbles etc.) of atoms of gaseous elements implanted into solids can be considered also as part of the same general field of study, and this has attracted much interest recently in connection with the formation of gas bubbles and the likely consequential "blistering" of the metal walls of the gas plasma chambers of possible future thermo-nuclear reactors; the subject of blistering and bubble formation has been recently reviewed by Roth [144]. Various aspects of implantation metallurgy including gas bubble formation were previously considered by Nelson [145].

One of the major aspects discussed in the articles by Nelson [145] and Picraux [77] is that of the formation of new phases and precipitates. Picraux notes that the inherent intimate mixing of the implanted atoms with those of the matrix (the separation often being less than 10 Å) means that diffusion-limited reactions can occur in times more than 10^4 shorter than would be needed for standard metallurgical alloying methods; new phases stable only at low temperatures may thus be producable by ion-implantation. He notes also that metastable substitutional alloys, for example of tungsten in copper, can be produced under conditions at low temperatures where the implanted tungsten can, perhaps as a result of collision cascades, move into substitutional lattice sites but not be able to diffuse far enough to form tungsten precipitates.

Of very great importance therefore is the possibility of forming and using at room temperature, alloys, of thermodynamically unstable compositions and structures, which are effectively stable because of the insignificant mobilities of the component atoms. Vogel [146] has shown that implantation of carbon ions (of 38 keV energy) into iron at room temperature produces the metastable martensitic iron-carbon phase of tetragonal symmetry, and that heating at about 330°C or above was required before the usual iron carbide phase of carbon steels began to form as precipitate particles.

As examples of enhanced diffusion two pieces of work concerned with the implantation of nickel ions are now mentioned. Enhanced diffusion in this connection is the increased diffusion rate of an impurity (or isotope) in a solid resulting from the presence of irradiation-induced lattice defects. Most commonly the effect occurs when the extra lattice vacancies aid the diffusion of substitutional impurities. In work at Harwell, Turner et al. [147] measured the concentration/depth profiles of 40-keV Ni^+ ions implanted into polycrystalline titanium at temperatures of about 300–600°C. (The profiles were determined from the energy spectra of back-scattered 1.5–3.5 MeV He^+ ions.) They found that for high implantation dose-rates ($\gtrsim 40 \ \mu A/cm^2$) the nickel profiles extended to depths (of 0.5–1 μm) which were an order of magnitude greater than expected from the combined effects of LSS implantation range and ordinary thermal diffusion, and they suggested that the results could be compatible with the occurrence of radiation-enhanced diffusion. The studies by van Wyk and Smith [148] also used 40-keV Ni^+ ions, but implanted at 60–70 $\mu A/cm^2$ into copper and silver targets at temperatures between 76 and 673 K. Their measurements showed that the amounts of implanted nickel retained in the targets were greater at all temperatures than the amounts expected on the basis of LSS depth profiles together with continuous nickel-ion-induced erosion (sputtering) of the target surfaces. They concluded that very significant radiation-enhanced diffusion had occurred, and they emphasized the importance of such effects, in the applications of ion-implantation to metals, for increasing both the concentrations and depths of the implanted atoms.

7. PROPERTY CHANGES AND APPLICATIONS FOR SEMICONDUCTORS

7.1. General aspects

Over the past ten years it has been the potential and actual doping of semiconductors, particularly silicon, by means of ion beams that has given the major encouragement to the development of the whole field of ion implantation to the techniques, the theory and the equipment. Various books and review articles devoted to the general consideration of ion implantation of semiconductors have been published [16, 17, 21, 23, 24, 26, 90–94] and many more reviews discuss the different kinds of semiconductor (see Table I); in addition, there are many hundreds of individual research papers on and pertinent to this subject. The present article can attempt only to outline important considerations and developments and to review some recent significant results and aspects.

Of very great importance for this field of physical science and technology were the two factors that the basic principles of the doping of semiconductors (during crystal growth and by thermal diffusion), for incorporating electrically active substitutional acceptor and donor impurities, were well known, and that the flourishing semiconductor industry was finding that these established techniques were nevertheless not sufficiently controllable and versatile for many requirements. Ion implantation of silicon is now widely used on a commercial scale as one of the essential processes in the fabrication of silicon integrated-circuits and of some discrete silicon devices; it is described without hesitation in recent physical electronics text books as a standard and very convenient device fabrication process. In contrast, the use of ion implantation for doping III–V compounds such as gallium arsenide and other semiconductor compounds is still at present in the research or research and development stages. The following sections consider the ion implantation of the various semiconductor materials; the emphasis is placed upon the electrical and optical properties since these have the greatest technological importance at the present time.

7.2. Electrical properties and devices of silicon

The use of ion implantation to change the electrical properties of silicon has been considered in a number of reviews [16, 17, 21, 90, 94], and the way in which this enables silicon electronic device structures to be fabricated has also been well described and discussed [91–102]. It is worth noting especially a fairly recent and excellent article by Lee and Mayer [92] which gives a comprehensive account of why this method of doping silicon is now so valuable, and of its application for making field-effect and bipolar transistors; the article shows also in graphical form compiled experimental data for the projected ranges and projected range stragglings of 10–1000 keV $^{11}B^+$, $^{31}P^+$ and $^{75}As^+$ ions incident on silicon.

For silicon, the basic background result is that ions of group III (such as especially boron) of the periodic table and of group V (such as phosphorus and arsenic) implanted into silicon take up substitutional lattice sites and become electrically active as acceptors (giving p-type silicon) and donors (giving n-type silicon), respectively, when the implanted material has been thermally annealed at a temperature of about 600°C. It has been found that if the ion implantation dose used produces enough radiation damage that the implanted layer becomes amorphous, then a subsequent anneal at 550–600°C gives layer recrystallization leading to a large proportion of the implanted dopant atoms taking up electrically active substitutional sites. Heating at up to 1000°C can be used if necessary to produce almost 100% substitutional fractions, and to anneal most lattice defects so as to produce the larger electron and hole mobilities characteristic of bulk single-crystal silicon.

Nicholas [149] has recently presented a simplified theoretical method for predicting the sheet resistances (in ohms per square) of silicon surfaces implanted with boron. It is assumed that all the boron atoms are electrically active and have a gaussian spatial distribution, and these correspondingly imply anneals at 900–1000°C at which temperatures no diffusion should occur. He finds that the sheet resistance R should be given by the expression

$$R = 1.0 \times 10^{11} \, D^{-0.7} \, s^{-0.3} \quad \text{ohms per square}$$

where D is the implanted boron ion dose per cm^2, s is the projected standard deviation of the implant in cm, and the numerical constant 1.0×10^{11} is deduced from experimental data. He shows that this expression gives values of R in good agreement (better than ± 10%) with experimental results for doses of $\sim 10^{13}$–10^{15} B$^+$/cm^2. The expression for R can be employed for predicting the results of implantation through silicon dioxide or other insulating layers by using appropriate values of s. Figure 7, from the same paper, shows the expression for R in the form of a nomogram valid for bare silicon, the value of s being set by the boron ion energy. It can be seen that the sheet resistance obtainable is fairly insensitive to the implanting energy. Presumably the same kind of theoretical treatment could be used for other implants in silicon provided that the assumption of 100% electrical activity is a useful one.

As mentioned previously, it is metal-oxide-semiconductor (MOS) technology for field-effect transistors (FET) that has very much benefitted from the application of ion implantation. To a large extent this is due to the geometrical planar structure of field-effect devices and of integrated circuits containing them. This structure is admirably suited to the large area coverage but smallish penetration of ion beam doping. An important example, considered first, is the use of ion implantation in MOSFET fabrication to give accurate alignment of the edges of source and drain electrodes with the edges of the metal gate electrode; this is

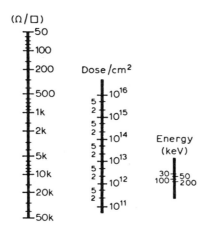

FIG. 7. *A nomogram for boron implantation of silicon relating sheet resistance obtainable after annealing at 900–950°C to the $^{11}B^+$ dose and energy used. From Nicholas [149].*

accomplished by initial formation of the gate by evaporation followed by implantation to produce the other electrodes, the gate structure acting as a mask ("auto-registration"). Gate-source and gate-drain capacitances are thus minimized. The accuracy of the electrode alignment is however limited by lateral spreading of the implanted ions (Section 3.2.2). A further very valuable implantation technique is used to reduce the threshold voltage V_{th} needed on the gate of an enhancement-mode FET to give type-inversion of the substrate beneath the gate so as to switch on the device; the majority of such enhancement devices are of the p-channel kind where the substrate is n-type and where V_{th} is negative. Boron ion doses in the substrate reduce its n-type character near the silicon surface and V_{th} is reduced in magnitude. The reduction of V_{th} is proportional to the boron dose, at a rate of about 1 V per 3×10^{11} B^+/cm^2 [97]. Since the initial, pre-reduction magnitude of V_{th} is usually 2–3 V a very significant reduction, with many circuit benefits, is obtained with modest boron doses.

Considering doping of silicon more generally, the great advantage of ion implantation over diffusion used by itself is that the ion charge incident on the semiconductor during implantation can be measured accurately by standard electronic equipment. Thus, the amount of dopant can be well controlled, and the reproducibility and reliability of the devices and circuits are very much enhanced. This is so important that low-energy implants (up to ~ 50 keV) are often used to provide sources of diffusant, the implant being carried out at room temperature and the "drive-in" diffusion at 1100–1200°C. For some purposes even the initial, silicon substrates, as cut from grown crystals or as deposited, have too much resistivity variability, and it is then advantageous to dope the substrates themselves sufficiently by ion implantation to make the initial variability insignificant.

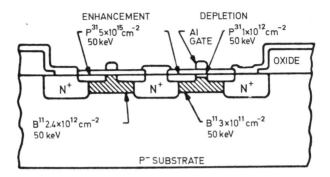

FIG. 8. *A composite enhancement and depletion mode MOSFET structure in a silicon fast logic circuit fabricated by electron-beam lithography and by boron and phosphorus implantations; the device has 1-μm gate widths. From Fang et al.* [150].

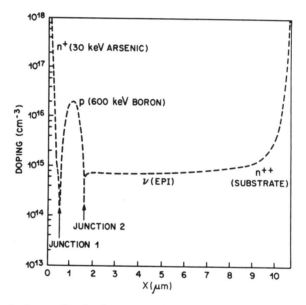

FIG. 9. *The doping profile of a silicon BARITT microwave-oscillator diode made by implantations of 600-keV 1 × 10^{12} B^+/cm^2 and of 30-keV 1 × 10^{15} As^+/cm^2 into a 6.5 ohm-cm n-type epitaxial layer on a 10^{-3} ohm-cm n-type substrate. From Eknoyan et al.* [152].

As examples of these techniques for silicon we now note briefly two particular device structures each formed by a sequence of implantations. Figure 8 (from Ref. [150]) shows a combination of n-channel enhancement-mode and depletion-mode MOSFETS on a lightly doped p-type silicon substrate; the drain and source extension electrodes were made by the 5 × 10^{15} cm^{-2} phosphorus dose, and the other phosphorus and boron implants served to give precise enhancement and depletion device-characteristics. In this structure (Fig. 8) the drain-to-source separations (gate widths) were only 1 μm, this being achieved by using electron-beam lithography to produce the silicon dioxide/metal gate masking pattern. The use of very small gate widths $\lesssim 1$ μm is essential in integrated circuits so as to produce high-speed devices and to aid the achievement of high device-packing density per unit area. In a forward-looking paper, Keyes [151] has recently discussed silicon integrated circuit miniaturisation including various aspects of lithographic processes as a prelude to the ion implantation and diffusion. He notes that 30 000 components can now be fabricated upon each silicon chip of size about 5 mm × 5 mm, and that ion-beam lithography may be a way of further decreasing the size of each individual component.

Figure 9 (from Eknoyan et al. [152]) shows the depth structure of a silicon BARITT (barrier-injection and transit-time) device formed by successive implantations of boron and arsenic into a lightly n-type epitaxial layer (ν) on a highly n-type substrate (n^{++}). (BARITT devices are used as the active components of microwave oscillators giving rather low power but also lower noise than for example IMPATT oscillators.) Eknoyan et al. found that the efficiency (microwave power out per DC power in) of the structure of Fig. 9 was about 5% which they state to be the highest ever reported for any BARITT device; they associate the high efficiency with the special n^+-intrinsic-p-ν-n multi-layer structure which can be very conveniently made by ion-implantation.

This section has aimed to outline the important and more recent aspects of ion implantation of silicon. This is now an established technique in industrial silicon-device technology, and the situation has come about because of the inherent accuracy and versatility (including the possibility of doping through thin passivating oxide layers) of ion implantation doping in the making of transistors, capacitors, resistors and other components especially in integrated circuit form. The editor of Solid State Technology has written [153] that "ion implantation is superior to other techniques because it allows precise control of charges and depth distribution of dopant profile as well as excellent reproducibility". Morehead and Crowder [93] note that ion implantation machines are available that allow 400 silicon slices to be loaded at a time and implanted in two hours to a uniformity of 1%.

7.3. Electrical properties and devices of diamond and germanium

This section considers ion implantation effects in diamond and germanium, which constitute the semiconducting solids immediately before and after silicon in group IV of the periodic table of elements. As for silicon, the atoms of groups III and V are expected to be acceptors and donors respectively when in substitutional sites. Compared with the situation for silicon, the present practical applications of implanted diamond and germanium are very few, and the number of individual research papers is also fairly small.

Ion implantation of diamond has been considered in two fairly recent review articles [109, 110]; a previous research paper [154] reviewed the results of various investigations to 1970. An early experimental difficulty in studying diamond was the presence of impurities and defects in natural material; synthetic samples of good quality and size are now available. Semiconducting diamond of p-type conductivity and blue colour exists both as natural crystals ("IIb" variety) where the acceptor impurity is believed to be boron [155], and as synthetic crystals grown with boron-doping; the acceptor ionisation energy is 0.37 eV [155]. Synthetic diamond grown with aluminium content is only slightly conducting and material containing nitrogen (Group V) is usually highly insulating.

Davidson et al. [154] studied the properties of diamond implanted with boron, phosphorus and nitrogen ions at room temperature. They concluded that the observed implantation-induced changes in electrical conductivity were probably due to radiation damage effects; heating at 950°C annealed the lattice disorder produced by phosphorus implantation but they observed no phosphorus-associated conductivity. Clark and Mitchell [156] pointed out the strong possibility that graphite or amorphous carbon layers of reasonable conductivity could be produced by ion bombardment, especially in experiments involving thermal annealing treatments, and Brosious et al. [157] indeed showed the presence in ion-implanted diamond of amorphous carbon, most of which could be recrystallized back to the diamond structure by careful annealing at 1400°C. However, Vavilov [110] has reviewed various sets of results giving evidence that elements of groups III and V can, under suitable conditions, be implanted to give electrically active acceptors and donors. He reports that the p-type conductivity of boron-implanted diamond increased with thermal annealing (presumably due to defect annealing and incorporation of the boron into active sites) and then remained constant even with heating at 1200°C; hole mobilities as high as 400–700 $cm^2 \cdot V^{-1} \cdot s^{-1}$ at 300 K could be obtained with doses of $\sim 5 \times 10^{14}$ B^+/cm^2 followed by annealing at 1350°C, and the acceptor activation energy was ~ 0.3 eV. Layers exhibiting stable n-type conduction could be made by hot (600°C) implantation of phosphorus, and p-n rectifying junctions could be formed by successive boron (with anneal) and hot phosphorus implants. Rectifying junctions could be formed also by successive boron and hot (800°C) antimony implantations followed by an anneal at 1400°C (see Fig. 10); the layers implanted with antimony at 800°C and annealed at 1400°C were found however to have very low carrier mobilities (less than 1 $cm^2 \cdot V^{-1} \cdot s^{-1}$) and the conduction was ascribed by Vavilov and co-workers to a charge-hopping mechanism.

In principle, because of its large energy band gap (about 5.4 eV), diamond could be of value as a high-temperature electronic device material; the research and development needed to hope to reach the present state of the technological use of silicon would however be enormous. In recent work, Blanchard et al. [158] have shown that available experimental electronic energy loss rates for boron ions

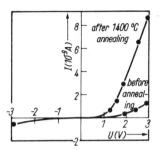

FIG.10. *The current-voltage characteristics of a p-n diode formed in diamond by room-temperature implantation of boron and 800°C implantation of antimony, before and after subsequent annealing at 1400°C. From Vavilov* [110].

in carbon used in LSS/moment calculations (see Section 3.2.2) give depth distribution profiles in reasonable agreement with their experimental data for 40–250 keV boron implanted into diamond.

In the case of germanium there have been a few detailed implantation studies, and applications concerning the fabrication of gamma-ray and particle counters have been described. Meyer [159] reported on investigations of the implantation of 27 elements (from groups III, V and others) and showed that annealed (300°C) B^+-, In^+- and Ga^+-implanted layers in n-type germanium were strongly p-type. He found that Li^+ implanted at 20°C into p-type germanium gave a low resistivity n-type layer without annealing treatments; implants of group V elements required heating at 300–600°C to remove acceptor-type lattice defects before n-type activity was produced. Gallium implantation at 12 keV and boron implantation at 15 keV have been used respectively by Dearnaley et al. [104] and by Ponpon et al. [106] to produce p^+ layers for germanium radiation detectors. Germanium has an advantage over silicon for detecting gamma rays and high-energy charged particles because of its higher attenuation coefficient and stopping power respectively for such radiation.

Detailed studies of the electrical properties and damage states of implanted germanium have been described for implants of B^+, Ga^+, P^+ and As^+ [105], of B^+ [106] and of Al^+ [160], for B^+ concerning radiation-enhanced diffusion during implantation [161], for B^+, C^+, N^+ and P^+ [162] and for B^+ and C^+ [108]. The general picture of the results for the group V elements is that implantations of phosphorus and arsenic in germanium lead to n-type activity upon thermal annealing at 400–500°C, and that nitrogen implantation gives some slight n-type conductivity upon annealing at 600–700°C. For implants of gallium and aluminium (of group III) annealing at 300–500°C produces strong p-type activity.

In contrast to gallium and aluminium, the implantation of boron into germanium at 20°C produces strong p-type activity which seems to be due to

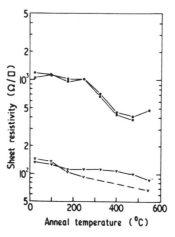

FIG.11. *The sheet resistances of the p-type surface layers of n-type germanium implanted at $20°C$ with 60-keV $^{12}C^+$ (filled circles and squares) or with 60-keV $^{11}B^+$ (triangles) to doses of 1.0×10^{15} ions/cm², before and after successive 30-min thermal annealing treatments. From MacDonald and Palmer [108].*

electrical activity of the implanted atoms without the need for thermal annealing treatment [105, 106, 108, 162]. Figure 11, from Ref. [108], shows the very different sheet resistances obtained by 60-keV C^+ and B^+ implantations of germanium at room temperature (without subsequent heating) for very similar implantation dose-rates. Both implanting ions produced p-type layers; but since these ions would create almost identical initial concentrations of lattice defects, the high conductivity of the B^+-implanted layer must have been due to considerable electrical activity of the implanted boron atoms (presumably on substitutional sites), together with significant defect annealing during the implant. Nuclear depolarization measurements involving the β-decay of ^{12}B have indeed indicated [163] that a substantial fraction of boron atoms implanted into germanium can become immediately substitutional in the lattice.

Various aspects of the fabrication of ion-implanted germanium transistors have been considered by Schmid et al. [107].

7.4. Electrical and optical properties of compound semiconductors and their device applications

The term "compound semiconductors" refers to a variety of compounds having electronic energy gaps of several tenths of eV to several eV, of which the most well known are binary III–V and II–VI materials such as, respectively, GaAs and CdS, and the IV–IV compound SiC. Ion implantation results for compound semiconductors have been reviewed in various articles [111–115]; the recent valuable reviews by Hemment in 1975 [114] and Donnelly in 1976 [115] concentrate on GaAs, which is of much technological importance, and contain many references. The conference and symposium proceedings [12, 13, 164, 165] include additional research papers relating to ion implantation in GaAs and the fabrication of devices by this means.

Just as for the elemental semiconductors, impurity atoms in the compound materials can produce p-type or n-type electrical conductivity if they take up substitutional sites in the crystalline lattice. Thus, in III–V compounds, atoms of group II elements, such as zinc and cadmium, can occupy the sites of the group III element and give p-type activity by becoming acceptors. Similarly, group VI impurity atoms, such as sulphur and selenium, can occupy group V positions and become donors. The methods by which effective doping can be achieved in GaAs by ion-implantation are now known [114, 115], although the physical reasons why they work are certainly not understood in detail. It has been found that p-type electrical activities approaching 100% for lowish doses (up to $\sim 10^{14}$ ions/cm^2) can be achieved in GaAs by implantation of Be$^+$, Zn$^+$ or Cd$^+$ ions at room temperature followed by thermal annealing of the implanted material at 800–900°C. For high ion doses some diffusion of the implanted element usually occurs during the annealing treatment, the diffusion rate probably being affected (enhanced or reduced) by the presence of implantation-induced lattice defects. The production of n-type conductivity in GaAs by implantation of the group VI ions S$^+$, Se$^+$ or Te$^+$ requires implantation at 200–400°C, followed also by an annealing at 800–900°C; it is thought that the hot implantation requirement is related to a known radiation-damage annealing stage at about 150°C. As an example of n-type doping, Fig. 12 (from [166]) shows the measured sheet electron concentration as a function of the temperature of implantation of 10^{14} S$^+$/cm^2 into GaAs preceding an anneal at 825°C; for implant temperatures of 200 to about 450°C an electrical activity of about 40% was obtained. For lowish doses of Se$^+$ and Te$^+$, activities of up to 80% can be produced [115]. Some diffusion of the group V implants usually occurs during the annealing treatment. Implantations of the group IV elements silicon [167, 168] and tin [169] into GaAs also produce n-type electrical conduction, and this donor activity implies that the silicon and tin must replace gallium in the lattice. The Si$^+$ implantation method of forming n-type layers is advantageous in that room-temperature implantation (following by anneal at 800–900°C) leads to high electrical activity; i.e. the more difficult, hot implantation is not needed.

Since all of these implantations of GaAs require heat treatments at 800–900°C to anneal lattice disorder and to produce the highest possible electrical activities and carrier mobilities, an encapsulant layer is needed around the GaAs to prevent its decomposition and the out-diffusion of the implanted element; it has been found (see discussion and references in [114] and [115]) that Si$_3$N$_4$ is very

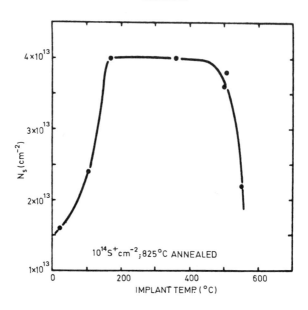

FIG. 12. The effect of the temperature at which 1-MeV sulphur ions are implanted into GaAs (annealed subsequently at 825°C) upon the sheet electron concentration N_s of the implanted surface. From Davies et al. [166].

effective for this purpose. Welch et al. in 1974 [170] described how the use of a layer of Si_3N_4 allowed successful annealing of S^+-implanted GaAs at 900°C and the fabrication of an n-channel depletion-mode, the Schottky-barrier field-effect transistor (metal-semiconductor FET or MESFET) having excellent characteristics.

It is clear that GaAs devices incorporating ion-implanted regions will form a very important part of future semiconductor technology. Firstly, the electron mobility in n-GaAs is very much larger than that in n-Si, and this gives the possibility of n-channel FET's working to higher frequencies as amplifiers, and with shorter switching times in logic circuits. Hunsperger and Hirsch in 1975 [171] described the fabrication by ion-implantation and the characteristics of an n-channel depletion-mode GaAs MESFET, with a 2 μm gate length, for microwave amplification up to 20 GHz. In 1976, Liechti [172] of the Hewlett-Packard Corporation described n-channel GaAs MESFET structures formed by ion implantation and having switching times of only 50–100 ps; he notes that all the required GaAs processing techniques including ion implantation are now available for GaAs fast-logic technology. Secondly, it is likely that ion-implantation will be valuable in fabrication of light-emitting diodes and lasers of GaAs and of other binary and ternary III–V compounds [173–176]. Thirdly, there are the very important microwave-producing devices such as the IMPATT diode; Donelly [115] discusses and gives references to several ion-implanted forms of

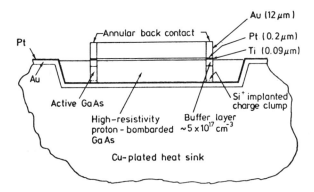

FIG.13. *The structure of a Pt-GaAs Schottky-barrier IMPATT diode in which a very efficient Lo-Hi-Lo n-type doping profile is obtained by Si^+ implantation; the annular active region is then defined by masking the device by means of the ring-shaped back contact during a series of 50–1400 keV proton irradiations. From Murphy et al. [177].*

this. Figure 13, from Murphy et al. in 1976 [177], shows a GaAs Schottky-barrier IMPATT diode of novel and ingenious annular design where proton irradiation, masked by the platinum-ring Schottky layer, was used finally to produce a high-resistivity [114, 115, 178] inactive region, after Si^+ implantation from the other side had been employed to form a high-efficiency n-type doping profile ("Lo-Hi-Lo"). The special annular shape improves heat dissipation in the device during operation as a microwave generator, and Murphy et al. report 7.4 W power output at 3 GHz with the very good conversion efficiency of 35%.

The conclusion to be drawn from the ion-implantation results for GaAs reviewed in the previous paragraphs is that this is an important and fast-developing field of semiconductor technology. Donelly in 1976 [115] expressed the belief that this technology should soon start to move from the research laboratory to the production line.

Finally in this section, let us not dismiss compound semiconductors other than GaAs without further mention. General discussion and references for ion implantation in these materials can be found in the articles by Allen [111] and Stephen [112]. The compounds such as SiC, GaP, GaN, ZnO, ZnS, ZnSe, ZnTe and CdS, with electronic band gaps greater than about 2 eV, are of special interest in connection with photoluminescence and electroluminescence.

Luminescence properties and lattice disorder related to ion implantation in ZnS, ZnSe and CdS were the subjects of six research papers at the Yorktown Heights Conference [9], and sharp cathodoluminescence lines due to implanted ytterbium in ZnS and several other II–VI compounds have been observed by Bryant and Fewster [179]. Bontemps et al. in 1974 [180] studied (by the channelling method) the lattice disorder produced in ZnTe by Ar^+ implantation and suggest that the saturation of the disorder before an amorphous state is reached is due to the strongly ionic nature of ZnTe. For the same compound, Pautrat et al. in 1976 [181] showed that ion implantation can produce an electrically compensated insulating region over a depth of several micrometres related to the long-range migration of defects; on the basis of an observed anomalously large penetration of ^{65}Zn implanted into ZnTe, it has been suggested [182] that these defects are interstitial zinc atoms. In earlier work [183], chlorine-implanted ZnTe annealed at 520°C was found to have n-type electrical activity, and n-p junctions and electroluminescence could be observed in implanted p-ZnTe. In ZnO, implantations of P$^+$ and V$^+$ ions have been observed [184] to give n-type conductivity, not due to lattice disorder but possibly due to impurity-band conduction. The doping of II–VI compounds by implantation was investigated initially on the grounds that high-temperature growth and diffusion methods of incorporating impurities always led in these ionic materials to electrical compensation by the formation of charged lattice defects; at the present time effective, practical ion-implantation of these materials seems also to be hindered by the interactions of defects during the implantation itself and during the annealing treatments.

As a valuable recent example of the use of ion-implantation to change the luminescence properties of III–V compounds we can note that Pankov and Hutchby in 1976 [185], in a study of the photoluminescence emission spectra of the wide-band-gap material GaN after implantation with each of thirty-five elements and anneal at 1050°C, found that the elements Mg, Zn, Cd, Ca, As, Hg and Ag produce characteristic emission bands between 1.5 and 3.4 eV. This could certainly be of value in the fabrication of light-emitting diodes to give outputs of various colours (e.g. the zinc emission band is blue).

8. PROPERTY CHANGES AND APPLICATIONS FOR GLASSES AND OTHER INSULATORS

The properties of glasses and other insulators depend, as do those of metals and semiconductors considered in Sections 6 and 7, upon their compositions and structures, and it is to be expected therefore that ion implantation can alter these properties. The experimental studies that have been made of ion implantation effects in these materials have been those where near-surface changes in properties or special surface patterns of implanted impurities could have value for device purposes. Three particular kinds of investigation have concerned changes in the refractive index in glasses, electro-optic information storage possibilities and lattice expansion effects in magnetic bubble garnets, again in connection with information storage. Each of these is briefly considered here.

Experimental investigations of how ion implantation can affect the refractive index of glassy materials have almost all concerned silica glass SiO$_2$. The subject has been discussed in reviews by Brown in 1971 [186], Dearnaley et al. in 1973 [116] and more recently by Townsend in 1975 [117]. In summary, the findings are that the refractive index of the SiO_2 can be increased by about 1% or more by ion-implantation but that the change is due to an irradiation-damage-induced compaction (increase of density) of the glass and is not ion-specific; the refractive index returns to its pre-implantation value upon thermal annealing. There has been much consideration of the possible use of the implantation-produced refractive index changes for fabricating optical light-guides [117, 186, 188, 189, 204], but the optical absorption associated with implantation-induced defects corresponds to a power loss of ~ 1 dB/cm, whereas optical communication guides require losses of $\lesssim 5$ dB/km [196]. Investigations of the disorder, chemical and other properties of implanted silica glass have also been made [190−192]. Mattern et al. in 1976 [191] suggested that SiO_2 could be useful as a protective layer for structural components subject to gaseous ion implantation in thermonuclear fusion reactors, since it seems that SiO_2, unlike other glasses, oxides and metals, would not flake or blister under such conditions.

The fabrication of electro-optic information storage devices (i.e. "memories") by ion implantation of insulators has been considered in the review article by Townsend [117]. He notes that both the electronic and optical properties of impurity centres can be important in giving information-storage possibilities, the writing in and reading out method being by a scanned electron or light beam; ion implantation doping of the insulator has the advantage of allowing the formation of an optimum depth profile of the impurity near the surface and of a fine-mesh impurity distribution pattern in the plane of the surface. He cites $LiNbO_3$, $BaTiO_3$, MgO and ZnO as being among insulators of interest. The storage of information in alkali halides by use of the anisotropic polarization-dependence of optical absorption by certain colour centre defects has been considered for a number of years (e.g. Schneider et al. in 1970 [193]). A recent paper by Magee and Lehmann [194] has described studies of the M-centre information storage properties of NaF implanted with Li^+, B^+, C^+, O^+ or F^+ ions. Polarized light (which orients the M-centres) in a spatially-modulated pattern provided the information for the NaF, and erasure of the information could be accomplished by illumination with unpolarized light. It was found that NaF implanted with the lightest ions (Li^+ and B^+) had memory storage and retrieval properties, and that of these the Li^+-implanted material was the more effective medium, allowing high-contrast, non-fading (in $2\frac{1}{2}$ years) storage of detailed holographic images. It seems that this whole area of electro-optical pattern and information storage, aided by the use of ion-implantation, could be of much future technological importance.

Finally, in this section we note the use of ion-implantation in connection with the properties of magnetic-bubble materials (Brown in 1971 [186], North and Wolfe in 1972 [118, 195] and Dearnaley et al. in 1973 [187]). Magnetic bubbles are cylindrical magnetic domains that can be formed in thin layers of the cubic ferrimagnetic insulators called iron garnets which have the general formula $R_3Fe_5O_{12}$ where R is an element of the lanthanum series or yttrium, and where some of the iron may be replaced by gadolinium or aluminium. It has been found that the lateral stress produced in an implanted surface region of an iron garnet crystal as a result of lattice defects induced by the implantation can, through magnetostrictive effects, change the direction of easy magnetization; this can allow magnetic bubble domains to be formed with axes perpendicular to the surface by the influence of smallish magnetic fields. The absence or presence of a bubble in a particular region can be used for binary information storage; field-induced bubble motion can in principle allow sequential read-out. Ion implantation can be used also [118, 195] to suppress "hard bubbles" which are magnetic bubble domains that require fields of up to $\sim$ 100 Oe for their formation instead of the usual value of about 75 Oe. It remains to be seen whether this technique of information storage can be developed to compete with the established and still progressing metal-insulator-semiconductor device methods.

9. LATTICE LOCATIONS OF IMPLANTED ATOMS

It is the lattice location of an implanted impurity that very often determines the effect that the impurity has upon the properties of the material, and the factors that influence the lattice site and the use of ion-channelling and other methods for determining the site are important aspects of ion-implantation. However, lack of space prevents a detailed discussion of this topic here and only references to relevant publications can be given. Suffice it then to say that review descriptions of the theoretical and experimental aspects of these can be found in a number of books (Mayer et al. [16], Townsend et al. [20], Carter and Grant [21]) and in recent articles (Davies [197], general features of the channelling method; de Waard and Feldman [198], channelling and hyperfine-interaction methods and results for metals; Picraux [199], the channelling method and results for metals and semi-conductors). Subsequent individual research papers include those of Beloshitsky et al. [200] on detailed studies of boron in silicon, Borders and Poate [201] concerning various implanted impurities in copper, silver, nickel and palladium, Ligeon and Guivarc'h [65] on hydrogen in silicon, Bugeat et al. [202] on hydrogen in aluminium, and Nojiri et al. [203] for boron-12 in nickel by use of a hyperfine-interaction nuclear technique. The papers by Ligeon et al. [65, 202] describe the very interesting use, in a channelling experiment, of the $^1H(^{11}B, \alpha)$ reaction induced by a $^{11}B^+$ beam to determine the lattice location of hydrogen.

10. CONCLUSIONS

This paper has aimed to give a comprehensive outline of the field of research known as ion implantation and to describe the very great variety of technological applications actually in current use or being investigated. For metals there are ion implantation effects of much potential technological importance in connection with surface hardness and wear, catalysis, corrosion resistance and super-conductivity, and the use of ion beams in the simulation of neutron irradiation damage is now well established. Ion implantation is employed as a production-line process in silicon integrated-circuit technology, and there are strong indications that it will soon be used commercially in the fabrication of field-effect and microwave devices from GaAs and related compound semiconductors; effects in other semiconducting materials are being investigated. It seems likely that ion-implantation will also be of definite value for making pattern-storage devices in insulating optical materials. An essential background to all such studies is formed by the LSS-based calculations of the spatial distributions of implanted ions and of the energy depositions into elastic collision and electronic excitation processes, these however still being the subjects of current experimental and theoretical study.

Ion implantation relates to many aspects of physics, chemistry, materials science and electronics; it continues to be an important, exciting and still developing field of basic and applied research.

ACKNOWLEDGEMENTS

The author would like to thank Dr. J.C. Combasson for valuable discussions in connection with various ion-range concepts, and also to express his gratitude to Miss L. Loveday for her careful typing of the paper.

REFERENCES

[1] Proc. Int. Conf. III on Atomic Collision Phenomena in Solids, University of Sussex, UK, 1969 (PALMER, D.W., THOMPSON, M.W., TOWNSEND, P.D., Eds), North Holland Publishing Co. (1970).

[2] Proc. Int. Conf. IV on Atomic Collisions in Solids, Gausdal, Norway, 1971 (ANDERSEN, S., BJÖRKQVIST, K., DOMEIJ, B., JOHANSSON, N.G.E., Eds), Gordon and Breach (1972); also in Radiat. Effects **12** and **13** (1972).

[3] Proc. Int. Conf. V on Atomic Collisions in Solids, Gatlinburg, USA, 1973 (DATZ, S., APPLETON, B.R., MOAK, C.D., Eds), Plenum Press (1975).

[4] Proc. Int. Conf. VI on Atomic Collisions in Solids, Amsterdam, 1975 (SARIS, F.W., Van der WEG, W.F., Eds), North Holland Publishing Co. (1976); also as Nucl. Instrum. Methods **132** (1976).

[5] Proc. Int. Conf. on the Applications of Ion Beams to Semiconductor Technology, Grenoble, France, 1967, Ophrys, Paris (1968).

[6] Proc. Int. Conf. on Ion Implantation in Semiconductors, Thousand Oaks, California, 1970 (EISEN, F.H., Ed.), Gordon and Breach (1971); also in Radiat. Effects 6 (1970) and 7 (1971).

[7] Proc. Eur. Conf. on Ion Implantation, Reading, UK, 1970, Peregrinus Ltd., Stevenage, UK (1971).

[8] Proc. Int. Conf. on Ion Implantation in Semiconductors, Garmisch-Partenkirchen, Fed. Republic of Germany, 1971 (RUGE, I., GRAUL, J., Eds), Springer-Verlag, Berlin (1971).

[9] Proc. Int. Conf. on Ion Implantation in Semiconductors and Other Materials, Yorktown Heights, USA, 1972 (CROWDER, B.L., Ed.), Plenum Press (1973).

[10] Proc. Int. Conf. on Applications of Ion Beams to Metals, Albuquerque, USA, 1973 (PICRAUX, S.T., EERNISSE, E.P., VOOK, F.L., Eds), Plenum Press (1974).

[11] Proc. Int. Conf. on Ion Implantation in Semiconductors and Other Materials, Osaka, Japan, 1974 (NAMBA, S., Ed.), Plenum Press (1975).

[12] Proc. Int. Conf. on Applications of Ion Beams to Materials, University of Warwick, UK, 1975 (CARTER, G., COLLINGON, J.S., GRANT, W.A., Eds), Institute of Physics, London, Conf. Series 28 (1976).

[13] Proc. Int. Conf. on Ion Implantation in Semiconductors and Other Materials, Boulder, USA, 1976 (CHERNOW, F., Ed.), Plenum Press (1977).

[14] NELSON, R.S., The Observation of Atomic Collisions in Crystalline Solids, North Holland Publishing Co. (1968).

[15] CARTER, G., COLLINGTON, J.S., Ion Bombardment of Solids, Heinemann, London (1968).

[16] MAYER, J.W., ERICKSSON, L., DAVIES, J.A., Ion Implantation in Semiconductors, Academic Press (1970).

[17] DEARNALEY, G., FREEMAN, J.H., NELSON, R.S., STEPHEN, J., Ion Implantation, North Holland Publishing Co. (1973).

[18] MORGAN, D.V., Ed., Channeling, Wiley (1973).

[19] WILSON, R.G., BREWER, G.R., Ion Beams – with Applications to Ion Implantation, Wiley (1973).

[20] TOWNSEND, P.D., KELLY, J.C., HARTLEY, N.E.W., Ion Implantation, Sputtering and their Applications, Academic Press (1976).

[21] CARTER, G., GRANT, W.A., Ion Implantation of Semiconductors, Arnold Ltd, London (1976).

[22] ZIEGLER, J., Ed., New Uses of Ion Accelerators; Plenum Press (1975).

[23] DEARNALEY, G., Ion bombardment and implantation, Rep. Prog. Phys. 32 (1969) 405.

[24] DEARNALEY, G., Ion implantation, Annu. Rev. Mater. Sci. 4 (1974) 93.

[25] GRANT, W.A., WILLIAMS, J.S., The modification of surface layers by ion implantation, Sci. Prog., Oxf. 63 (1976) 27.

[26] PALMER, D.W., "Ion-irradiation, implantation and channelling effects in solids", Utilization of Low Energy Accelerators, Unpriced document 171, IAEA, Vienna (1975) 23.

[27] AGAJANIAN, A.H., Ion implantation – an annotated bibliography, Radiat. Eff. 23 (1974) 73.

[28] PLUNKETT, J.C., STONE, J.L., A selected bibliography on ion implantation in solid state technology, Solid State Technol. Dec. (1975) 49.

[29] DEARNALEY, G., in Ref. [17] p. 9.

[30] LINDHARD, J., K. Dan. Vidensk. Selsk., Mat.-Fys. Medd. **28** 8 (1954).
[31] SCHIØTT, H.E., Stopping power and range-energy relations, Proc. C. Erginsoy Memorial Meeting, Istanbul, 1972.
[32] THOMPSON, M.W., NEILSON, G., Phys. Lett., **A49** (1974) 151.
[33] FEUERSTEIN, A., KALBITZER, S., OETZMANN, H., Phys. Lett., A **51** (1975) 165.
[34] ANDERSEN, H.H., BØTTIGER, J., JØRGENSEN, H.W., Appl. Phys. Lett. **26** (1975) 678.
[35] BARAGIOLA, R.A., CHIVERS, D., DODDS, D., GRANT, W.A., WILLIAMS, J.S., Phys. Lett., A **56** (1976) 371.
[36] GRANT, W.A., DODDS, D., WILLIAMS, J.S., CHRISTODOULIDES, C.E., BARAGIOLA, R.A., CHIVERS, D., in Ref. [13].
[37] COMBASSON, J.L., FARMERY, B.W., McCULLOCH, D., NEILSON, G.W., THOMPSON, M.W., (University of Sussex), 1977, to be published in Radiation Effects.
[38] BOHR, N., LINDHARD, J., K. Dan. Vidensk. Selsk., Mat.-Fys. Medd. **28** 7 (1954).
[39] LINDHARD, J., SCHARFF, M., SCHIØTT, H.E., K. Dan. Vidensk. Selsk., Matt.-Fys. Medd. **33** 14 (1963).
[39a] GOMBÁS, P., Die Statistische Theorie des Atoms; Springer-Verlag, Vienna (1949).
[40] LINDHARD, J., SCHARFF, M., Phys. Rev. **124** (1961) 128.
[41] SCHIØTT, H.E., Radiat. Eff. **6** (1970) 107.
[42] BETHE, H.A., Ann. Phys. (Leipzig) **5** (1930) 325.
[43] BLOCH, F., Ann. Phys. (Leipzig) **16** (1933) 285.
[44] SCHIØTT, H.E., Can. J. Phys. **46** (1968) 449.
[45] SCHIØTT, H.E., K. Danske Vidensk. Selsk., Mat.-Fys. Medd. **35** 9 (1966).
[46] SIGMUND, P., SANDERS, J.B., in Ref. [5] p. 215.
[47] BRICE, D.K., Radiat. Eff. **11** (1971) 227.
[48] WINTERBON, K.B., Radiat. Eff. **13** (1972) 215.
[49] BRICE, D.K., in Ref. [9] p. 171.
[50] JOHNSON, N.L., KOTZ, S., Continuous Univariate Distributions, Vol. II, Houghton-Mifflin, New York (1970).
[51] MYLROIE, S., GIBBONS, J.F., in Ref. [9] p. 243.
[52] GIBBONS, J.F., JOHNSON, W.S., MYLROIE, S.W., Projected Range Statistics: Semiconductors and Related Materials, Dowden, Hutchinson and Ross Inc. (1975) (distributed by Halstead Press, a Division of John Wiley and Sons Inc.).
[53] FURUKAWA, S., MATSUMURA, H., ISHIWARA, H., Jap. J. Appl. Phys. **11** (1972) 134.
[54] FURUKAWA, S., MATSUMURA, H., in Ref. [9] p. 193.
[55] HVELPLUND, P., FASTRUP, B., Phys. Rev. **165** (1968) 408.
[56] CHESHIRE, I., DEARNALEY, G., POATE, J.M., Proc. R. Soc. (London), Ser. A **311** (1969) 47.
[57] LAND, D.J., BRENNAN, J.G., in Ref. [4] p. 89.
[58] FIRSOV, O.B., Sov. Phys. JETP **9** (1959) 1076.
[59] EISEN, F.H., Canad. J. Phys. **46** (1968) 561.
[60] MOAK, C.D., et al. in Ref. [4] p. 95.
[61] BETZ, G., ISELE, H.J., RÖSSLE, E., HORTIG, G., Nucl. Instrum. Methods **123** (1975) 83.
[62] GRAHMANN, H., KALBITZER, S., in Ref. [4] p. 119.
[63] NORTHCLIFFE, L.C., SCHILLING, R.F., Nucl. Data Tables A **7** (1970) 254.
[64] ZIEGLER, J.F., CHU, W.K., Atomic and Nucl. Data Tables **13** (1974) 463.
[65] LIGEON, E., GUIVARC'H, A., Radiat. Eff. **27** (1976) 129.
[66] LANGLEY, R.A., BLEWER, R.S., in Ref. [4] p. 109.
[67] BRICE, D.K., Phys. Rev., A **6** (1972) 1791.

[68] BRICE, D.K., Radiat. Eff. **6** (1970) 77.
[69] VOOK, F.L., Radiation Damage and Defects in Semiconductors (Proc. Int. Conf. Reading, UK, 1972), Institute of Physics, London and Bristol, Conf. Series 16 (1973) 60.
[70] KINCHIN, G.H., PEASE, R.S., Rep. Prog. Phys. **18** (1955) 1.
[71] SIGMUND, P., Appl. Phys. Lett. **14** (1969) 114.
[72] BRICE, D.K., Ion Implantation Range and Energy Deposition Distributions, Vol. 1, Plenum Press (1975).
[73] PALMER, D.W., Radiation Effects in Semiconductors (Proc. Int. Conf. Dubrovnik, 1976), Institute of Physics, London and Bristol, Conf. Series 31 (1977) 144.
[74] DEARNALEY, G., in Ref. [7] p. 162.
[75] FREEMAN, J.H., in Ref. [12] p. 340.
[75a] THOMPSON, D.A., ROBINSON, J.E., in Ref. [4] p. 261.
[76] THOMPSON, M.W., in Ref. [7] p. 109.
[77] PICRAUX, S.T., in Ref. [12] p. 183.
[78] NELSON, R.S., in Ref. [10] p. 221.
[79] DEARNALEY, G., in Ref. [22] p. 283.
[80] DEARNALEY, G., in Ref. [10] p. 63.
[81] GRANT, W.A., in Ref. [12] p. 127.
[82] HARTLEY, N.E.W., in Ref. [12] p. 210.
[83] MEYER, O., in Ref. [22] p. 323.
[84] STRITZKER, B., in Ref. [12] p. 160.
[85] NELSON, R.S., MAZEY, D.J., HUDSON, J.A., J. Nucl. Mater. **37** (1970) 1.
[86] NORRIS, D.I.R., Radiat. Eff. **14** (1972) 1 and **15** (1972) 1.
[87] KULCINSKI, G.L., in Ref. [10] p. 613.
[88] JOHNSTON, W.G., ROSOLOWSKI, J.H., in Ref. [12] p. 228.
[89] ENGLISH, C.A., in Ref. [12] p. 257.
[90] GLOTIN, P., in Ref. [7] p. 46.
[91] BEALE, J.R.A., in Ref. [7] p. 81.
[92] LEE, D., MAYER, J., Proc. IEEE **62** (1974) 1241.
[93] MOREHEAD, F.F., CROWDER, B.L., Sci. Am. **228** 4 (1973) 64.
[94] KIMERLING, L.C., POATE, J.M., Lattice Defects in Semiconductors (Proc. Int. Conf. Freiburg, 1974), Institute of Physics, London and Bristol, Conf. Series 23 (1975) 126.
[95] DILL, H.G., BOWER, R.W., TOOMBS, T.N., in Ref. [6] and Radiat. Eff. **7** (1971) 45.
[96] MACRAE, A.U., in Ref. [6] and Radiat. Eff. **7** (1971) 59.
[97] DILL, H.G., TOOMBS, T.N., BAUER, L.O., in Ref. [8] p. 315.
[98] MACRAE, A.U., in Ref. [8] p. 329.
[99] PRUSSIN, S., et al., J. Electrochem. Soc. (US) **121** (1974).
[100] SHACKLE, P.W., et al., J. Vac. Sci. Technol. (US) **10** (1973) 1090.
[101] SOPIRA, M.M., et al., J. Vac. Sci. Technol. (US) **10** (1973) 1086.
[102] TOKUYAMA, T., in Ref. [13].
[103] STEPHEN, J., in Ref. [17] p. 685.
[104] DEARNALEY, G., HARDACRE, A.G., ROGERS, B.D., Nucl. Instrum. Methods **71** (1969) 86.
[105] HERZER, H., KALBITZER, S., in Ref. [8] p. 307.
[106] PONPON, J.P., et al., in Ref. [8] p. 420.
[107] SCHMID, K., et al., Phys. Status Solidi A **23** (1974) 523.
[108] MACDONALD, P.J., PALMER, D.W., Lattice Defects in Semiconductors (Proc. Int. Conf. Freiburg, 1974), Institute of Physics, London and Bristol, Conf. Series 23 (1975) 504.

[109] BOURGOIN, J.C., WALKER, J., Diamond Res. 24 (1975).
[110] VAVILOV, V.S., Phys. Status Solidi A 31 (1975) 11.
[111] ALLEN, R.M., in Ref. [7] p. 127.
[112] STEPHEN, J., in Ref. [17] p. 689.
[113] EISEN, F.H., in Ref. [11] p. 3
[114] HEMMENT, P.L.F., in Ref. [12] p. 44.
[115] DONNELLY, J.P., in Ref. [165] p. 166.
[116] DEARNALEY, G., et al., in Ref. [17] p. 724.
[117] TOWNSEND, P.D., in Ref. [12] p. 104.
[118] NORTH, J.C., WOLFE, R., in Ref. [9] p. 505.
[119] DEARNALEY, G., GOODE, P.D., MILLER, W.S., TURNER, J.F., in Ref. [9] p. 405.
[120] ANTILL, J.E., BENNETT, M.J., DEARNALEY, G., FERN, F.H., GOODE, P.D., TURNER, J.F., in Ref. [9] p. 415.
[121] NAGUIB, H.M., KRIEGLER, R.J., DAVIES, J.A., MITCHELL, J.B., J. Vac. Sci. Technol. 13 (1976) 396.
[122] ASHWORTH, V., BAXTER, D., GRANT, W.A., PROCTOR, R.P.M., in Ref. [11].
[123] GRENNESS, M., THOMPSON, M.W., CAHN, R.W., J. Appl. Electrochem. 4 (1974) 211.
[124] KANAYA, K., KOGA, K., TOKI, K., J. Phys., E (London) 5 (1972) 541.
[125] GABOVICH, M.D., BUDERNAYA, L.D., PORITSKII, V.Y., PROTSENKO, I.M., Proc. Kiev Meeting on Ion Beam Physics (1974).
[126] PAVLOV, A.V., PAVLOV, P.V., ZORIN, E.I., TETEL'BAUM, D.I., Proc. Kiev Meeting on Ion Beam Physics (1974).
[127] HARTLEY, N.E.W., DEARNALEY, G., TURNER, J.F., in Ref. [9] p. 423.
[128] HARTLEY, N.E.W., DEARNALEY, G., TURNER, J.F., SAUNDERS, J., in Ref. [10] p. 123.
[129] THOMPSON, M.W., Defects and Radiation Damage in Metals, Cambridge University Press (1969).
[130] DEERY, M., GOH, K.H., STEPHENS, K.G., WILSON, I.H., Thin Solid Films 17 (1973) 59.
[131] WILSON, I.H., GOH, K.H., STEPHENS, K.G., in Ref. [10] p. 269.
[132] BARDEEN, J., COOPER, L.N., SCHRIEFFER, J.R., Phys. Rev. 108 (1957) 1175.
[133] POATE, J.M., TESTARDI, L.R., STORM, A.R., AUGUSTYNIAK, W.M., in Ref. [12] p. 176.
[134] STRITZKER, B., BUCKEL, W., Z. Phys. 257 (1972) 1.
[135] STRITZKER, B., BECKER, J., Phys. Lett., A 51 (1975) 147.
[136] SKOSHIEWICZ, T., Phys. Status Solidi A 11 (1972) K123.
[137] MEYER, O., in Ref. [12] p. 168.
[138] Proc. Int. Conf. on Fundamental Aspects of Radiation Damage in Metals, Gatlinburg, 1975, USERDA, NSF and ORNL, CONF – 751006 P1 and P2.
[139] VOOK, F.L., et al., Rev. Mod. Phys. 47 Suppl. No. 3 (1975), Report to the American Physical Society by the study group on physics problems relating to energy technologies: radiation effects on materials.
[140] VOOK, F.L., Physics Today, Sep. (1975) 34.
[141] HUDSON, J.A., J. Nucl. Mater. 60 (1976) 89.
[142] BULLOUGH, R., PERRIN, R.C., in Proc. Int. Conf. on Radiation-Induced Voids in Metals (CORBETT, J.W., IANNIELLO, L.C., Eds), Albany, NY, USA, 1971; AEC Symposium Series CONF – 710601, page 769.
[143] MARWICK, A.D., J. Nucl. Mater. 55 (1975) 259.
[144] ROTH, J., in Ref. [12] p. 280.

[145] NELSON, R.S., in Ref. [17] p. 226.
[146] VOGEL, F.L., Thin Solid Films **27** (1975) 369.
[147] TURNER, J.F., TEMPLE, W., DEARNALEY, G., in Ref. [9] p. 437.
[148] VAN WYK, G.N., SMITH, H.J., Radiat. Eff. **30** (1976) 91.
[149] NICHOLAS, K.H., Radiat. Eff. **28** (1976) 177.
[150] FANG, F., HATZAKIS, M., TING, C.H., J. Vac. Sci. Technol. **10** (1973) 1082.
[151] KEYES, R.W., Science **195** 4283 (1977) 1230.
[152] EKNOYAN, O., YANG, E.S., SZE, S.M., Solid State Electronics **20** (1977) 291.
[153] MARSHALL, S., Solid State Technol. **17** 11 (1974) 29.
[154] DAVIDSON, L.A., CHOU, S., GIBBONS, J.F., JOHNSON, W.S., Radiat. Eff. **7** (1971) 35.
[155] COLLINS, A.T., WILLIAMS, A.W.S., J. Phys., C (London) **4** (1971) 1789.
[156] CLARK, C.D., MITCHELL, E.W.J., Radiat. Eff. **9** (1971) 219.
[157] BROSIOUS, P.R., CORBETT, J.W., BOURGOIN, J.C., Phys. Status Solidi A **21** (1974) 677.
[158] BLANCHARD, B., COMBASSON, J.L., BOURGOIN, J.C., Appl. Phys. Lett. **28** (1976) 7.
[159] MEYER, O., IEEE NS – **15** (1968) 232.
[160] ITOH, T., OHDOMARI, I., Jap. J. Appl. Phys. **10** (1971) 1002.
[161] GUSEVA, M.I., MANSUROVA, A.N., Radiat. Eff. **20** (1973) 207.
[162] MACDONALD, P.J., D. Phil. Thesis, University of Sussex, UK (1974).
[163] McDONALD, R.E., McNAB, T.K., Phys. Rev. **13** (1976) 39.
[164] Proc. Sixth Int. Symp. on Gallium Arsenide and Related Compounds, Edinburgh, 1976 (HILSUM, C., Ed.), Institute of Physics, Bristol and London, Conf. Series 33a (1977).
[165] Proc. Sixth Int. Symp. on Gallium Arsenide and Related Compounds, St. Louis, 1976 (EASTMAN, L.F., Ed.), Institute of Physics, London and Bristol, Conf. Series 33b (1977).
[166] DAVIES, D.E., ROOSILD, S., LOWE, L., Solid State Electronics **18** (1975) 733.
[167] SANSBURY, J.D., GIBBONS, J.F., Radiat. Eff. **6** (1970) 269.
[168] HARRIS, J.S., in Ref. [8] p. 157.
[169] WOODCOCK, J.M., SHANNON, J.M., CLARK, D.J., Solid State Electronics **18** (1975) 267.
[170] WELCH, B.M., EISEN, F.H., HIGGINS, J.A., J. Appl. Phys. **45** (1974) 3685.
[171] HUNSPERGER, R.G., HIRSCH, N., Solid State Electronics **18** (1975) 349.
[172] LIECHTI, C.A., in Ref. [164] p. 227.
[173] ONO, Y., SAITO, K., SHIRAKI, Y., SHIMADA, T., Jap. J. Appl. Phys. **14** (1975) 1489.
[174] ITOH, T., OANA, Y., Appl. Phys. Lett. **24** (1974) 320.
[175] CHATTERJEE, P.K., VAIDYANATHAN, K.V., McLEVIGE, W.V., STREETMAN, B.G., Appl. Phys. Lett. **27** (1976) 567.
[176] BARNOSKI, M.K., HUNSPERGER, R.G., LEE, A., Appl. Phys. Lett. **24** (1974) 627.
[177] MURPHY, R.A., in Ref. [165] p. 210.
[178] DAVIES, D.E., KENNEDY, J.K., HAWLEY, J.J., in Ref. [12] p. 81.
[179] BRYANT, F.J., FEWSTER, R.H., Radiat. Eff. **20** (1973) 239.
[180] BONTEMPS, A., LIGEON, E., DANIELOU, R., Radiat. Eff. **22** (1974) 195.
[181] PAUTRAT, J.L., BENSAHEL, D., KATIRCIOGLU, B., PFISTER, J.C., REVOIL, L., Radiat. Eff. **30** (1976) 107.
[182] LIGEON, E., PAUTRAT, J.L., BOURIANT, M., Phys. Lett., A **59** (1976) 307.
[183] MARINE, J., in Ref. [7] p. 153.
[184] THOMAS, B.W., WALSH, D., J. Phys., D (London) **6** (1973) 612, and personal communication.

[185] PANKOVE, J.I., HUTCHBY, J.A., J. Appl. Phys. **47** (1976) 5387.
[186] BROWN, W.L., in Ref. [8] p. 430.
[187] DEARNALEY, G., et al., in Ref. [17] p. 720.
[188] BAYLEY, A.R., Radiat. Eff. **18** (1973) 111.
[189] NAMBA, S., ARITOME, H., NISHIMURA, T., MASUDA, K., TOYODA, K., J. Vac. Sci. Technol. **10** (1973) 936.
[190] EERNISSE, E.P., J. Appl. Phys. **45** (1974) 167.
[191] MATTERN, P.L., THOMAS, G.J., BAUER, W., J. Vac. Sci. Technol. **13** (1976) 430.
[192] WEBB, A.P., HOUGHTON, A.J., TOWNSEND, P.D., Radiat. Eff. **30** (1976) 177.
[193] SCHNEIDER, I., MARRONE, M., KABLER, M.N., Appl. Optics **9** (1970) 1163.
[194] MAGEE, T.J., LEHMANN, M., in Ref. [12] p. 112.
[195] WOLFE, R., NORTH, J.C., Bell Syst. Tech. J. **51** (1972) 1436.
[196] Electronic Engineering (UK), May (1977) 5.
[197] DAVIES, J.A., in Ref. [18] p. 391.
[198] DE WAARD, H., FELDMAN, L.C., in Ref. [10] p. 317.
[199] PICRAUX, S.T., in Ref. [22] p. 229.
[200] BELOSHITSKY, V.V., DIKII, N.P., KUMAKHOV, M.A., MATYASH, P.P., SKAKUN, N.A., Radiat. Eff. **25** (1975) 167.
[201] BORDERS, J.A., POATE, J.M., Phys. Rev., B **13** (1976) 969.
[202] BUGEAT, J.P., CHAMI, A.C., LIGEON, E., Phys. Lett., A **58** (1976) 127.
[203] NOJIRI, Y., HAMAGAKI, H., SUGIMOTO, K., Phys. Lett., A **60** (1977) 77.
[204] TOWNSEND, P.D., J. Phys., E (London) **10** (1977) 197.

TRACK FORMATION

Principles and applications

M. MONNIN
CNRS, Laboratoire de Physique Corpusculaire,
Université de Clermont,
Clermont-Ferrand, Aubière,
France

Abstract

TRACK FORMATION: PRINCIPLES AND APPLICATIONS.
　　The principles and technical aspects of track formation in insulating solids are first described. The characteristics of dialectic track detection are discussed from the technical point of view: the nature of the detectors, the chemical treatment, the sensitivity and the environmental conditions of use. The applications are reviewed. The principle of each type of applied research is described and then the applications are listed. When used as a detector, nuclear tracks can provide valuable information in a number of fields: element content determination and wrapping, imaging, radiation dosimetry, environmental studies, technological uses and miscellaneous other applications. The track-formation process can also be used for making well-def defined holes; this method allows other applications which are also described. Finally, some possible future applications are mentioned.

1. INTRODUCTION

　　The continuous damage trail created by fission fragments in a dielectric solid was first observed by E.C.H. Silk and R.S. Barnes, and the etching enlargement technique was first described by D.A. Young. But their papers remained comparatively unnoticed and it was only the remarkable work of P.B. Price, R.L. Fleischer and R.M. Walker that put the now well-known track-formation processes in the spotlight. Their long list of discoveries began in 1962. They showed that the tracks could be etched and enlarged so as to become visible under the optical microscope and that most insulating solids exhibited this property; they discovered the sensitivity energy threshold and later the fossil tracks in uranium-bearing natural rocks; they proposed models for the track-formation mechanism. At the early stage of development of this technique they described how what was already known as "solid-state nuclear detectors" could benefit a lot of research areas including nuclear physics, solid-state physics, cosmic-ray studies, extraterrestrial material studies, geology and so on. Not only did Price,

Fleischer and Walker carry out pioneering work, but they also created a broad new discipline. Their work was followed by many laboratories all over the world and now hundreds of research teams are involved in track studies.

A comprehensive review of the subject entitled Nuclear Tracks in Solids was published in 1975 by Fleischer, Price and Walker [1] and contains most of the data available to-day. In this paper, the opportunities offered by the track technique are outlined and some of the most recent literature references are given.

2. TRACK FORMATION: PRINCIPLES

2.1. The basic fact

"The passage of heavily ionizing particles through most insulating solids creates narrow paths of intense damage on an atomic scale. These damage tracks may be revealed and made visible in an ordinary optical microscope by treatment with a properly chosen chemical reagent that rapidly and preferentially attacks the damaged material. It less rapidly removes the surrounding undamaged matrix in such a manner as to enlarge the etched holes that mark and characterize the sites of the original, individual damaged region" [1]. In the following text, the term "tracks" will always refer to such a revealed channel etched along the path of a heavy ionizing particle.

2.2. Heavy ions interaction with matter

The density of damage created along the trajectory of any charged particle can be roughly correlated with the so-called "linear energy transfer" (LET). The charged particles can be classified into two main groups: the low LET particles (electrons, muons, pions, elementaries, and the high LET particles (ions). The low LET particles are unable to create sufficient damage to give rise to a continuous etched track. Therefore, we will only deal with ions ranging from alpha particles to uranium ions, with a possible but infrequent extension to protons.

When an ion is travelling through a solid it primarily interacts with the electrons belonging to the atoms or molecules within the medium. In this process, electrons are either excited to a higher energy level or ejected from their parent body and then act themselves in the same manner, i.e. by interacting with other electrons (delta rays). In the process, the ionizing particle alternatively loses some of its own electrons and captures electrons from the medium. This purely Coulombian interaction is responsible for the slowing down of the moving particles, for the fact that an effective average charge Z^* can be attributed to a given ion travelling in a solid of known composition and, finally, for the damages induced in the bombarded medium.

The rate of energy loss as well as the effective charge can be calculated from various mathematical functions (see, for example, Refs [2–6]). For practical purposes, energy loss rate and range can be found in tables such as Northcliff's [7] and Benton's [8]. From these data, one notices that the energy transfer from ions to a medium is more than a thousand times higher than for electrons, even for the lightest ion (alpha particles). Furthermore, recent calculations [8, 9], already corroborated by experiments, have shown that the energy released to the medium is confined within a very narrow region around the ion path; in addition, this energy is deposited in a very short period of time. Accordingly, not only is the energy density extremely high (thousands of J/g) but the power impulse is tremendous (of the order of 100 GW/g). Therefore, it is not surprising that fast moving ions will induce a lot of phenomena which will result ultimately in a trail of high-density specific damages. Various mechanisms have been proposed to explain the ways by which these damages are created both in organic and inorganic media and, similarly, the chemical etching itself has been thoroughly investigated. A complete review of the current state of the problem can be found in Ref. [1].

Let us just remember that the passage of a moving ion through an insulating solid results in a high concentration of damage along the ion trajectory and that this damage trail can be chemically etched and enlarged so as to become visible under an optical microscope (see Figs 1 and 2).

We will now look at the practical characteristics of the insulating solids that exhibit preferential etching after ion penetration. These sensitive media are also called "solid-state nuclear track detectors" (SSNTD) or "dielectric track detectors".

2.3. Characteristics of solid-state track detectors

2.3.1. Nature

One will notice first that the track detectors belong to the organic world as well as to the inorganic one. High polymers and plastics, minerals and glasses do exhibit these dual properties. Nevertheless, they have to be dielectric media to store the damages properly and a lower limit of 2000 $\Omega \cdot$ cm is a minimum value for their electrical resistivity. The nature of the most commonly used detectors, together with their registration sensitivity (see later), are given in Table I.

2.3.2. Etching techniques

The etching procedure for a solid that properly records and stores tracks can be defined by:

  (a) the composition and concentration of the chemical reagent,
  (b) the temperature of the etching bath, and
  (c) the duration of the chemical attack.

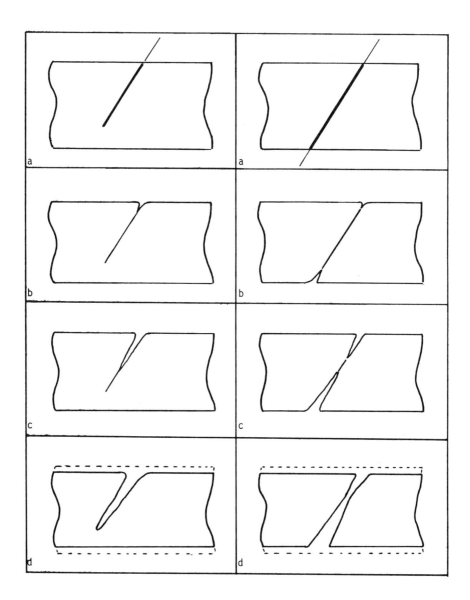

FIG.1. Nuclear-track formation by an ion stopping within the detector (left) and by an ion travelling through the detector (right).
(a) Energy deposition by the ion.
(b) and (c) Track formation during etching.
(d) Final shape of the track.

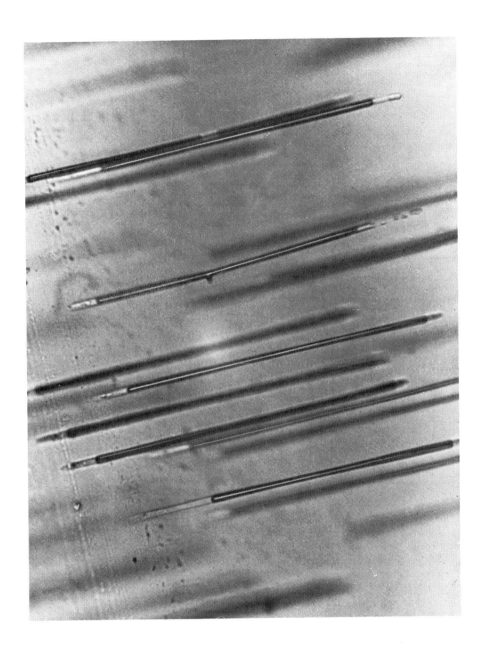

FIG.2. Iron-ion tracks in polycarbonate.

TABLE I. REGISTRATION SENSITIVITY [1]

	Organic detectors	
Detector	Atomic composition	Least ionizing ion seen
Amber	$C_2H_3O_2$	Full-energy fission fragments
Phenoplaste	C_7H_6O	
Polyethylene	CH_2	Fission fragments
Polystyrene	CH	
Polyvinylacetochloride	$C_6H_9O_2Cl$	42 MeV ^{32}S
Polyvinylchloride-polyvinylidene chloride copolymer	$C_2H_3Cl + C_2H_2Cl_2$	42 MeV ^{32}S
Polyethylene terephthalate (Cronar, Melinex)	$C_5H_4O_2$	
Polyimide	$C_{11}H_4O_4N_2$	36 MeV ^{16}O
Ionomeric polyethylene (Surlyn)		36 MeV ^{16}O
Bisphenol A-polycarbonate (Lexan, Makrofol)	$C_{16}H_{14}O_3$	0.3 MeV ^{4}He
Polyoxymethylene (Delrin)	CH_2O	28 MeV ^{11}B
Polypropylene	CH_2	1 MeV ^{4}He
Polyvinylchloride	C_2H_3Cl	
Polymethylmethacrylate (Plexiglas)	$C_5H_8O_2$	3 MeV ^{4}He
Cellulose acetate butyrate	$C_{12}H_{18}O_7$	
Cellulose triacetate (Cellit, Triafol-T, Kodacel TA-401 unplasticized)	$C_3H_4O_2$	
Cellulose nitrate (Daicell)	$C_6H_8O_9N_2$	0.55 MeV ^{1}H

TABLE I (Cont.)

	Organic detectors	
Hypersthene	$Mg_{1.5}Fe_{0.5}Si_2O_6$	100 MeV ^{56}Fe
Olivine	$MgFeSiO_4$	
Labradorite	$Na_2Ca_3Al_8Si_{12}O_{40}$	
Zircon	$ZrSiO_4$	
Bronzite	$Mg_{1.7}Fe_{0.3}Si_2O_6$	
Enstatite	$MgSiO_3$	
Diopside	$CaMg(SiO_3)_2$	170 MeV ^{56}Fe
Augite	$CaMg_3Fe_3Al_2Si_4O_{19}$	170 MeV ^{56}Fe
Oligoclase	$Na_4CaAl_6Si_{14}O_{40}$	4 MeV ^{28}Si
Bytownite	$NaCa_4Al_9Si_{11}O_{40}$	4 MeV ^{28}Si
Orthoclase	$KAlSi_3O_8$	100 MeV ^{40}Ar
Quartz	SiO_2	100 MeV ^{40}Ar
Phlogopite mica	$KMg_2Al_2Si_3O_{10}(OH)_2$	
Muscovite mica	$KAl_3Si_3O_{10}(OH)_2$	2 MeV ^{20}Ne
Silica glass	SiO_2	16 MeV ^{40}Ar
Flint glass	$18SiO_2:4PbO:1.5Na_2O:K_2O$	2–4 MeV ^{20}Ne
Tektite glass (Obsidian similar)	$22SiO_2:2Al_2O_3:FeO$	
Soda lime glass	$23SiO_2:5Na_2O:5CaO:Al_2O_3$	20 MeV ^{20}Ne
Phosphate glass	$10P_2O_5:1.6BaO:Ag_2O:2K_2O:2Al_2O_3$	

Usually, only a few chemical reagents are suitable for etching a particular detector. In contrast, temperature, time and concentration can be varied within a rather large range, according to the particular purpose of the experiment or the experimentor's personal habits or preference. Table II gives the etching conditions for revealing fission fragment tracks in various detectors. For other particles, the etching conditions can be derived from this table after a few preliminary tests (according to the charge and speed of the particle one seeks to get tracks of) like increasing or decreasing the temperature, or concentration, or both, or by combining time, temperature and concentration accordingly.

The best results are obtained with simple equipment. A temperature-controlled bath contains a vessel with the etching solution which is occasionally stirred. The detector is allowed to stay in the vessel in such a way that it is entirely surrounded by the etching solution. After etching, the detector is removed, carefully washed and dried. In some cases, when only a quick and/or approximate result is wanted, the procedure can be simplified to the extreme: heat up the reagent in a beaker, drop the detector in it, wait, remove the detector, wipe it and examine it under the microscope. This rusticity in the use of track detectors is not the least of their advantages, especially for practical applications.

2.3.3. Sensitivity

Not all the detectors exhibit the same sensitivity to heavily ionizing radiation. Some of them are able to record fission fragments only, while others record alpha particles or even low-energy protons. Generally, plastic detectors are more sensitive than minerals. There has been a lot of argumentation among the scientific community to decide which criterion should be used to characterize the sensitivity of a particular recording solid. Actually, there is only one definite way to do it, that is to state the limit in energy (or speed) above which a particular ion is no longer recorded in the considered detector etched with a well-defined reagent concentration at a fixed temperature. And this should be done for all the possible ions. An alternative way is to determine the fastest and lightest ion recorded (Table I), but this is not very easy to use either because it does not allow one to decide a priori whether the particular ion one is interested in will be recorded or not. This is why we would like to go back to the old "dE/dx criterion" proposed at the early stage of development of this technique by Price, Fleischer and Walker. It states that if an ion A loses a certain amount of energy per unit of path, an ion B losing the same or more energy per unit of path will be recorded equally. Therefore, for each detector a $(dE/dx)_t$ limit value can be attributed as being the threshold under which one cannot get any track whatsoever (Fig.3). It must be pointed out that this criterion does not fit the data perfectly, particularly for high-energy ions up to the relativistic region. Nevertheless, for the sake of simplicity, it can be considered a useful tool.

2.4. Environmental effects

The solid-state nuclear track detectors are amongst the least sensitive detectors to environmental effects. However, they are not totally inert and it must be kept in mind that, sometimes, external constraint can affect their recording ability.

2.4.1. Thermal effect

"By far the most pervasive and useful environmental effect on particle tracks has been the thermal alteration of tracks" [1]. This effect is exhibited by both

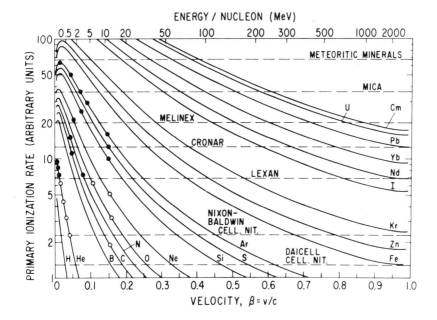

FIG.3. Sensitivity thresholds (after Ref.[1]).

organic and inorganic solids, even though it is much more striking in organic media than in minerals or glasses. In general, an irradiated detector stored at high temperature will lose some of the information recorded in it. If the temperature is high enough, or if the storage time is long enough, this effect may even cause a total loss of information. Otherwise, if the storing conditions are not too severe, only a fraction of the recorded tracks will be lost, or only a fraction of their etchable length will be erased. This track annealing effect is summarized in Table III. Advantage can be taken of this effect in some particular cases. As a matter of fact, low Z particle tracks are more easily annealed than high Z particle tracks. Therefore if, for any reason, the two types of particle are recorded and if one is interested only in the heavy ones, the light particle tracks can be preferentially removed. The annealing properties can also be used as a technique for dating natural or man-made samples (see later).

2.4.2. Gases

Most, if not all, of the inorganic detectors are not affected by the presence or absence of a gas. On the other hand, the characteristics of organic detectors may be altered, depending on their structure, if they are bombarded by ions in vacuum or in the presence of gases, if they record tracks while oxygen is present or not, and so on. Extensive studies have been performed on this topic and

TABLE II. ETCHING CONDITIONS OF FISSION FRAGMENT TRACKS FOR VARIOUS MATERIALS (after Ref. [1])

Material	Etching conditions
Feldspar (Albite, $NaAlSi_3O_8$)	3 g NaOH: 4 g H_2O, 85 min, boiling
Feldspar (Anorthite, $CaAl_2Si_2O_8$)	3 g NaOH: 4 g H_2O, 14 min, boiling
Feldspar (Microcline, Orthoclase, $KAlSi_3O_8$)	5 g KOH: 1 g H_2O, 80 min, 190°C
Fluorite (CaF_2)	98% H_2SO_4, 10 min, 23°C
Autunite ($Ca(UO_2)_2P_2O_8 \cdot 8H_2O$)	10% HCl, 10–30 sec, 23°C
Mica (Biotite, $K(Mg,Fe)_3AlSi_3O_{10}(OH)_2$)	20% HF, 1–2 min, 23°C
Mica (Lepidolite; Zinnwaldite, $K_2Li_3Al_4Si_7O_{21}(OH,F)_3$)	48% HF, 3–70 sec, 23°C
Mica (Muscovite, $KAl_3Si_3O_{10}(OH)_2$)	48% HF, 10–40 min, 23°C
Mica (Phlogopite; Lepidomelane, $KMg_2Al_2Si_3O_{10}(OH)_2$)	48% HF, 1–5 min, 23°C
Quartz (SiO_2)	KOH(aq), 3 h, 150°C, or 48% HF, 24 h, 23°C
Lead phosphate glass	1 ml 70% HNO_3 : 3 ml H_2O, 2–20 min
Phosphate glass	48% HF, 5–20 min
Silica glass (fused quartz; Vycor; Libyan Desert Glass)	48% HF, 1 min
Soda-lime (microscope slide; cover slip; window glass)	48% HF, 5 sec (better: 5% HF, 2 min) 24% HBF_4 : 5% HNO_3 : 0.5% acetic acid, 1 h
Cellulose acetate (Kodacel; Triafol T; Cellit)	1 ml 15% NaClO : 2 ml 6.25N NaOH, 1 h, 40°C 25 g NaOH : 20 g KOH : 4.5 g $KMnO_4$: 90 g H_2O, 2–30 min, 50°C
Cellulose acetate butyrate	6.25N NaOH, 12 min, 70°C
Cellulose nitrate (Diacell; Nixon-Baldwin)	6.25N NaOH, 2–4 h, 23°C
Cellulose triacetate (Kodacel TA401, unplasticized; Bayer TN)	1 ml 15% NaClO : 2 ml 6.25N NaOH, 1 h, 40°C
HBpaIT (Polyester, $C_{17}H_9O_2$)	6.25N NaOH, 8 min, 70°C
Ionomeric polyethylene (Surlyn)	10 g $K_2Cr_2O_7$: 35 ml 30% H_2SO_4 1 h, 50°C

TABLE II (Cont.)

Material	Etching conditions
Polyamide (H-film)	$KMnO_4$ (25%, aq), 1.5 h, 100°C
	6N NaOH solution
Polyimide	$KMnO_4$ in H_2O
Polycarbonate (Lexan; Makrofol; Merlon; Kimfol)	6.25N NaOH, 20 min, 50°C
	6.25N NaOH + 0.4% Benax, 20 min, 70°C
Polyethylene terephthalate (Mylar, Chronar, Melinex, Terphane)	6.25N NaOH, 10 min, 70°C
	$KMnO_4$ (25%, aq), 1 h, 55°C
Polystyrene	sat. $KMnO_4$, 2.5 h, 85°C
	10 g $K_2Cr_2O_7$, 35 ml, 30% H_2SO_4 3 h, 85°C
Silicone-polycarbonate copolymer	6.25N NaOH, 20 min, 50°C

TABLE III. THERMAL EFFECT [1]

Material	1-hour annealing temperature (°C)		
	Total fading	50% track loss	Start of track loss
Apatite	530	336	400
Aragonite ($CaCO_3$)	150	-	130
Cellulose acetate (Cellit-T)	165	160	100
Cellulose nitrate	147	140	110
Feldspar (Anorthite)	680	550	350
Feldspar (Bytownite)	790	750	690
Glass (Borosilicate, Pyrex)	380	302	-
Glass ($V_2O_5 \cdot 5P_2O_5$)	-	95	-
Mica (Muscovite)	540	510	450
Polycarbonate	185	-	-
Pyroxene (Diopside)	880	850	820
Quartz (SiO_2)	1050	-	1000

useful data may be found in the literature [1]. However, the general and more important features of this effect are that the absence of oxygen during or after irradiation decreases the detector sensitivity and sometimes even precludes any recording, whereas electron donor gases increase the sensitivity. The desensitization due to the lack of oxygen can be more or less avoided if the sample is stored in oxygen at a pressure of about 100 atmospheres.

2.4.3. Radiation effect

Solid-state detectors are normally not sensitive to radiation other than to ions. In other words, one should not expect to see a track of electrons and it is not possible to use the detectors to detect "gamma" radiation, for instance. This is an extremely useful feature since it is possible to handle the detectors in the open daylight and to use them even in a high intensity radiation background. However, irradiation with ultraviolet and higher energy photons and with electrons or other particles can have significant and sometimes profound effects on the properties of track detectors. No general rule can be made and if one uses a particular detector under some of the above-mentioned conditions it is best to first make sure that the corresponding effect is acceptable [1].

2.5. The reading of the tracks

The more widely used piece of equipment in track scanning is the optical microscope which allows the sample to be viewed either in transmitted or reflected light. Some care should be taken in preparing the sample; in particular it should be dried thoroughly. Various attempts have been made to enhance the contrast (see Ref.[1]). In a recent paper [11], it has been shown that by using the detector foil as a light guide the contrast can be increased and the unwanted background decreased.

The optical microscope allows one to count the tracks, locate them with great accuracy, measure their dimensions, etc., but its numerous advantages are somewhat counterbalanced by the fact that optical microscope scanning is a time-consuming, delicate and often tedious operation. This is why several techniques have been proposed that do not require the optical microscope [175]. These techniques are reviewed in Nuclear Tracks in Solids [1] which also contains descriptions of the very fruitful automatic spark scanning devices. A novel and interesting procedure has been devised: the energy spectrum of a monoenergetic particle beam transmitted through the detector is measured. The particles that have been transmitted at a place in the detector where a track is located merge with an appreciably higher energy than those transmitted through the bulk detector. After proper calibration, the ratio of the high-energy-to-low-energy peaks can be expressed in terms of the number of tracks [168].

2.6. Available detectors

The numerous solids that can be used as detectors were not primarily considered for this purpose. Most of the minerals are of natural source with the exception of some artificial micas, quartz and diamonds. The same is true for the glasses and plastics which are available from private companies as slabs, sheets, or foils, but since they are produced for industrial use their properties are not always constant and laboratory tests should be performed on each batch before being used as a detector. To our knowledge, Kodak-Pathé (Vincennes, France) is the only company that especially devised and produces plastic sheets for track detection. The products are the now well-known LR-115 high-contrast alpha-particle-sensitive cellulose nitrate, the CA-80-15 and the neutron-sensitive foils.

2.7. Other techniques

Tracks can be obtained in plastics by other means than chemical etching. The grafting technique is under development. It consists of growing a polymer along the trail produced by particle damage. This different polymer is then dyed and the track is made visible [13–17].

The use of amorphous semiconductors has been proposed as a possible way of producing tracks [12], but so far no practical applications have been developed.

2.8. Summary of the track-formation characteristics

Solid-state nuclear track detectors are:

insulating solids
able to record ion trajectories
easy to etch
available in large quantities
inexpensive
mostly insensitive to radiation and light.

Their recording properties are as follows:

there is an energy threshold above which they record ions with a 100% efficiency
they do undergo some thermal fading but it is far less severe than for any other known integrating detector
when chosen according to the planned experiment they can be used under extreme environmental conditions
their reading can be automatized.

3. TRACK FORMATION: APPLICATIONS

Right from the beginning, almost immediately after their initial discovery, solid-state track detectors (SSTD) have been widely used in an ever-increasing number of applications relevant to almost any field of scientific research and technology. Many opportunities have been offered by their particular properties and a considerable amount of work has been done in fundamental research. An extensive survey of the subject is given in Ref.[1]. Here, interest is focused on the numerous applications in which an SSTD is considered to be a very simple and easy to use tool; that is to say when extremely careful etching conditions are not required, when multiple corrections are not necessary, when side effects are not of prime importance, etc. Applications that involve particle charge or energy identification or both are omitted. With regard to applications, solid-state nuclear track detectors can be used as particle detectors as well as finely perforated sheets. In the first case, the sample to be studied emits heavily ionizing particles or modifies their initial number or energy spectrum, and the particles are then recorded by the detector which can be the sample itself. In the second case, the detector is used as a wall with a fixed number of "passages" or "holes" acting as a filter to let "something" pass through.

Here, instead of mentioning all the possible applications, information is given about the different techniques and some of the applications relevant to each technique are listed.

3.1. Element mapping

"Solid state track detectors have unique capabilities for measuring the concentration and spatial distribution of certain elements. In principle, any isotope is capable of being studied if it emits heavy nuclear particles, either directly because of its natural radioactivity, or, more important, as a result of specific nuclear reactions when bombarded in an accelerator or nuclear reactor. Certain isotopes with large cross sections for specific reactions are, of course, more suitable than others. This area of research began with the measurement of uranium via the detection of fission fragments in samples irradiated with thermal neutrons. As time has gone on, however, more and more nuclides have been studied and the number of diverse problems that can be approached by track methods is enlarging rapidly" [1].

3.1.1. Uranium determination

The uranium concentration of a given sample can be determined by means of track detectors by recording the number of fission events induced by a known flux of thermal neutrons. Since only ^{235}U undergoes thermal neutron fission, the isotopic ratio of ^{235}U to ^{238}U has to be known. One usually assumes that it is equal to that of the natural abundance but, in some cases, it has to be determined independently. Absolute measurement of the thermal neutron flux is generally avoided because it is rather difficult to do and it can be a major source of error. Instead, one uses solids of known uranium concentration (usually glasses). Preferably, a glass with a uranium content close to that of the unknown sample is chosen. The standard glass is irradiated simultaneously with the studied sample so that they receive exactly the same number of thermal neutrons. Before irradiation an auxiliary detector is placed in close contact with the glass. The ratio of induced tracks in this detector to those measured coming from the sample itself allows the uranium content of the sample to be calculated. Two methods can be used. In the external method, an auxiliary detector is placed in close contact with the unknown sample and the standard glass. In the internal method, an auxiliary detector is placed against the standard glass but the sample itself is used as a detector in order to measure the number of fission events induced within the sample. It must be pointed out that this relative measurement of uranium content can be greatly affected by the range of the fission fragments both in the sample and in the standard glass. If, for instance, the standard glass has an elementary composition with an average Z much lower than that of the unknown sample, the range of fission fragments within the standard glass can be much larger than their range within the sample. Consequently, even if the number of induced fission events is the same, the number of tracks recorded in the detector in contact with the glass will be larger than the number of fission fragments emerging from the sample. Therefore, when possible, the standard solid must have an average Z close to that of the unknown sample. With the external method, a uranium concentration as low as 50 ng·g^{-1} can be measured, whereas with the internal method, measurements down to a 10^{-14} atom fraction can be made. Of course, when the internal method is used the unknown sample has to be heated sufficiently to remove any possible fossil track provided by spontaneous fission of ^{238}U during the "life" of the sample.

The technique that has been schematically summarized can be used for measuring the uranium content of samples of uniform concentration. This is not likely to be true for most samples. Heterogeneity adds special complications in determining a bulk, average uranium concentration. Two possible ways of removing the heterogeneities are by dissolving the sample and measuring the uranium concentration in the dried solute product [18] or by crushing the sample to a scale that is much finer than the scale of the heterogeneities [19]. Another way is to move the sample away from the detector and to irradiate in vacuum [20]. If the sample is a liquid or a suspension, a uniform solid target must be made. This can be done either by evaporation or collection on a filter, for instance. It is also possible to add known amounts of uranium standards to several samples of the same origin. The track densities are measured for each case and then extrapolated to zero addition to get the initial uranium content of the sample.

3.1.2. Uranium mapping

It is often very important to measure the detailed spatial uranium distribution. With dielectric track detectors it is easy to measure the uranium distribution on a micrometre scale [174]. Basically, the technique is always the same. A fission-fragment track detector is tightly applied to the flat surface of the sample in which one wants to determine the uranium distribution. This "pack" is subjected to thermal neutron irradiation. The external detector is then etched out and the uranium-rich regions of the sample are revealed. They appear like small sea urchins or tiny stars. There are, of course, many possible ways to map uranium but the principle of the technique is the same. The main difficulty is to relate accurately the fission stars seen on the detector and the actual position of uranium grains in the sample. If one decides to keep the detector and the sample together, thus assuring a good spatial resolution, the problem of etching arises. This problem is overcome by the use of a detector that is thinner than the fission-fragment range. Then it is possible to etch the detector from the outside, without separating it from the sample, but this is not an easy method. Most micromapping has been done by separating the detector from the material after irradiation. The etching is easier than with the previous method but locating the uranium-bearing regions is a little more complicated. In one way or another some sort of fiducial marks have to be made. For instance, a striking feature of the sample can be connected to a scratch or a grid engraved on the surface of the detector and a photograph of the montage taken; or, the sample itself produces a print of its structure on the surface detector [25].

3.1.3. Fission-fragment mapping of other elements

Fission of elements other than uranium can be induced by fast-moving particles and high-energy gamma rays. The technique is generally the same as for uranium. The only added difficulty is to have a detector or a target thin enough to allow the impinging particle to reach the fissionable material. Thorium mapping is achieved by bombardment with 30-MeV alpha particles or 85-MeV protons [21,22]. Since uranium is associated with thorium, the uranium content is measured first by the thermal neutron method. Then, the fast-moving particle irradiation induces both uranium and thorium fission and the thorium content is eventually determined by subtraction. To measure the Th/U ratio in samples with a low uranium content, fast neutrons can be used too [23, 24]. Other elements which undergo fission when hit by fast-moving particles are also likely to be mapped. These include Pb and Bi, which could be measured at the ng/g level, and Au when bombarded by energetic mesons; photo-fission of U, Th and Bi can be induced; elements as light as Ho can be split by energetic oxygen ions [1].

3.1.4. Alpha particles and proton mapping

The lightest ions are alpha particles and protons. They can be recorded, at low energies, by solid-state track detectors. Consequently, any element that emits these particles, either naturally or because of an n,α-induced nuclear reaction, can be mapped by techniques similar to uranium mapping. As far as spontaneous emission is concerned, one deals with the "alpha autoradiography" technique. This technique was already used with nuclear emulsion but it has been greatly improved by using track detectors. Isotopes of uranium, plutonium, thorium, lead (via the ^{208}Pb (α, 2n) Po reaction), polonium and other heavy elements can be used as alpha emitters for autoradiography. Elements that are not naturally radioactive can be mapped too. Boron, for instance, has an n,α cross-section of 760 barns and it is the element that is the more easily mapped in this way. Boron can be detected to a level of μg/g. The minimum detectable particle size is 10^{-16} g and the short range of the alpha particle and Li fragment makes the spatial resolution extremely good (0.3 μm). Other elements like ^{6}Li and ^{17}O can be determined by neutron irradiation and subsequent alpha-particle emission. Proton bombardment provides a way to measure ^{18}O and ^{15}N, while ^{14}N content can be measured by neutron irradiation leading to the emission of recorded protons.

3.2. Applications of element mapping

3.2.1. Fission-track dating

The first application of the measurement of the average content of uranium that should be mentioned is the dating of samples of various origin. If a sample by itself is a fission-fragment detector (or if it contains grains of such minerals) and if, in addition, it does contain some uranium, it can be dated with a good accuracy. Immediately after the "birth" of the sample (the cooling down of the melt for a rock, the solidifying process for a glass, the firing for a piece of pottery, etc.), the uranium contained in the sample starts to undergo spontaneous fission. Each fission fragment is recorded by the surrounding material and its track is stored in it. The spontaneous fission half-life of ^{238}U is known (4.51×10^9 years). Therefore, if both the number of spontaneous fission events that occurred since the birth and the uranium content are known, the age of the sample can be measured. Typically, the studied sample is first etched to reveal and count the fossil tracks and then the uranium content is measured by neutron irradiation. This technique has been applied to many different samples: rocks (terrestrial and extraterrestrial), glasses (natural and man-made), and various minerals. Of course, some care should be taken when handling and measuring the sample. Particularly the thermal history of the sample is of great importance (for details see Ref. [1]). The fission-track dating method has proved extremely successful in solving problems in geophysics, geochronology and archaeology. It is impossible to list here all the innumerable applications. Among the more spectacular findings has been the information obtained from dating lunar samples, evidence for the spreading of the ocean bed and the knowledge acquired about archaeological fields like the one at the Olduvai gorge.

As soon as Apollo XI returned from the moon, the lunar samples were submitted to fission-track investigations and the age of the moon samples was successfully measured by the technique described here [1].

Spreading of the ocean bed can be measured by dating geological samples taken from the bottom of the ocean at various distances from the mid-ocean ridge. Since these samples were hot when they emerged at the surface of the submerged earth crust, they started recording tracks at this time only. Therefore, if there is any spreading of the ocean bed, the samples found at a greater distance from the median valley should exhibit greater ages. Fission-track dating provided clear results that fully conform with the ideas of ocean-bed spreading [1, 26–28].

The K-Ar age (1.75×10^6 years) of the archaeological field at the Olduvai gorge seemed surprising to the experts, but when the track dating of volcanic pumice from this location gave a cross-check of 2.0×10^6 years, the site became considered as one of the best-dated of the anthropological sites that are too ancient for carbon dating [29].

3.2.2. Biology

Most of the elements that can be measured or mapped are fairly ubiquitous and can be found in many biological samples. They can induce several transformations within the biological material (radiation effect) or they combine with molecular products of prime biological importance. Biologists were quick to recognize the usefulness of solid-state track detectors in these studies. In addition to the sensitivity of the detectors, their relative innocuity to biological media offered a further advantage.

The content of uranium has been measured in, for instance, blood cells [30], seaweeds and plants [31] and bones [32, 33], and the content of plutonium has been determined in bones and bioassays of various origins [34, 35]. Both uranium and plutonium distributions have been determined in bones [44–47], liver, kidney and animal blood that has previously been injected with these radionuclides [36]. Such studies indicate the processes by which harmful nuclides diffuse within the body and which organs are more likely to be affected in an accidental or chronic internal contamination.

Boron distribution has been determined in a variety of biological materials [37] including bone [38] and leaves [39]. It can be seen that boron is not homogeneously distributed within the leaf and strong depletions of boron are observed along the veins of the leaf. Boron can also be used for labelling a compound of biological importance to follow its biochemical processes.

The problem of lithium determination in biological material has been discussed by Carpenter [40].

Oxygen-17 can be used as a tracer in, for instance, the preparation of labelled citric acid and in the study of oxygen uptake and transport in the brain. An animal is given ^{17}O-enriched air to inhale before it is killed and subsequent ^{17}O mapping of the brain sections shows the oxygen-rich regions in relation to the brain structure [1].

Similarly, nitrogen has been determined in several biological materials at the National Bureau of Standards [41–43] by means of the n,p reaction on ^{14}N.

"The localization of deuterium and tritium via the D(T,n) reaction has been demonstrated by Geisler and his colleague ([44–46] in Ref.[1]) who have emphasized the potential biomedical applications of the method. In particular they have measured lymphocytes labelled with deuterated thymidine and have suggested the use of this method in human cancer detection" [1].

Koechel and Kalff [70] have used autoradiography to determine the productivity of phytoplankton species, and other biological applications have been suggested [71].

3.2.3. Geology and geochemistry

Apart from the remarkable studies on the lunar and meteoritic samples — to which an entire paper could be devoted (see Ref. [1]), "terrestrial" geology has benefited greatly from the track method. The average content of uranium has been measured in terrestrial samples like petroleum [48], in manganese nodules found at the bottom of the ocean [49], in sediments [50, 51], and in almost any imaginable rock. A particularly interesting example of uranium mapping is offered by the technique of plastic prints initiated by Kleeman and Lovering [25]. A plastic sheet is placed in contact with a polished section of the rock. After irradiation by thermal neutrons (fission fragment print) or after an exposure time long enough (alpha autoradiography), the detecting material gives a "print" that shows the repartition of uranium or thorium. This technique was also used by Chantret [52, 53] to study radionuclide distribution in uranium ores.

The distribution of alpha-particle emitters in minerals and rocks (like Li) has been discussed [54] and the distribution of boron was determined [55] in diamond and shown to bear a relation to the colour and other optical properties as well as the electrical properties of the sample.

For more details on geological applications the reader is referred to Refs [56–66] and Ref. [178].

3.2.4. Environmental studies

Either naturally or as a result of man's activities, radioactive nuclides are present in our biosphere. They are a potential hazard to human, animal and plant life. It might be of vital importance to know the size and abundance of the particles that contain radioactive nuclides present in the air and water. The size of plutonium oxide particles has been measured by means of track detectors [1]. Uranium as an air and water contaminant can be traced too [68, 82], and plutonium has been detected in various environmental samples [67, 69].

3.2.5. Technological applications

Some man-made materials contain elements that can be detected or mapped by the track technique. The measurement of their distribution by this technique has cast some light on the understanding of their behaviour and structure. For instance, boron can be found in various alloys and can affect the metallurgical properties; its distribution is of great importance in understanding steel composition [73]. Boron can be responsible for crack formation and swelling [72]. Boron mapping shows that this element is preferentially distributed along crystal boundaries and in inclusions in silicon-iron steel [1]. Boron distribution has been rather extensively determined for alloys and steel structure studies [74–76, 173]. Similarly, some experiments have been performed to elucidate the diffusion of trace amounts of uranium in metallic crystals [1] and in steel [77] and to measure the thickness of the reaction zone in Nb-Sn superconducting wires [78].

Numerous applications of track etching have been developed to study the internal structure of materials [79]; this is particularly true for atomic energy technology. Solid-state track detectors have been used for controlling the thickness of electrodeposited radioactive sources [80], for testing reactor materials [81], for measuring the isotopic enrichment of reactor fuel [83] and for determining contamination of nuclear targets [84].

3.3. Imaging or internal structure probing

We will now discuss non-destructive techniques for studying the internal structure of a body. Basically, these techniques are based on the fact that when charged particles are travelling through a condensed medium, they undergo a lot of interactions with the constitutive electrons, atoms or molecules of the medium. As a result, the characteristics of a particle beam entering the medium will be drastically changed so that the characteristics of the out-going beam can be used as an indication of the internal structure. For instance, the particles can be scattered or preferentially absorbed by some of the constituents only, or their energy can be differentially decreased. Obviously, the probing particles are generated by an external source and their initial energy must be high enough for at least some of them to travel through the sample and be detected at the other side. Two types of particles are used: neutrons and ions (alpha particles or heavier ions) that produce an image to be analysed [169].

3.3.1. Neutron radiography

As neutrons are non-charged particles they cannot be detected directly by solid-state detectors. Therefore an active material, which can convert neutrons into detectable particles, is placed between the sample and the detector. When using thermal neutrons, one takes advantage of the n,α reaction on boron or of the n,f reaction on uranium because of their large cross-section. With fast neutrons, detectable charged particles can be induced by recoil reactions within the detector itself by the (n,n')3α reaction on carbon or by fission reactions of various nuclides such as ^{237}Np or ^{232}Th [1].

Neutron radiography is particularly useful when X-ray radiography is less suitable, for instance with samples made of light elements like organic materials. Berger [85] has made a survey of the work on track etch radiography so let it suffice here to state that neutron radiography has four special qualities. It can be performed in a high radiation environment; it has a high resolution, only limited by track length; radiographs can be taken with a low exposure to neutron beam and their sensitivity to thickness differences is high ($\sim$ 1%) [85, 86, 168]. The practical applications of neutron radiography are numerous and are not limited to small-sample studies. In fact, G. Farny, at the CEN-Fontenay, France, has developed neutron radiography for inspecting irradiated reactor fuel elements on a large scale. Recently, a complete survey of neutron radiography was published in Atomic Energy Review (Vol. 15, No.2, 1977).

3.3.2. Charged-particle radiography

"Whereas neutron radiography is sensitive to variations of certain chemical elements, proton and heavy radiography are sensitive to small changes in density or mass thickness along the beam. The general principle is to shoot a monoenergetic beam of charged particles through the object to be imaged and to bring the beam to rest in a stack of plastic sheets" [1]. "Where the specimen is thicker or has a higher stopping power the particles stop earlier in the stack; where it is thinner or has a low stopping power, the particles stop in a layer of the stack which is further downstream" [87]. The principles of this technique were developed by Tobias and Benton in 1972. Since then, they have added further improvements and developed it into a routine method [87–89]. The detector can be "read out" by photographing each layer individually; it is also possible to synthesize the images with the aid of a computer. This technique is extremely powerful and makes it possible to detect very small abnormalities that would hardly be detected with X-ray techniques, particularly in soft tissue. For instance, mass-thickness changes as small as 0.03 g/cm^2 can be detected. However, it should be mentioned that this sensitivity is achieved with particles of charge greater than 6 for thick objects and greater than 2 for thin objects (less than 1 cm). The technique is clearly the most sensitive tool available at present for the measurement of internal density distribution of objects. The striking opportunities offered by the method for diagnostic radiology will undoubtedly add to the interest in building heavy-ion medical accelerators. Nevertheless, the greater availability of protons and alpha beams has urged scientists to study the possibilities of using such beams for radiography [90, 91].

Low-energy alpha particles from a radioactive source can be used to study thin metallurgical samples and to detect defects such as grain boundaries, twin boundaries and the accumulation of dislocations [1, 92–94]. The sample under study is placed between the alpha-particle source and the detector. The energy of the particles is adjusted by means of an absorber-collimator so that non-scattered particles will just reach the detector and be recorded. On the other hand, scattering will reduce the flux detected. The resulting picture allows information to be obtained on the internal structure of the sample or measurements of activation energies of metallurgical processes [96]. The method is not limited to alpha-particle probing and the study of heavy metals like platinum and tungsten is carried out with fission fragments [95].

When charged particles move between atomic planes or atomic rows of a crystallized sample, they can travel farther than in other directions. This phenomenon is known as "channelling". Consequently, if charged particles enter a sample with an energy a little less than needed to get out, only the channelled particles will emerge and be recorded on an adjacent track detector. With this technique one can measure crystal structure, symmetries and orientation [97, 98] in the same way as one would with an X-ray Laue pattern.

3.4. Radiation dosimetry

Some of the unique properties of solid-state track detectors make them extremely attractive for radiation dosimetry and it is not surprising that the first neutron fluence measurements were performed as far back as 1963 [99]. Since then the development of the method has been astonishing and hundreds of papers have been published on the subject. It is not possible to make a detailed survey of this field of application of solid-state track detectors here. Instead, the principles of the method are outlined and a few salient literature references are given.

Research in radiation dosimetry began in the United States of America and in Europe at about the same time [100–104] and was mainly centred on thermal neutron dosimetry. The scope of interest soon widened to include other particle dosimetry (for a review see Ref.[105]).

Application of solid-state track detectors in radiation dosimetry extends from thermal and fast neutron fluence measurements to alpha-particle and heavy ion dosimetry. In contrast, beta-ray, gamma-ray, and X-ray doses cannot be measured by this technique.

3.4.1. Thermal neutrons

The method for thermal neutron dosimetry is based on a similar arrangement as described for the imaging technique. Some form of converter has to be used to induce the emission of detectable particles whose number would be in relation to the neutron flux. There are two possibilities, the n,f and the n,α reactions.

With fission fragment registration, the targets that are more likely to be used are natural uranium, ^{235}U-enriched foils and ^{239}Pu, but because of the chemical toxicity of plutonium it is preferable to use uranium converters. Besides, by varying the concentration of uranium of the target, an enormous range of neutron fluence can be measured (from 10^3 to 10^{22} n/cm^2). The detectors themselves are chosen according to the environmental conditions, or the etching or reading technique one intends to use. They can be mica or glass sheets or plastic foils. Various possibilities are offered for the reading of the dosimeter. Optical and electron microscopes can be used but, more often, scientists have tried to replace manual counting by automatic or integrating devices. Most of these devices have been developed by groups concerned with radiation dosimetry because research in dosimetry will ultimately lead to application on a large scale and an increasing number of dosimeters have to be read repeatedly and in a short time.

The advantages of solid-state dosimeters are [1]:

No need for specialized electronic counting equipment
Ease of processing and evaluation
No need for gram quantities of fissile materials
Insensitivity to β-, X- and γ-radiation which makes track counting in nuclear emulsion impossible at high doses
No fogging, fading or other storage problems at ambient temperatures
Huge range of doses amenable to study
Ease of activation or inactivation by separating fissile and detector foils.

When the thermal neutron flux is small one has to look for alpha-induced particles. The 3840-barn and 950-barn cross-sections of the ^{10}B(n,α) and ^{6}Li(n,α) reactions are significantly higher than the thermal neutron fission cross-section of ^{235}U. These two elements are used in conjunction with an alpha-sensitive detector in order to measure low thermal neutron fluxes. The sensitivity that can be achieved is high (0.013 track/neutron) [106] with a 0.2-cm-thick boron sheet. This high sensitivity was proven useful for several special applications besides practical dosimetry. For instance, low fluxes of thermal neutrons produced by cosmic rays over the earth's surface can be measured [106] and the first direct measurement of the distribution of thermal neutrons with depth in the lunar soil was carried out in 1973 [107].

Special mention should be made of neutrons in the energy range of about 20 to 100 keV. In this energy interval, the cross-section for the n,α reaction on Li or B is too low to be useful and recoils of heavier nuclei in the detector (see later) would have too short a range to be detectable. It is then possible to detect proton recoil directly induced in a hydrogen-rich organic detector [108, 109].

3.4.2. Fast neutrons

Fast neutron dosimetry is a bit more complicated than thermal neutron dosimetry because one should get information not only on the neutron fluence but also on the energy spectrum of neutrons. Fission can be induced in various elements such as ^{232}Th, ^{238}U, ^{231}Pa and ^{237}Np. By using several such fissile materials with different thresholds for induced fission it is possible to measure fluence from approximately 10^7 to 10^{19} neutrons per cm^2, and to obtain a rather good idea of their energy spectrum. The first particle personnel dosimeter on this principle was made by Baumgartner and Brackenbush [110].

Neptunium-237 is particularly attractive for neutron dosimetry. Its fission threshold is well defined at 0.4 MeV, it has a high cross-section and at the same time a low sensitivity to high-energy gamma rays. It has been extensively studied [111] and its sensitivity has been as good as 6×10^{-4} tracks/(neutron/cm^2).

With ^{232}Th, the measurable dose range lies between 20 mrad and 20 krad [112]. This raises the question of converting neutron fluence to biological doses. There is no simple relation between the energy of neutrons and their efficiency for inducing damage in the human body. This underlines the importance of getting information on the neutron energy spectrum. Some attempts have been made to optimize this measurement [113, 114].

In addition to fission fragments in fissile materials, fast neutrons can be detected by observing the recoil nuclei that were directly hit by neutrons. Light nuclei have high elastic collision cross-sections and are the most favourable. They can be initially located within the detector itself (H,C,N,O) or they can be contained in a separate screen in contact with the detector (He, Be) [115, 116]. The sensitivity to fast neutrons ranges from about 3×10^{-6} to 2×10^{-5} tracks/neutron depending on the plastic and the neutron energy.

3.4.3. Alpha particles

The major sources of potentially dangerous alpha particles are airborne radon gas and aerosol particles containing radon daughters, such as ^{218}Po and ^{214}Po.

Radon can be found almost everywhere, even at high altitudes in the atmosphere, though generally at trace levels. It is present at appreciable concentrations in the air in uranium mines where it is a serious hazard to miners and must be monitored. Because of the warm, humid atmosphere in the mines and the necessity for an opaque wrapping to prevent the 5.5-MeV alpha particles from being detected, nuclear track emulsions are unsatisfactory as dosimeters. On the other hand, alpha-sensitive plastic detectors do not have to meet these requirements and are very suitable for miners' personal dosimeters [1, 117, 118]. Their sensitivity is sufficiently high, but some practical problems have still to be solved. For instance, dirt tends to deposit on the detector surface and alter its response.

Radon dosimeters have been used in mines as well as to study the distribution of radon in the atmosphere [119].

3.4.4. Highly charged particles

Man may suffer from highly charged particle bombardment only when placed under special conditions. Highly energetic and charged particles are not present in our usual biosphere, but they can be found at high altitude (SST and space flights) in the cosmic rays, and on the earth around heavy ion accelerator facilities when subjecting patients to heavy ion therapy or diagnosis. Since heavy ions are extremely powerful projectiles that can induce permanent damage in cells it is of prime importance to be able to monitor them. This can be done for fluence as well as charge and energy measurement by means of plastic detectors [120–122].

For recent data on dosimetry see Refs [123–133].

3.5. Miscellaneous applications of the charged-particle detectors

This section describes some more situations, not covered in the previous sections, in which solid-state track detectors can be profitably used.

3.5.1. Tracers

Small particles containing atoms detectable by track detectors can be used to trace the flow of a fluid, whether it is a gas or a liquid. Again, fission fragments and alpha particles are the available alternatives. In the first case, 0.5-μm uranium oxide particles or any other fissile material like thorium, gold and platinum are suspended in the fluid flow. The fluid is then filtered, the filter is pressed against a detector and neutron or high-energy particle irradiated to produce fission events that can be recorded [134, 135]. The measured concentration in "tracer" is related to the fluid flow characteristics. Similarly, boron, lithium and ^{17}O can be used with alpha-sensitive detectors [136]. It is also possible to code individually the tracer particles by specifying, for example, the uranium-to-boron ratio [1].

The tracer technique has many attractive features; it can be used to study pollution and water networks, to follow atmospheric and sea stream behaviour, for environmental surveys, in medicine, etc.

3.5.2. Nuclear technology

In addition to the applications already mentioned, solid-state track detectors have been used in nuclear power plants and nuclear research centres. For instance, they have been used for accurate mapping of the neutron beam structure from a reactor [137, 170] and for measuring the reactors' static and dynamic characteristics [171, 172] such as fission density distribution inside fuel elements, near the interface and along the reactor holes, as well as fractional burn-up [1, 138–142, 148].

Another interesting application in connection with nuclear energy is the localization of uranium ore deposits. "The method depends on the fact that one member of the ^{238}U decay chain, ^{222}Rn, is a gas with a half-life of 3.8 days, which is long enough for radon to be transported to the earth's surface from ores buried at considerable depths, provided the permeability of the overlying rock and soil is sufficiently high. Plastic cups, each containing a piece of cellulose nitrate, are inverted and buried at shallow depths where they record alpha decays from ^{222}Rn and its daughters throughout a burial time of two to four weeks" [1]. The points of high radon concentration are selected for exploratory drilling. This procedure has now located previously unknown ore bodies at depths of up to 300 ft and obtained clear signals from known deposits at depths of 450 ft [143–145].

3.5.3. Barometric measurements

Since the stopping power of a gas is a direct function of its pressure, the range of charged particles will also be a function of the pressure. The lower the pressure, the larger the range. Therefore, a radioactive source and a detector can be made into a pressure gauge. This idea was put to practical use by Gustafson et al. [146] who measured the maximum altitude of birds. They devised an altimeter made out of a polonium source and a cellulose acetate detector tilted at 45° to the particle from the source and found that homing pigeons fly at altitudes ranging from 300 to 1700 m and that swifts fly at a height of 1400 to 3600 m.

3.5.4. Polymer stability

The characteristics of track storage and etching have been used to determine the technical thermal stability of polymers [147].

3.6. Application of track detectors as hole-filled materials

"Many of the uses of the particle track derive from the simple fact that an etched particle track is a hole. By proper control of the spatial distribution of the track forming particles and of the etching conditions, structures can be produced that have unique, useful geometries" [1].

3.6.1. Filters

By exposing a thin detector to a collimated beam of particles that produce tracks across the entire thickness of the sheet and by subsequent etching, one can produce the required number of holes with a known geometry [149]. The diameter of the holes can be as small as 50 Å. Continuously adjustable holes can be obtained by perforating a stretchable plastic such as a silicon-polycarbonate film [150]. Filters are commercially available from the Nuclepore Corporation. The first applications of such filters were of biomedical interest. Seal [151] demonstrated that when malignant cells were present in blood they could be separated and recognized and that their number was related to the stage of development of the tumour. The same technique has been used to detect cancer cells in spinal fluid and lung material sampled by needle aspiration biopsy [152, 153]. Nuclepore filters have also been used in immunology [154] and in the study of metabolic interactions [155].

Environmental studies have also benefited from track-etch filters. Airborne particles, radioactive aerosols, cloud nuclei and fine particulate matter from suspension in ocean water can be sampled by this method [1, 156–158].

Even every-day commodities can be improved by Nuclepore filters. They can be used to clarify and stabilize wine and beer by removing the bacteria, sediment and yeast. In contrast to thermal sterilization this technique does not change the taste of the beverage [1].

Track-etch filters are also used for cleaning gases that are used for removing the dust from electronic components before encapsulation [1].

3.6.2. Virus and bacteria counter

The DeBlois-Bean counter employs a single etched track to count and measure small particles in an electrolyte [159]. The single perforated membrane is placed between two electrodes. The resistance between the electrodes depends on the conducting path through the hole. When a charged and insulating particle enters the hole the resistance increases proportionately to the size of the particle, and the velocity of the particle moving through the hole is a measure of its charge. Thus, from a knowledge of the resistance change between the electrodes and its time behaviour it is possible to count, characterize and identify the particles.

The method has been successfully applied to virus and bacteria counting [160]. An added bonus of the technique is that individual pulses with harmonic oscillation are sometimes observed when elongated particles pass through the hole. From the distribution of amplitudes it is possible to infer the shape of such particles [161].

3.6.3. Diverse applications

Since holes created by ion bombardment and etching can be filled with appropriate substances, new materials of technological interest can be produced.

Fleischer et al. [1] demonstrated that hot, liquid gallium can be forced into micrometresized holes in mica where it solidifies. Similarly, photosensitive silver chloride can be made to precipitate in such holes. Lead iodide fixed in the same manner can be transformed into superconducting lead by electromagnetic or electron irradiation at 180°C. Superconductivity can be studied in fine wires produced by electroplating tin into fine holes in etched mica. Wires 400 Å in diameter and 15 μm long can be produced from tin, zinc and indium.

Porous muscovite mica produced by track etching has been used as a model system to test theories of flow and ideas of the structure of water and of superfluid flow [162, 163]. Some of the ideas that anomalous water exists as an unusual ice-like structure on the interior of fine capillaries have been refuted by convincing experiments on fine holes [164–166].

It has also been shown that etched tracks filled with a metal by chemical reduction can serve as anchoring points for a continuous electroplated layer on top of a plastic object. Adhesion strengths of more than 50 kg/cm² can be achieved [176].

CONCLUSIONS

A large number of practical applications has emerged from the development of solid-state nuclear track detectors. Some applications are routine methods (e.g. radiation dosimetry) and others are still laboratory techniques. The use of solid-state track detectors has made major contributions to our knowledge in many scientific disciplines.

Although it is dangerous to predict the future development of a method, a conservative extrapolation on utilization of nuclear-track formation in some tentative, possible, potential and not too exotic applications of solid-state track detectors is presented.

Since environmental factors can affect the registration characteristics of tracks, solid-state track detectors could perhaps be used to study or measure these factors. The rate of fading of registered tracks provides a method to determine temperature, vacuum can be measured from the loss of efficiency of registration, and pressure from travelling distances of detectable particles.

Obviously, the possible applications depend primarily on the ingenuity of the experimentator. One possible future for solid-track detectors is the generalization of their use for monitoring environmental conditions. Detectors can be activated without any power consumption and offer advantages for a long-term experiment or measurement. Such detectors migh be placed in the atmosphere or in waters (sea, river or lake) to measure the content of radioactive elements, to detect possible accidental contamination or to follow the evolution of a pollutant in the biosphere. This type of permanent control might be useful particularly near nuclear power plants or fuel-treatment factories.

Solid-state track detectors have been used in conjunction with tracer elements. Such applications will increase in number and importance. In this particular case we are in the position of having answers and being able to seek questions.

There is a good chance that the micro-hole filters will be of great assistance in solving problems in various fields of science and technology. As an example, consider the catalytic sheets made of large plastic foils with a catalytic agent inserted. This method would allow the separation of the catalytic agent from the reagents and final products at low cost [177].

The actual surface area of a solid in which tracks have been etched is much larger than its apparent surface. One could take advantage of this surface property for catalytic effects in a similar way as zeolites are used.

Plastics can be electroplated after irradiation and etching. By this method, large and cheap electrodes could be produced for the electrolysis of water or salt solutions.

By putting appropriate compounds into the holes or by using the grafting technique it might be possible to make fluorescent screens or digital display devices.

A single hole made in a piece of detector (like in the virus counter) might provide a way to obtain an extremely well-defined point source of light and several holes in a suitable geometrical distribution might lead to the formation of light-scattering screens.

As a far-reaching but possible technological development, one can think of successive deposition of several different chemicals (metals, semiconductors, isolators) along an etched long track (made by heavy ions). Extremely micro-sized electronic components could be manufactured in this manner.

Addresses of companies commercially involved in track detectors

Detectors

Kodak-Pathé, 30 rue des Vignerons, 94300 Vincennes, France

Filters

Nuclepore Corporation, 7035 Commerce Circle, Pleasanton, California, USA
Nomura Micro Science Co, Ltd., Japan

Uranium exploration

Terradex Corporation, 1900 Olympic Blvd, Walnut Creek, California 94596, USA

REFERENCES

(1) FLEISCHER, R.L., PRICE, P.B., WALKER, "Nuclear Tracks in Solids", University of California Press, Berkeley, USA (1975)
(2) BLOCH, F., Stopping power of atoms with many electrons, Z. Physik, 81 (1933) 363
(3) LINHARDT, J., SCHARFF, M., Energy dissipation by ions in the keV region, Phys. Rev., 124 (1961) 128
(4) FLEISCHER, R.L., PRICE, P.B., WALKER, R.M., HUBARD E.L. Criterion for registration in dielectric track detectors, Phys. Rev., 156 (1967) 353
(5) WHALING, W., Handbuch der Physik, Springer, Berlin (1968)
(6) HECKMAN, H.H., PERKINS, B.L., SIMON, W.G., SMITH, F.M., BARKAS, W., Range and energy loss processes of heavy ions in emulsion, Phys. Rev., 117 (1960) 544
(7) NORTHCLIFF, L.C., SCHILLING, R.F., Range and stopping power tables for heavy ions, Nuclear data tables, 7 (1970) 233
(8) BENTON, E.V., Charged Particle Tracks in Polymers, USNRDL-TR-67-80, US Nav. Rad. Def. Lab., San Francisco, Calif. (1967)
(9) CHATTERJEE, A., MACCABEE, H., TOBIAS, C.A., Radial cut-off LET and radial cut-off dose calculations for heavy charged particles, Rad. Res., 54 (1973) 419
(10) FAIN, J., MONNIN, M., MONTRET, M., Spatial energy distribution around heavy-ion path, Rad. Res., 57 (1974) 399
(11) DIXON, G.P., WILLIAMS, J.G., Track revelation in solid state track recorders utilizing the light guide principle, Nucl. Instr. and Meth., 135, 2 (1976) 293
(12) MACKOWSKI, J., Detection of charged particles with amorphous track detectors, Onde Electrique, 56, 2 (1976) 75
(13) MONNIN, M., BLANFORD, G.E., Detection of charged particles by polymer grafting, Science, 181, 4101 (1973) 743
(14) MAYBURY, P.C., LIBBY, W.F., Nature, 254 (1975) 209
(15) GOURCY, J., MONNIN, M., Charged Particles Detection : The Graft and Dye Method, 9th Int. Conf. on Solid State Track Detectors, Munich (1976)
(16) SOMOGYI, G., Processing of plastic track detectors, Nucl. track detection, 1, 1 (1977) 3
(17) BESANT, C.B., QAQISH, A., Detection Efficiency and Range Measurement of Alpha Particles and Protons in Cellulose Nitrate, 9th Intern. Conf. on Solid State Track Detectors, Munich (1976)
(18) FLEISCHER, R.L., LOVETT, D.B., Uranium and boron content of water by particle track etching, Geochim. Cosmochim. Acta, 32 (1968) 1126

(19) FISCHER, D.E., Homogeneized fission track determination of uranium in whole rock geologic samples, Analyt. Chem., 42 (1970) 414
(20) GEISLER, F.H., SHIRCK, J., WALKER, R.M., A New Method of Average U-Determination in Heterogeneous Samples, unpublished report (1974)
(21) BIMBOT, R., MAURETTE, M., PELLAS, P., A new method for measuring the ratio of the atomic concentrations of thorium and uranium in minerals and natural glasses, Geochim. Cosmochim. Acta, 31 (1967) 263
(22) HAIR, M.W., KAUFHOLD, K., MAURETTE, M., WALKER, R.M., "Th microanalysis using fission tracks", Rad. Effects, 7 (1971) 285
(23) BERZINA, I.G., GURVICH, M.Y., KHLEBIKOV, G.E., Determination of the concentration and spatial distribution of thorium in minerals and rocks according to the tracks from fission fragments, Nucl. Sci. Abstracts, 23 (1969) 3625
(24) CROZAZ, G.C., BURNETT, D., WALKER, R.M., Uranium and thorium distribution in meteorites, Meteoritics
(25) KLEEMAN, J.D., LOVERING, J.F., Uranium distribution in rocks by fission tracks registration in lexan plastic, Science, 156 (1967) 512
(26) FLEISCHER, R.L., VIERTL, J.R., PRICE, P.B., AUMENTO, F., Mid-atlantic ridge : age and spreading rate, Science, 161 (1968) 1339
(27) AUMENTO, F., The mid atlantic ridge near 45° N. Fission track and manganese chronology, Can. J. Earth Sci., 6 (1969) 1431
(28) AUMENTO, F., LONGAREVIC, B.D., ROSS, D.I., Hudson geotraverse : geology of the mid atlantic ridge, Phil. Trans. Roy. Soc., A268 (1971) 623
(29) FLEISCHER, R.L., PRICE, P.B., WALKER, R.M., Application of Fission Tracks and Fission Dating to Anthropology, Proc. Seventh Int. Congress on Glass, Brussels, 224 (1965) 1
(30) CARPENTER, B.S., CH CHEEK trace determination of uranium in biological material by fission track conting, Analyt. Chem., 42 (1970) 121
(31) ABDULAEV, K.H., ZAKHVATAEV, B.B., PERELYGIN, V.P., Determination of uranium concentration in plants from tracks of uranium fission fragments, Radiobiologiya, 8 (1968) 765
(32) OTGONSUREN, O., PERELYGIN, V.P., CHULTEM, D., Build up of uranium in animal bones, Atomnaya energiya, 29 (1970) 301
(33) STORR, M.C., HOLLINS, J.G., An improved method for fission fragment radiography of Pu in rat bones, Int. J. Appl. Radiat. and Isot., 26, 12 (1975) 708
(34) BRACKENBUCH, L.W., BAUMGARTNER, W.V., Detection of Plutonium in Bioassay Programs Using Nonelectrical Fission Fragment Track Detectors, AEC Rept., No BNWL-SA-58, (1965) 1
(35) BLEANEY., B., The radiation dose rates near bone surface in rabbits after intravenous or intramuscular injection of ^{239}Pu, Brit. J. Radiol., 42 (1969) 51
(36) HAMILTON, E.I., The registration of charged particle in solids an alternative in autoradiography in life sciences, Int. Journ. Appl. Rad. and Isotp, 19 (1968) 159
(37) CARPENTER, B.S., Quantitative application of the nuclear track technique, Microscope, 20 (1973) 175
(38) JEE, W.S., DELL, R.B., MILLER, L.G., High resolution neutron induced autoradiography of bone containing ^{239}Pu, Health Phys., 22 (1972) 761
(39) STINSON, R.H., Boron autoradiography of botanical specimens, Can. J. Botany, 50 (1972) 245
(40) CARPENTER, B.S., Lithium determination by the nuclear track technique, J. Radioanal. Chem., 19 (1974) 233
(41) CARPENTER, B.S., LAFLEUR, P.D., Observing proton tracks in cellulose nitrate, Int. J. Appl. Rad. Isotopes, 23 (1972) 157

(42) CARPENTER, B.S., LAFLEUR, P.D., Nitrogen determination in biological samples using the nuclear track technique, Am. Nucl. Soc. Trans., 15 (1972) 118
(43) LYON, W.S., ROSS, H.H., Nucleonics, Anal. Chem., $\underline{48}$, 5 (1976) 96
(44) BECKER, K., JOHNSON, D.R., Nonphotographic alpha autoradiography and neutron induced autoradiography, Science, 167 (1970) 1370
(45) COLE, A., SIMMONS, D.J., CUMMINS, H., CONGEL, F.J., KASTNER, J., Application of cellulose nitrate film for alpha autoradiography of bone, Health Physics, 19 (1970) 55
(46) SIMMONS, D.J., FITZGERALD, K.T., Application of Cellulose Nitrate Film for Alpha Autoradiography of Bone, Argonne Nat. Lab. Rept. ANL/7760 PtII, (1970) 208
(47) SCHLENKER, R.A., OLTMAN, B.G., Fission Track Autoradiography, ANL8060, Radiological and environmental research division annual report, Part II, (1973) 163
(48) BERZINA, I.G., POPENKO, D.P., SHIMELEVICH, Y.S., Determination of trace amount of uranium in petroleums from the fission fragment tracks, Geokhimiya, 8 (1969) 1024
(49) YABUKI, H., SHIMA, M., Measurement of uranium, thorium in manganese nodule using fission tracks and alpha particle tracks, Rept. Inst. Phys. and Chem. Res., 47 (1971) 27
(50) RYDELL, H., FISHER, D.E., Uranium content of caribbean core, P6304-9, Bull. of Marine Sci., 21 (1971) 787
(51) LAHOUD, J.A., MILLER, D.S., FRIEDMAN, G.M., Relationship between depositional environment and uranium concentration of molluskan shells, J. Sed. Petrology, 36 (1966) 541
(52) CHANTRET, F., Un example d'application des détecteurs solides de particules atomiques à l'étude du déséquilibre radiatif des minerais d'uranium, Bul. Soc. Fr. Mineral. Cristallogr., 96, (1973) 223
(53) CHANTRET, F., Etude de la Répartition de l'Uranium dans les Minerais, Journées d'Etude des Applications des Détecteurs de Trace, Fontenay-aux-Roses, CEA (1976) (To be published)
(54) BERZINA, I.G., BERMAN, I.B., NAZAROVA, A.S., Detection of the spatial distribution and determination of lithium concentration in minerals and rocks, Dokl. Akad. Nauk SSSR, 201 (1971) 686
(55) CHRENKO, R.M., Boron content and profiles in large laboratory diamond, Nature, 229 (1971) 165
(56) COPPENS, R., RICHARD, P., BASHIR, S., Utilization of α-Autoradiography of Rocks in the Investigation of the Radioactive Equilibrium 9th Int. Conf. on SSNTD, Munich, (1976)
(57) LIEHU, A.E., Geologic Analysis by Track Etch Method, ibid.
(58) JENSEN, T.U., ENGE, W., ERLENKENSER, H., WILLKOM, H., Age Determination of Sediments by Pb-210 Using a Plastic Detector Technique, ibid.
(59) DANIS, A., On the Nature and Distribution of the Fissionable Element Impurities in Minerals and Soils, ibid.
(60) AMIEL, A.J., MILLER, D.S., FRIEDMAN, G.M., Uranium distribution in coordonate sediments of a hypersaline pool, Gulf of Eilat, Red. Sea., Israël J. Earth Sci., 21 (1972) 187
(61) AUMENTO, F., Uranium content of mid-oceanic basalts, Earth Planet. Sci. Lett., 11 (1971) 90
(62) BIGAZZI, C., RINALDI, G.F., Variazione del rapporto U/$CaCO_3$ nelle concrezioni di grotta, Atti Soc. Tosc. Sc. Nat. Mem., A LXXV (1968) 647
(63) CHALOV, P.I., MAMYROV, V., MUSIN, Y.A., Comparative study of some methods for measuring the relative ^{235}U content in natural uranium samples, Izv. Akad. Nauk Kirg-SSR, 5 (1970) 13

(64) FISHER, D.E., Achonditric uranium, Earth Planet. Sci. Lett., 20, (1973) 151
(65) GAVSHIN, V.M., Uranium concentration in natural stratified alumosilicates, Dokl. Akad. Nauk SSSR, 705 (1972) 956
(66) SEITZ, M.G., Uranium and thorium diffusion in diopside and apatite Carnegie Institution year, Book, 72 (1973) 586
(67) FLEISHER, A.L., RAABE, O.G., Dissolution of Respirable Pu-Dioxyd Particles in Water by Alpha Decay, 9th Int. Conf. on SSNTD, München (1976)
(68) HAMILTON, E.I., The concentration of uranium in air from contrasted natural environments, Health Phys., 19 (1970) 511
(69) SAKANOVE, M., NAKAURA, M., IMAI, T., The Determination of Pu in Environmental Samples, Int. Symp. on Rapid Methods for Measurement of Radioactivity in the Environment, München (1971)
(70) KOECHEL, R., KALFF, J., Track autoradiography for the determination of phtytoplankton species productivity, Limnol. Oceanogr., 21, 7 (1976) 590
(71) JEAMMARIE, L., BALLADA, J., Testing Solid State Track Detectors with a View to their Use in Radiotoxicological Laboratories, Assesment Radioactive Contam. Man. Proc. Symp. 1971, 267 (1972) 81
(72) HARRIES, D.R., Neutron irradiation embrittlement of austenitic stainless steels and nikel-base alloys, J. Brit. Nucl. Energy Soc., 5 (1966) 74
(73) ROSENBAUM, H.S., ARMIJO, J.S., Fission track etching as a metallographic tool, J. Nuc. Mat., 22 (1967) 115
(74) GARNISH, J.D., HUGHES, J.D., Quantitative analysis of boron in solids by autoradiography, J. Mat. Sci., 7 (1972) 7
(75) KAWASAKI, S., HISHIMNUMA, A., NAGASAKI, R., Behaviour of boron in stainless steel detected by fission track etching method and effect of radiation on tensile properties, J. Nucl. Materials, 39 (1971) 166
(76) ELEN, J.D., GLAS, A., Precipation of trace amounts of boron in AISI 340L and AISI 316L, J. Nucl. Materials, 34 (1970) 182
(77) VOBECKY, M., LEDROVA, E., Local microanalysis of uranium in steels by means of solid state track detectors, Radiochem. Radioanal. Lett., 24, 3 (1976) 151
(78) FLEISHER, R.L., ALTER, H.W., FURMAN, S.C., PRICE, P.B., WALKER, R.M., Technological applications of Science : the case of particle track etching, Science, 178 (1972) 255
(79) CICHOWSKA, K, WALIS, L., STVERAK, B., Track autoradiography in material studies, Radioisotopy, 13, 3 (1976) 434
(80) HASHIMOTO, T., Electrodeposition of americium and observation of the surface by means of α- particle tracks on cellulose nitrate film, J. Radioanal. Chem., 9 (1971) 251
(81) JOSEFOWICZ, K., Application of track detectors to reactor studies, Zh. Teplo energ., 2U (1976) 110
(82) HASHIMOTO, T., Determination of uranium content in sea water by a fission track method with condensed aqueons solutions, Analyt. Chim. Acta, 56 (1971) 347
(83) CHAPUIS, A.M., FRANCOIS, H., GERARD-NICODEME, N., Controle de l'enrichissement de l'uranium métallique à l'aide de nitrate de cellulose, Rad. Effects, 5 (1970) 91
(84) CHUNG, A., DIAMOND, W., LITHERLAND, A.E., A Sensitive Method for Determing U and Th Contamination in Nuclear Targets, Rept. AECL 5503 (1975)
(85) BERGER, H., Track etch radiography : alpha, proton and neutron, Nuclear Technology, 19 (1973) 188

(86) BERGER, H., LAPINSKI, N.P., Improved sensitivity and contrast. Track etch thermal neutron radiography, Trans. Am. Nucl. Soc., 15 (1972) 123
(87) BENTON, E.V., TOBIAS, C.A., HENKE, R.P., CRUTY, M.R., Heavy particle radiography with plastic nuclear track detectors, IXth Inter. Conf. on Solid State Nuclear Track Detectors, München (1976)
(88) BENTON, E.V., HENKE, R.P., TOBIAS, C.A., Heavy particle radiography, Science, 182 (1975) 474
(89) BENTON, E.V., HENKE, R.P., TOBIAS, C.A., CRUTY, M.R., Radiography with Heavy Particles, LBL-2887 (1975), Lawrence Berkeley Lab., California
(90) BERGER, H., LAPINSKI, N.P., BEYER, N.S., Proton Radiography : a Preliminary Report, Proc. 8th Symposium on non Destructive Evaluation in Aerospace, Weapons systems, Nuclear Applications, San Antonio, Texas (1971)
(91) SOMOGYI, G., SRIVASTAVA, D.S. Alpha radiography with plastic track detectors, Int. J. Appl. Rad. Isotopes, 22 (1971) 289
(92) MORY, J., DEGUILLEBON, D., DELSARTE, G., Measure of mean path length of fission fragments with the mica detector. Influence of the crystalline texture, Rad. Effects, 5 (1970) 37
(93) JOUSSET, J.C., MORY, J., QUILICO, J.J., Etude des Figures de Canalisation dans les Cristaux, Proc. Inter. Conf. Nucl. Track Registration in Insulating Solids, Clermont-Ferrand, France, IX-47 (1969)
(94) MEDVECKY, L., SOMOGYI, G., Induzierte Radiographie, Atomi Koslem., 12 (1970) 191
(95) MORY, J., DELSARTE, G., La canalisation des fragments de fission, Rad. Effects, 1 (1969) 1
(96) DELSARTE, G., JOUSSET, J.C., MORY, J., QUERE, Y., Dechanneling of Fast Transmitted Particles by Lattice Defects in Atomic Collision Phenomena in Solids (1970)
(97) MORY, J., Mesure du coefficient de canalisation par un défaut d'empilement dans l'or, Rad. Effects, 8 (1971) 139
(98) DELSARTE, G., DESARMOT, G., MORY, J., Determination par canaligraphie de l'orientation de microcristaux, Phys. Stat. Sol., 5 (1971) 683
(99) WALKER, R.M., PRICE, P.B., FLEISCHER, A versatile disposable dosimeter for slow and fast neutrons, Appl. Phys. Letters, 3 (1963) 28
(100) PROVO, P., DAHL, R.E., YOSHIKAWA, Thermal and fast neutron detection by fission-track production in mica, J. Appl. Phys., 35 (1964) 2636
(101) BAUMGARTNER, W.V., BRACKENBUSH, L.W., UNRUH, C.M., A New Neutron and High Energy Particle Dosimeter for Medical Dosimetry Applications, Symp. on Solid State and Chem. Rad. Dosimetry in Medicine and Biology, I.A.E.A., Vienne (1966)
(102) DEBEAUVAIS, M., MAURETTE, M., MORY, J., WALKER, R., Registration of fission fragment tracks in several substances and their use in neutron dosimetry, Int. J. Appl. Rad. Isotopes, 15 (1964) 289
(103) MEDVEDEZKY, M., SOMOGYI, G., Fast neutron flux measurement by means of plastics, Atomki Kozlem, 8 (1966) 226
(104) BECKER, K., Nuclear track registration in dosimeter glasses for neutron dosimetry in mixed radiation fields, Health Physics, 12 (1966) 769
(105) BECKER, K., Dosimetric Applications of Track Etching Topics in Radiation Dosimetry, Suppl. 1, (1972) 79, Academic Press, N.Y.
(106) ROBERTS, J.H., PARKER, R.A., GONGEL, F.G., KASTNER, J., OLTMAN, B.G., Environmental neutron measurements with solid state track recorders, Rad. Effects, 3 (1970) 283
(107) WOOLUM, D.S., BURNETT, D.S., BAUMAN, C.A., The Linear Neutron Probe Experiment, Apollo 17 preliminary science report, NASA pub. 330, 18 (1973) 1

(108) VARNAGY, M., CSIKAI, J., SZEGEDI, S., NAGY, S., Observation of proton tracks by a plastic detector, Nucl. Inst. Methods, 89 (1970) 27
(109) STERN, R.A., PRICE, P.B., Charge and energy information from heavy ion tracks in lexan, Nature Phys. Sci., 240 (1972) 82
(110) BAUMGARTNER, W.V., BRACKENBUSH, L.W., Neutron Dosimetry Using the Fission Fragment Damage Principle, Rept. BNWL-332, Battelle Northwest, Richland. Wash. (1966)
(111) SOHRABI, M., BECKER, K., Fast neutron personnel monitoring by fission fragment registration from ^{257}Np, Nucl. Inst. and Methods, 104 (1972) 409
(112) MONNIN, M., ISABELLE, D.B., Les détecteurs solides de traces et leurs applications en biologie, Ann. Phys. Biol. et Med, 4, (1970),95
(113) BURGER, G., GRUNAUER, F., PARETZKE, H., The Applicability of Track Detectors in Neutron Dosimetry, Pap. SM-143.17, Proc. Symp. on New Radiation Detectors, I.A.E.A., Vienne (1970)
(114) REMY, G., RALAROSY, J., TRIPIER, J., DEBEAUVAIS, M., STEIN, R., Dosimetrie et spectrométrie approchées de neutrons de fission et thermo-nucléaires à l'aide de détecteurs visuels plastiques, Rad. Effects, 5 (1970) 221
(115) BECKER, K., Direct Fast Neutron Interactions with Polymers, ORNL-Rept. 4446-226 (1969)
(116) FRANK, A.L., BENTON, E.V., Development of a high energy neutron detector, Defense Nucl. Agency Rept. 2918F (1972)
(117) BECKER, K., Alpha particle registration in plastics and its applications for radon and neutron personnel dosimetry, Health Physics, 16 (1969) 113
(118) FRANK, A.L., BENTON, E.V., A diffusion chamber radon dosimeter for use in mine environment, Nucl. Inst. Methods, 109 (1973) 537
(119) ANNO, J., BLANC, D., TEYSSIER, J.L., Collection of Rn Daughters on a Filter, Proc. 7th Int. Conf. Corpuscular Phot., Barcelona, 543 (1970)
(120) BENTON, E.V., COLLVER, M., Registration of heavy ions during the flight of Gemini VI, Health Physics, 13 (1967) 495
(121) COMSTOCK, G.M., FLEISCHER, R.L., GIARD, W.R., HART Jr., H.R., NICHOLS, G.E., PRICE, P.B., Cosmic ray tracks in plastics : Apollo helmet dosimetry experiment, Science, 172 (1971) 154
(122) ALKOFER, O.C., ENGE, W., HEINRICH, W., ROHRS, H., Preliminary Results of Measurements of Heavy Primaries in the Region of Supersonic Transport Using Plastic Stacks, Int. Conf. on Protection against Accelerator and Space Radiation, Geneva (1971)
(123) SPURNY, F., Neutron dosimetry based on recoils and alpha particle registration by solid state track detectors, Jad. Energy, 22, 2 (1976) 49
(124) SINGER, J., TROUSIL, J., PROUZA, Z., Track detectors in personnel neutron dosimetry, Radioisotopy, 17, 3 (1976) 371
(125) KNOWLES, H.B., RUDDY, F.H., TRIPARD, G.E., WEST, G.M., KLIGERMAN, M.M., Status Report : Direct Track Detector Dosimetry in Negative Pion Beams, 9th Int. Conf. on SSNTD, München (1976)
(126) KHAN, M.A., AHMAD, A.R., BUKHARI, K.M., SADDARUDIN, A., The Measurement of Radon and Thoron by Solid State Track Detectors, Ibid (1976)
(127) HASSIB, G.M., TUYN, J.W.N., DUTRA, J., On the electrochemical etching of neutron tracks in plastics and its application to personnel neutron dosimetry, Ibid (1976)
(128) SCHRANBE, H., PARETZKE, H.G., Neutron Fluence Measurement with Track Detectors, Ibid (1976)

(129) SPURNY, F., KAREL, T., Neutron Dosimetry by Means of Different Solid State Nuclear Track Detectors, Ibid (1976)
(130) DUTRANNOIS, J., TUYN, J.W.V., Application of Solid State Nuclear Track Detectors for Personnel Monitoring Around High Energy Accelerator, Ibid (1976)
(131) SHORABI, M., Amplification of recoil particle tracks in polymers and its applications in fast neutron personnel dosimetry, Health Physics, 27, 6 (1974) 598
(132) GOMAA, M.A., EL-KOLALY, M.A., Usability of track detectors in mixed fields of neutrons and gamma-rays, Nucl. Inst. and Methods, 134, 2 (1976) 253
(133) PIESCH, E., BURGKHARDT, B., Comparison of Albedo and Nuclear Track Detectors for Neutron Monitoring, Proc. Int. Congr. Int., Radiat. Prot. Assoc. 3rd, 1973 (pub. 1974)
(134) WEIDENBAUM, B., LOWETT, D.B., Development of a New Technique for Bulk Water Tracing, Vallecitos Nuclear Center Rept NEDC 12015-5 (1971)
(135) LOWETT, D.B., Track etch analysis of uranium particles as a tracer system, Trans. Am. Nucl. Soc., 15, 1 (1972) 121
(136) CARPENTER, B., SAMUEL, D., WASSERMAN, I., Quantitative applications of ^{17}O tracer, Rad. Effects, 19 (1973) 59
(137) DEBEAUVAIS, M., REMY, G., TRIPIER, J., Determination of Isoflux Curves of a 14 MeV-beam, 1st ASTM Euratom Symposium on Reactor Dosimetry, Petten, Holland, Spl. (1975)
(138) POPE, P., De COSTER, M., LANGELA, D., Burn-up determination nuclear fuels by high resolution gamma spectroscopy, track formation in solid state detectors and neutron dose measurements, Nucl. Appl. Technol., 9 (1970) 755
(139) TUYN, J.W.N., On the use of solid state nuclear track detectors in reactor physics experiments, Rad. Effects, 5 (1970) 75
(140) De COSTER, M., LANGELA, D., Accurate absolute determination of fission densities in fuel rods by means of solid state track recorders, Nucl. Appl. Technol., 9 (1970) 229
(141) LYCOS, T., BESANT, C.B., Measurement of fast reactor reaction rates using solid state track detectors, Microscope, 24, 3 (1976) 199
(142) JOSEFOWICZ, K., Application of track detectors for determining parameters of reactors, Zh. Teploenerg., 2U (1976) 110
(143) WEIDENBAUM, B., LOWETT, D.B., CAPUTI, R.W., Track etch research and development at GE Vallecitos nuclear center, Trans. Amer. Nucl. Soc., 13 (1970) 528
(144) GRINGRICH, J.E., LOWETT, D.B., A track etch technique for uranium exploration, Trans. Amer. Nucl. Soc., 15 (1972) 118
(145) GRINGRICH, J.E., Uranium exploration made easy, Power Eng., Aug. (1973) 48
(146) GUSTAFSON, T., KINDKVIST, B., KRISTIANSSON, K., New method for measuring the flight altitude of birds, Nature, 244 (1973) 112
(147) CASNATI, E., MARCHETTI, M., TOMMASINO, L., Use of heavy ion for the evaluation of polymer stability, Int. J. Appl. Radiat. Isot., 25, 7 (1974) 307
(148) KUZNETSOV, V.A., MOGILLNER, A.I., KOROLEVA, V.P., Use of solid track detectors in reactor experiments, T. Fiz. Energ. Inst., 123 (1974) 31
(149) BEAN, C.P., DOYLE, M.V., ENTINE, G., Etching of submicron pores in irradiated mica, J. Appl. Phys., 41 (1970) 1454
(150) FLEISCHER, R.L., VIERTL, J.R.M., PRICE, P.B., Biological filters with continuously adjustable hole size, Rev. Sci. Inst., 43 (1972) 1708

(151) SEAL, S., A sieve for the isolation of cancer cells and other large cells from the blood, Cancer, 17 (1964) 637
(152) WERTTAKE, P.T., MARKOVETS, B.A., STELLAR, S., Cytology evaluation of cerebrospinal fluid with chemical and histology correlation, Acta Cytologica, 16 (1972) 224
(153) KING, E.B., RUSSEL, W.M., Needle aspiration biopsy of the lung, Acta Cytologica, 11 (1967) 319
(154) HORWITZ, D.A., GARRETT, M.A., Use of leukocyte chemotaxis in vitro to assay mediators generated by immune reaction, J. Immunology, 103 (1971) 649
(155) BARTZDORF, U., KNOX, R.S., POKRESS, S.M., KENNADY, J.C., Membrane partitioning of the rose-type chamber for the study of metabolic interaction between different cultures, Stain Technology, 44, (1969) 71
(156) SPURNY, K.R., LODGE, J.P., FRANCK, E.R., SHEESLEG, D.C., Aerosol filtration by means of Nuclepore filters : aerosol sampling and measurement, Environmental Sci. and Technol., 3 (1969) 464
(157) SPURNY, K.R., LODGE Jr., J.P., A note on the measurement of radioactive aerosol, Aerosol Sci., 3 (1972) 407
(158) TWOMEY, S., Measurements of the size of natural cloud nuclei by means of Nuclepore filters, J. Atmos. Sci., 29 (1972) 318
(159) DeBLOIS, R.W., BEAN, C.P., Couting and sizing of submicron particles by the resistive pulse technique, Rev. Sci. Inst., 41 (1970) 909
(160) DeBLOIS, R.W., BEAN, C.P., Virus Detection and Characterization by the Resistive Pulse Technique, G.E. Co preprint 73 CRD 188 (1973)
(161) GOLIBERSUCH, D.C., Observation of aspherical particle rotation in Poiseuille-flox via the resistance pulse technique, Biophys. J., 13 (1973) 265
(162) GAMOTA, G., Creation of quantized vortex rings in Superfluid Helium, Phys. Rev. Letters, 31 (1973) 517
(163) NOTARYS, H.A., Electric field suppresion of the lambda point in liquid helium, Phys. Rev. Letters, 20 (1968) 1131
(164) BEAN, C.P., The physics of porous membranes-neutral pores, Membranes-A series of advances, G. Eisenman Ed., 1 (1972) 1
(165) BECK, R.E., SCHULTZ, J.S., Hindrance of solute diffusion within membranes as measured with microporous membranes of known pore geometry, Biochem., Biophys. Acta, 255 (1972) 273
(166) ANDERSON, J.L., QUINN, J.A., Ionic mobility in microcapillaries, J. Chem. Soc. Faraday Trans., I $\underline{68}$, 4 (1972) 744
(167) FLEISCHER, R.L., RAABE, O.G., Dissolution of Respirable Pu-Dioxyd Particles in Water by Alpha Decay, 9th Int. Conf. on SSNTD, München (1976)
(168) DURANI, S.A., KHAN, H.A., Solid State Track Detectors for Neutron Imaging, Radiogr. Neutrons Conf., 71, (1975) 7
(169) HARTMAN, W.J., Directional effects in track etch imaging, Nucl. Inst. and Methods, $\underline{140}$, 1 (1977) 139
(170) KRIVOKHATSKII, A.S., KADYROV, V.N., ALEKSAMDROV, B.M., AGIEVSKI, D.A., Measuring the characteristics of the neutron field in the horizontal channel of a WWR-S reactor with track detectors, Metrol. Neitron. Isluch. Reaktorakh. Uskor, Tr. Vses Soveshch, 2 (1972) 74
(171) GERZHMANSKI, B.V., Application of track detector for determining the static parameters of reactors, Sostsyanye i Perspekt. Rabot po Sozdaniya AES s reaktorami na bystrykh neikronakh, 3 (1975) 406
(172) KHAN, H.A., Natural quartz : a useful detector for in-core measurements, Nucl. Inst. and Methods, $\underline{128}$, 2 (1975) 245

(173) YOKOSUKA, Y., SATOH, H., KOMATSU, J., KOIZUMI, M., Technical improvement on α-radiography on Pu-fuels, Tokai Jigyo-sho, Doryoku-do, Kaku-nenryo, Kaibatsu, Jigyo-dam, N881, 72, 03 (1972) 62

(174) BLEACH, J.L., BECKER, K., Improvement in spatial resolution of track etching microradiography, Nucl. Inst. and Methods, 130, 2 (1975) 499

(175) VARNAGY, M., GYARMATI, E., SCTARICSKAI, T., Automatic track counting in gamma exposed polymers foils with a jumping spark counter, Nucl. Inst. and Methods, 133, 2 (1976) 371

(176) GAVAULT, G., Métallisation des matières plastiques, Thesis, University of Clermont-Ferrand (1974)

(177) MONNIN, M., Applications diverses, Rept. n° IX, Int. Conf. on Nuclear Track Registration in Insulating Solids, Clermont-Ferrand (1969).

(178) FISCHER, D.E., Fission/X analysis of the U, Th families, Nature (London) **265** 5591 (1977) 227.

LIST OF PARTICIPANTS

H. Arriola   Centro de Estudios Nucleares, UNAM,
                    Circuito Exterior, CU,
                    México 20 DF

S. Bashkin   University of Arizona,
                    Tucson, AZ 85721,
                    United States of America

J.P. Blewett   Department of Physics,
                    Brookhaven National Laboratory,
                    Associated Universities Inc.,
                    Upton, NJ 11973,
                    United States of America

N. Clark   Escuela de Física,
                    Universidad de Costa Rica,
                    Ciudad Universitaria "Rodrigo Facio",
                    Costa Rica

R.L. Cohen   Bell Laboratories,
                    600 Mountain Avenue,
                    Murray Hill, NJ 07974,
                    United States of America

M. Gallardo   Universidad de Costa Rica,
                    Ciudad Universitaria "Rodrigo Facio",
                    Costa Rica

L. Haug   Universidad de Costa Rica,
                    Ciudad Universitaria "Rodrigo Facio",
                    Costa Rica

P. Hautojärvi   Helsinki University of Technology,
                    Department of Technical Physics,
                    SF-02150 Espoo 15,
                    Finland

LIST OF PARTICIPANTS

R. Jiménez	Universidad de Costa Rica, Ciudad Universitaria "Rodrigo Facio", Costa Rica
G. Kostorz	Institut Laue-Langevin, Avenue des Martyrs, 156X Centre de Tri, 38042 Grenoble Cedex, France
J.A. Lubkowitz	Chemistry Department, University of Georgia, Athens, GA 30602, United States of America and Instituto Venezolano de Investigaciones Científicas, Centro de Petróleo y Química, Apartado 1827, Caracas, Venezuela
M. Monnin	Laboratoire de physique corpusculaire, Université de Clermont, Clermont-Ferrand, 63170 Aubière, France
L. Moya	Universidad de Costa Rica, Ciudad Universitaria "Rodrigo Facio", Costa Rica
S.S. Nargolwalla	Nuclear Applications Research Laboratory, SCINTREX Ltd, 222 Snidercroft Road, Concord, Ontario L4K 1B5, Canada
D.W. Palmer	The University of Sussex School of Mathematical and Physical Sciences, Falmer, Brighton BN1 9QH, United Kingdom
T.B. Pierce	UKAEA Research Group, Instrumentation and Applied Physics Division, Building 148, AERE, Harwell, Oxfordshire OX11 0RA, United Kingdom

LIST OF PARTICIPANTS

A. Salazar	Universidad de Costa Rica, Ciudad Universitaria "Rodrigo Facio", Costa Rica
E.A. Schweikert	Texas A & M University, Dept. of Chemistry, College Station, TX 77843, United States of America
V. Valković	Rice University, Dept. of Physics, Houston, TX 77001, United States of America and Institut Ruđer Bošković, Zagreb, Yugoslavia
G.W. Wertheim	Bell Laboratories, 600 Mountain Avenue, Murray Hill, NJ 07974, United States of America

Scientific Secretary

J. Dolničar	Division of Research and Laboratories, International Atomic Energy Agency, Kärntner Ring 11, P.O. Box 590, A-1011 Vienna, Austria

HOW TO ORDER IAEA PUBLICATIONS

An exclusive sales agent for IAEA publications, to whom all orders
and inquiries should be addressed, has been appointed
in the following country:

UNITED STATES OF AMERICA   UNIPUB, P.O. Box 433, Murray Hill Station, New York, N.Y. 10016

In the following countries IAEA publications may be purchased from the
sales agents or booksellers listed or through your
major local booksellers. Payment can be made in local
currency or with UNESCO coupons.

ARGENTINA	Comisión Nacional de Energía Atómica, Avenida del Libertador 8250, Buenos Aires
AUSTRALIA	Hunter Publications, 58 A Gipps Street, Collingwood, Victoria 3066
BELGIUM	Service du Courrier de l'UNESCO, 112, Rue du Trône, B-1050 Brussels
C.S.S.R.	S.N.T.L., Spálená 51, CS-113 02 Prague 1
	Alfa, Publishers, Hurbanovo námestie 6, CS-893 31 Bratislava
FRANCE	Office International de Documentation et Librairie, 48, rue Gay-Lussac, F-75240 Paris Cedex 05
HUNGARY	Kultura, Bookimport, P.O. Box 149, H-1389 Budapest
INDIA	Oxford Book and Stationery Co., 17, Park Street, Calcutta, 700016
	Oxford Book and Stationery Co., Scindia House, New Delhi-110001
ISRAEL	Heiliger and Co., 3, Nathan Strauss Str., Jerusalem
ITALY	Libreria Scientifica, Dott. Lucio de Biasio "aeiou". Via Meravigli 16, I-20123 Milan
JAPAN	Maruzen Company, Ltd., P.O. Box 5050, 100-31 Tokyo International
NETHERLANDS	Martinus Nijhoff B.V., Lange Voorhout 9-11, P.O. Box 269, The Hague
PAKISTAN	Mirza Book Agency, 65, Shahrah Quaid-e-Azam, P.O. Box 729, Lahore-3
POLAND	Ars Polona-Ruch, Centrala Handlu Zagranicznego, Krakowskie Przedmiescie 7, Warsaw
ROMANIA	Ilexim, P.O. Box 136-137, Bucarest
SOUTH AFRICA	Van Schaik's Bookstore (Pty) Ltd., P.O. Box 724, Pretoria 0001
	Universitas Books (Pty) Ltd., P.O. Box 1557, Pretoria 0001
SPAIN	Diaz de Santos, Lagasca 95, Madrid-6
	Diaz de Santos, Balmes 417, Barcelona-6
SWEDEN	AB C.E. Fritzes Kungl. Hovbokhandel, Fredsgatan 2, P.O. Box 16358 S-103 27 Stockholm
UNITED KINGDOM	Her Majesty's Stationery Office, P.O. Box 569, London SE1 9NH
U.S.S.R.	Mezhdunarodnaya Kniga, Smolenskaya-Sennaya 32-34, Moscow G-200
YUGOSLAVIA	Jugoslovenska Knjiga, Terazije 27, POB 36, YU-11001 Belgrade

Orders from countries where sales agents have not yet been appointed and
requests for information should be addressed directly to:

Division of Publications
International Atomic Energy Agency
Kärntner Ring 11, P.O.Box 590, A-1011 Vienna, Austria